AF841667

WITHDRAWN

The Urban Book Series

Series Advisory Editors

Fatemeh Farnaz Arefian, University College London, London, UK
Michael Batty, University College London, London, UK
Simin Davoudi, Newcastle University, Newcastle, UK
Geoffrey DeVerteuil, Cardiff University, Cardiff, UK
Karl Kropf, Oxford Brookes University, Oxford, UK
Marco Maretto, University of Parma, Parma, Italy
Vítor Oliveira, Porto University, Porto, Portugal
Christopher Silver, University of Florida, Gainesville, USA
Giuseppe Strappa, Sapienza University of Rome, Rome, Italy
Igor Vojnovic, Michigan State University, East Lansing, USA
Jeremy Whitehand, University of Birmingham, Birmingham, UK

Aims and Scope

The Urban Book Series is a resource for urban studies and geography research worldwide. It provides a unique and innovative resource for the latest developments in the field, nurturing a comprehensive and encompassing publication venue for urban studies, urban geography, planning and regional development.

The series publishes peer-reviewed volumes related to urbanization, sustainability, urban environments, sustainable urbanism, governance, globalization, urban and sustainable development, spatial and area studies, urban management, urban infrastructure, urban dynamics, green cities and urban landscapes. It also invites research which documents urbanization processes and urban dynamics on a national, regional and local level, welcoming case studies, as well as comparative and applied research.

The series will appeal to urbanists, geographers, planners, engineers, architects, policy makers, and to all of those interested in a wide-ranging overview of contemporary urban studies and innovations in the field. It accepts monographs, edited volumes and textbooks.

More information about this series at http://www.springer.com/series/14773

Fatemeh Farnaz Arefian

Organising Post-Disaster Reconstruction Processes

Housing Reconstruction after the Bam Earthquake

Springer

Fatemeh Farnaz Arefian
Silk Cities, The Bartlett Development Planning Unit
University College London
London
UK

ISSN 2365-757X ISSN 2365-7588 (electronic)
The Urban Book Series
ISBN 978-3-319-70910-9 ISBN 978-3-319-70911-6 (eBook)
https://doi.org/10.1007/978-3-319-70911-6

Library of Congress Control Number: 2017958553

© Springer International Publishing AG, part of Springer Nature 2018
This work is subject to copyright. All rights are reserved by the Publisher, whether the whole or part of the material is concerned, specifically the rights of translation, reprinting, reuse of illustrations, recitation, broadcasting, reproduction on microfilms or in any other physical way, and transmission or information storage and retrieval, electronic adaptation, computer software, or by similar or dissimilar methodology now known or hereafter developed.
The use of general descriptive names, registered names, trademarks, service marks, etc. in this publication does not imply, even in the absence of a specific statement, that such names are exempt from the relevant protective laws and regulations and therefore free for general use.
The publisher, the authors and the editors are safe to assume that the advice and information in this book are believed to be true and accurate at the date of publication. Neither the publisher nor the authors or the editors give a warranty, express or implied, with respect to the material contained herein or for any errors or omissions that may have been made. The publisher remains neutral with regard to jurisdictional claims in published maps and institutional affiliations.

Printed on acid-free paper

This Springer imprint is published by the registered company Springer International Publishing AG part of Springer Nature
The registered company address is: Gewerbestrasse 11, 6330 Cham, Switzerland

"Some Hindus have an elephant to show.
No one here has ever seen an elephant.
They bring it at night to a dark room.

One by one, we go in the dark and come out
saying how we experience the animal.
One of us happens to touch the trunk.
A water-pipe kind of creature.

Another, the ear. A strong, always moving
back and forth, fan-animal. Another, the leg.
I find it still, like a column on a temple.

Another touches the curve back.
A leathery throne. Another, the cleverest,
feels the tusk. A rounded sword made of
porcelain.
He is proud of his description.

Each of us touches one place
and understands the whole in that way.
The palm and the fingers feeling in the dark
are how the senses explore the reality of the
elephant.

If each of us held a candle there,
and if we went in together, we could see it."

Elephant in Dark
Rumi, the 13th Century Persian poet and teacher of Sufism
From The Masnavi, Book III, Story V
Translation by Coleman Barks

To Maryam, Farangis and memory of Ali

Preface

The story of this book somehow mirrors a personal quest for understanding a contemporary challenging issue with global importance. The book draws on my Ph. D. research at the Bartlett Development Planning Unit (DPU), University College London, and my real-time involvement in the implementation of Bam reconstruction that triggered my doctoral studies and subsequent work leading to a collective 11 years of practice and research.

As a reflective practitioner who returned to academia to investigate the questions reached through practice, I hope publishing this book will contribute to bridging the gap between theory and practice in this field. It is a milestone in this personal journey to share learning opportunities we can extract from the Bam experience that are nevertheless helpful for other reconstruction cases, whether to manage them or to participate in planning and implementation. The evidence-based learning opportunities and research propositions advance the practice theory knowledge and contributes to bridging the existing theoretical gap. In particular, the book contributes to:

- Understand, from the literature on reconstruction, recovery and development, the expectations a reconstruction programme can be expected to fulfil.
- Understand the organisational characteristics of reconstruction programmes and how they influence the programme's ability to approach and achieve its stated objectives.
- Enable a deeper understanding and support better practice on organising reconstruction through the contribution of a multi-perspective conceptual framework for organisational configuration and management for post-disaster reconstruction programmes.

At the time of the Bam earthquake, I had my urban design and architectural consultancy company based in Tehran undertaking large-scale urban design, planning projects and affordable housing. The Bam earthquake was a national disaster in Iran and touched all. The earthquake destroyed 80% of a city, which was internationally famous for its ancient citadel (Arg-e-Bam) and historical urban identity. Immediate inspections showed that even newly-built buildings were destroyed because they did not comply with the existing national seismic regulations. I first visited Bam 40 days after the earthquake, at a time when local people were living in temporary accommodation. Soon after, a call was announced for identifying qualified interested consultancy companies to be engaged in housing reconstruction. Consultancies were requested to establish their local branch in Bam, be ready to work in difficult conditions, and to work within the overall programme framework. Each consultancy company would directly work with an estimated 2,000 disaster affected families on architectural and technical requirements for the reconstruction of their houses on an individual basis. Similar to many other companies, we applied and got involved. There was a sense of duty and desire to contribute to reconstruction beyond the difficult practicalities and financial matters. Later, with a few other consultancies, we were chosen to work on a few urban design projects. These projects were combined with the reconstruction of adjacent retail units and a collaborative effort with the respective owners on an individual basis. The first days started with having all consultancy representatives placed in a school, one of the few safe buildings in the area. Those extraordinary days included frequent meetings and efforts in Tehran and Bam to build the required project teams, trying internal and external mechanisms to get ready to face the unknown trend of demands and requests. Simultaneously we were developing broad housing typologies based on potential specifications of building plots, for example, direction, proportions, location and access as requested by the reconstruction executive body Housing Foundation of Islamic Revolution (HFIR) and The Bam Architectural and Urbanism Council (BAUC), which was established quickly after the earthquake to guide us. We later moved to a purpose-built construction bazaar, a one-stop-shop for disaster-affected people.

I was the project director for the housing reconstruction of 2,100 houses and the streetscape urban design project of the reconstruction of adjacent buildings during the Bam reconstruction period for my company. Over the course of 4 years in the Bam case, my company worked with local beneficiaries and reconstruction stakeholders and actors, dealing with the realities and emerging issues of the highly complex, traumatic and complex post-disaster condition. My reflections on my personal experience in Bam follow two distinct directions. The first was recognizing an insufficiency of traditional architectural professional routines and educational curriculums to balance other pressing issues in extreme traumatic situations. The second was the importance of the programme level operational framework and its organisational configurations for reconstruction activities that define the process of who does what and when, tying us all stakeholders together. While the first direction can be the subject of further examination, this book reflects on the latter. Reading international cases and noticing that there is indeed a gap or at least imbalance in disaster-development that leans towards *what* is expected from reconstruction than *how* to approach these expectations prompted a personal journey for understanding of the bigger picture on '*how*'.

This book, per se, is an advocacy of an interdisciplinary approach and outreaching to organisation theory to help us organising reconstruction activities that is more complex than construction management. As you might think, understanding the bigger picture required going beyond my disciplinary-self. Reflecting on this I undertook a post-graduate course on Strategic Management and Leadership offered by the Chartered Management Institute (CMI) and pursued my Prince 2 management qualification. So you will see the analytical approach to the subject and tone of the book reflects the '*wholeness*' of the delivery system of the programme organisation, instead of focusing on one system element (the actor position).

As you will see, this book presents debates on how reconstruction would address the Bam characteristics and how '*urbanising*' the existing know-how led to the introduction of three objectives: safeguarding historical urban identity, building earthquake resistant buildings, and mobilising people to participate. This book explores, within the case of Bam following the 2003 earthquake, how *organisational design* and *management* of the reconstruction programme have influenced the approach and achievement of the *objectives* of the reconstruction process. The housing reconstruction programme in Bam was a complex case, offering learning opportunities to understand *organisational design* and *management*. An examination of organisational characteristics of reconstruction programmes at the international level provides entry points to organisation theory for the examination of the Bam case. Furthermore, it provides organisational configuration of the delivery system and its evolution after being put into practice. Examples of the complementing and interconnected insights examine the following points: influential organisational attributes linked with consistencies and inconsistencies of the system; strategic characteristics of the case and the application of strategic thinking approach during the configuration and its implementation; the social characteristics of reconstruction and the very interesting construct of social learning during the housing reconstruction programme; and multi-organisational characteristics of reconstruction activities and related hidden threats and potentials. The book introduces visual analytical methods for understanding and analysing organisational configurations for reconstruction processes and programmes. The book offers a multi-perspective conceptual analytical framework for understanding and examining organisational aspects of post-disaster reconstruction programmes in order to organise related processes.

The completion of the original research and preparation of this book owes to generous support provided by many people. Thanks to Dr. Cassidy Johnson and Prof. Julio Davila, my research supervisors, whose generous support went well beyond the doctoral studies and extended to the Silk Cities initiative, which I founded as a doctoral researcher at DPU/UCL. Silk Cities is an independent, professional and academic initiative for contextual knowledge exchange, research and advocacy, and has an initial focus of the Middle East and Central Asia (www.silk-cities.org). I must also thank Dr. John Twigg and Tony Lloyd-Jones for their reading and the questions which shaped the original research. It was also heart-warming to see how key people from organisations involved in the Bam reconstruction, from the policy formulation and initial formation of the delivery system to putting the programme into practice, were supportive of an independent research that would be useful for the future of reconstruction activities in the country and beyond. I am thankful to Eng. Majid Joody and Dr. Pirooz Hanachi, Mr. Roohollah Abyari and Dr. Abbas Esmaeili, Eng. Mohammad Tofigh and Dr. Victoria Kianpour, Dr. Mohamoud Fatemi Aghda and all colleagues involved in the Bam reconstruction. I must also thank my good old friends for all their on-going supports. Therefore, in-depth and semi-structured interviews with key and knowledgeable people from involved organisations, other related organisations and beneficiaries became possible. I owe special thanks to all 59 interviewees for their insights and generosity. Reaching a deeper level of understanding of the bigger picture—*'the wholeness of the multi-organisational system'* —became possible because all those allocated time and shared their experiences and reflections in a depth beyond their formal positions. And last but not least, I am indebted to my family for all their support, especially my daughter Maryam, mother Farangis and my late father Ali for their unconditional love and patience. The original research was self-funded and the fieldwork expenses were partially covered by the DPU doctoral research bursary and UCL Institute for Risk and Disaster Reduction (IRDR).

I Hope You Enjoy the Book.

London, UK Fatemeh Farnaz Arefian

Contents

About the Author

Dr. Fatemeh Farnaz Arefian is a consultant, researcher and founding director of Silk Cities initiative (www.silk-cities.org) for research and knowledge exchange on contextual contemporary challenges of cities in the Middle East and Central Asia, associated with the UCL Bartlett Development Planning Unit. She combines academic research and education with extensive professional experience and has a background in leading her private sector consultancy for delivering large-scale urban design/planning and architectural projects in the Middle East. She has delivered international research and training on post-disaster reconstruction, disaster risk reduction and resilience. She served as a board member of the Society of Iranian Urban Planners, and is a fellow of Institute of Civil Protection and Emergency Management, Chartered Management Institute, and member of International Development Network at RTPI. She is invited and visiting lecturer at universities and international workshops. Farnaz has published academic and professional books and papers. Publication examples include: *Organising Post-Disaster Reconstruction Processes* (Springer 2018), *Getting Ready for Urban Reconstruction: Organising Housing Reconstruction in* Bam (Springer 2016), *Urban Change in Iran* (Springer 2016), co-author of *Research Roadmap, Disasters and the Built Environment* (CIB World 2014), Urban Design in the Middle East (Special topic, Urban Design Group 2012). Her past knowledge exchange activities received special attention from UNESCO Director General.

Acronyms

BAUC	Bam Architecture and Urbanism Council
BRCSR	Bureau for Research and Coordination of Reconstruction Affairs
BTID	Building Technical Identification Document
CMI	Chartered Management Institute
DAU	Department of Architecture and Urbanism
DM	Disaster Management
DRR	Disaster Risk Reduction
GFDRR	Global Facility for Disaster Reduction and Recovery
HDI	Human Development Index
HFA	Hyogo Framework for Action
HFIR	Housing Foundation of Islamic Revolution
IDNDR	International Decade for Natural Disaster Reduction
IFRC	International Federation of the Red Cross and Red Crescent Societies
IIEES	International Institute for Earthquake Engineering and Seismology
INSC	Iran National Statistical Centre
ISMN	Iran Strong Motion Network
KEO	Kerman Engineering Organisation
MDG	Millennium Development Goals
MHUD	Ministry of Housing and Urban Development
NBC	National Building Codes
NCEO	National Construction Engineering Organisation
NDMO	National Disaster Management Organisation
NHD	Natural Disaster Headquarters
OCHA	United Nations Office for the Coordination of Humanitarian Affairs
ODI	Oversees Development Institute
ODR	Owner Driven Reconstruction

ORA	Office of Reconstruction Affairs
PRP	International Recovery Platform
SBU	Shahid Beheshti University
Steering Committee	Steering Committee of Bam Reconstruction
UN	United Nations
UNDP	United Nations Development Programme
UNESCO	United Nations Educational, Scientific and Cultural Organization
UNISDR	The United Nations Office for Disaster Risk Reduction
WCDRR	UN World Conference on Disaster Risk Reduction

Part I
Setting the Scene

Chapter 1
Introduction

Abstract Contemporary advances in the intersection between disaster and development studies characterise recovery and reconstruction with a multiplicity of tasks and objectives. In practice, however organising such reconstruction programmes that can achieve development objectives and contribute to multi-dimension recovery has proven problematic. Despite the increasing recognition of problems and difficulties in organising reconstruction processes, there is an imbalance in the literature in disaster fields that lean towards '*what*' is expected from reconstruction with less attention to '*how*' to organise reconstruction towards the delivery of such expectations. This chapter deals with questions of why, what and how. A combination of factors identifies the Bam case study as a strategic case for this research that are based on information-oriented and contemporary theoretical values. The Bam case for permanent housing reconstruction programme in an urban environment had three objectives: building earthquake-resistant buildings, safeguarding historical urban identity and mobilising people to participate. It was a complex case, offering learning opportunities to understand organisational design and management of reconstruction programmes and processes. The evidence-based learning opportunities and the theoretical research propositions advance the practice theory knowledge contributes to bridging the aforementioned theoretical gap towards better organising reconstruction processes in practice.

1.1 Reconstruction Programmes: Optimistic, but Difficult to Organise

Contemporary advances in the intersection between disaster and development studies point out the importance of reconstruction roles for an integrated multidimensional recovery (e.g. Davis 2007; Wisner 2004; Cuny 1991; Blaikie et al. 1994). In practice, however, organising such reconstruction programmes, especially on a large scale, have proved problematic (e.g. Gaillard 2017; Lassa 2012; Lyons et al. 2010; Lizarralde et al. 2009; Comfort and Kapucu 2006; Barakat 2003). The current epistemological developments in disaster studies characterise recovery and

© Springer International Publishing AG, part of Springer Nature 2018

F. F. Arefian, *Organising Post-Disaster Reconstruction Processes*,
The Urban Book Series, https://doi.org/10.1007/978-3-319-70911-6_1

reconstruction, with a multiplicity of tasks and objectives, as the intersection for the two proactive and reactive entry points to disasters that are Disaster Risk Reduction (DRR) and post-disaster emergency relief and rehabilitation (e.g. Gaillard 2017; Baas et al. 2008; Davis 2007; Wisner 2004; Blaikie et al. 1994; Cuny 1991).

Despite the increasing expectations from post-disaster reconstruction and recognition of the existence of problems and difficulties in delivering them in practice, there is an imbalance in the literature in disaster fields that leans towards '*what*' is expected from reconstruction with less attention to '*how*' to organise reconstruction towards the delivery of such expectations.

My goal for working on this topic has been to find out the organisational attributes that influence organisational design and management of post-disaster reconstruction programmes in order to understand their formation and implementation. The Bam case presents a strategic choice for this research because of some underlying factors, which are discussed in the next section. It is acknowledged that in the case of a single case study, '*strategic choice of a case greatly add to the generalisability of a case study*' (Flyvbjerg 2006). Deep analytical generalisation then becomes the tool for generalisation and contributing to knowledge production and theory development. This is through rich descriptions and analysis, focusing on the depth of the case study that brings out the details of deep cultural phenomena and interactions (e.g. Eisenhardt 1989; Bryman 2008; Yin 2009). Thus, moving towards the above goal the research explores, in the case of Bam after its 2003 earthquake, how the achievement of the objectives of the reconstruction programme was influenced by organisational design and management approach of the programme. The book specifically contributes to: (a) Understand, from the literature on reconstruction, recovery and development, the expectations a reconstruction programme can be expected to fulfil. (b) Understand the organisational characteristics of reconstruction programmes and how they influence the programme's ability to approach and achieve its stated objectives, and (c) Enable a deeper understanding and support better practice on organising reconstruction through the contribution of a multi-perspective conceptual framework for organisational design and management for post-disaster reconstruction programmes.

A brief review shows that recent decades have been faced with growing catastrophic natural disasters with the escalating financial and human costs of disasters since 1960. Examples of a long list of recent examples include Amatrice earthquake, Italy 2016; Gorkha earthquake, Nepal 2015; Tohoku earthquake and tsunami, Japan 2011, Haiti earthquake 2010; L'Aquila earthquake, Italy 2009, Sichuan earthquake, China 2008; Peru earthquake and tsunami 2007; Kashmir earthquake, Pakistan 2005; Indian Ocean Tsunami 2004; Bam earthquake, Iran 2003; Gujarat earthquake, India 2001, Marmara earthquake, Turkey 1999, Kobe earthquake, Japan 1995, Alto Mayo earthquake, Peru 1990. As the Centre for Research on the Epidemiology of Disasters (CRED 2016) reports, between 1994 and 2013, there were 6,873 recorded natural hazards worldwide, which claimed 1.35 million lives or almost 68,000 lives on average each year. On average natural hazards affected 218 million people per annum during this 20-year period. The annual average of the number of disasters in decade 2005–2014 was 380 with their annual average numbers of 196.3 million victims and US$ 159.8 billion financial damages. In 2015 alone, 376 natural-triggered disasters were registered, which resulted in 22,765 deaths, 110.3 million victims worldwide, with estimates placing economic damages at US$ 70.3 billion. There were also 315 natural catastrophe events in 2016 that generated economic losses of US$ 210 billion. A review of the single geological hazards (earthquake, tsunami, volcanic activity, landslide) with over 1000 casualties during 1994–2017 shows 23 incidents occurred in 16 countries, causing 700,057 casualties. Nine countries of the top ten countries in terms of number of victims were low or lower middle-income countries. Considering all scales of casualties, earthquakes (including tsunamis) killed more people than all other types of disaster put together, claiming nearly 750,000 lives between 1994 and 2013 (AON Benfield 2017; AON Benfield 2015; CRED 2016; Guha-Sapir et al. 2016; UNDP Evaluation Office 2010; UN 2010).

The growing international concern about the increased frequency of large-scale catastrophic disasters has increased the international drive to reduce the destructive effects on the lives and livelihoods of individuals and communities. For example, the 1990s was known as the 'decade for reducing disaster risks'. The World Conference on Disaster Reduction, held in Kobe, Hyogo Prefecture, Japan in 2005 was a platform to share the momentum and concerns (UNISDR 2005). Moving the trend of tackling disastrous incidents at global level from after to before the event, the proactive entry point, DRR, as the Hyogo Framework for Action (HFA), 2005–2015 emphasised, is a shift from reactive emergency relief to proactive disaster risk reduction (e.g. Bosher 2008; Baas et al. 2008; Pearce 2003; UNDP 2004; International Federation of the Red Cross and Red Crescent Societies—IFRC 2004). The Third UN World Conference in Sendai, Japan, on 18 March 2015 led to the Sendai Framework for Disaster Risk Reduction 2015–2030 (Sendai Framework) which was the first major agreement with the post-2015 development agenda (UNISDR 2015). Disaster risk reduction is also related to development efforts as disasters are also manifestations of the failure of development programmes (e.g. Benson and Twigg 2007; e.g. Pelling 2003; Wisner 2004). As a result of the strong interrelation of disasters and development from DRR perspective, since the late 1990s, there has been increasing recognition of the need for mainstreaming disaster risk reduction into development (e.g. Bosher 2008, Benson and Twigg 2007; UNDP 2004; Wisner 2004).

In parallel, however, the reactive entry point of post-disaster management retains its importance because earthquakes and other extreme natural hazards do not wait until our cities get ready! Prevention following a proactive approach is better than a cure, but it has proved to be illusive in disaster research and response (Pelling 2003). In the 1990s over 90% of all funds for disaster mitigation went to disaster relief and reconstruction (Bender 1992). A similar pattern emerged from a joint review report by Kettlett and Caravani (2013) for the Global Facility for Disaster Reduction and Recovery (GFDRR) and Oversas Development Institute (ODI), in 2013. Although the trend for financial commitments related to disasters has been on the increase since the late 1990s, this is largely due to activities undertaken in the aftermath of events rather than by DRR, especially the emergency response. Disastrous events, mostly earthquakes, have been particularly important in the increase of overall disaster financing while also, due to the massive impact and media attention they have generated, often helping at least to put DRR on the agenda. With their very visible and sudden impact, such events generate significant attention, pushing up financing of both response and reconstruction activities. However, since 2003, financing of DRR has been roughly stable at about 10% of overall financing for disasters each year (Kettlett and Caravani 2013). Recovery and reconstruction form an important part of disaster management, which is concerned with the organisation and management of resources and responsibilities for dealing with all the humanitarian aspects of a disaster, including preparedness, and response and recovery in order to lessen the impact of future disasters (UNISDR 2013; IFRC n.d). Reflecting the advances in disaster studies, UNISDR defines recovery as '*The restoration and improvement where appropriate, of facilities, livelihoods and living conditions of disaster-affected communities, including efforts to reduce disaster risk factors*' (UNISDR 2009).

Such an approach provides longer-term prospects for recovery and puts a strategic emphasis on recovery for linking post-disaster rebuilding to development because the aim is that disaster-affected communities get back to their normal life with less vulnerability and more resilience (Wisner 2004). Parallel to shifts in the discourse of recovery and its links to development, the discourse of reconstruction broadened to reflect all the qualitative principles of successful disaster risk reduction, including, for example people's participation in recovery. The focus in reconstruction has shifted from replacing lost products into facilitating the process of producing products in order to avoid recreating the pre-existing situation and to improve the disaster resistance of housing stock (Spence 1992; Maskrey 1994). Reducing disaster risks for the future and a focus on the participation of people in the process of reconstruction are arguably two most commonly advocated strategies in reconstruction in recent years.

Current advances in disaster studies have raised expectations from reconstruction programmes, however, they, in turn, made organising such reconstruction programme more difficult. As Davis (2007) states, the advantages of a multidimensional recovery (physical, psychological, economic, social, and environmental) through a facilitative reconstruction, also entails multiplying complexities for organising reconstruction processes. In practice, however, practitioners and academics alike have reported reoccurring problems in reconstruction practice at various levels, from high-level policies to operational arrangements. Examples are exposed operational bottlenecks during implementations, imposing unforeseen setbacks and sudden changes in operations, already expressed by researchers and practitioners (e.g. Gaillard 2017; Lassa 2012; Lyons et al. 2010; Lizarralde et al. 2009; Lloyd Jones 2006; Özerdem and Jacoby 2006; Barakat 2003; Maskrey 1994; e.g. Lassa 2012; Comfort and Kapucu 2006; Haigh et al. 2006). Nevertheless there is optimism in the field. Although the realism of the full delivery of idealistic expectations has been questioned, there is an acknowledgment that some improvements by working within the existing systems are possible; failures in their delivery do not invalidate the causes (e.g. Christopolos et al. 2006; Wisner 2004). *It is rhetoric versus reality.*

There appears to be a gap in theory to help the practice of organising reconstruction programmes and processes, which are overarched by contemporary expectations from multidimensional recovery, for example between the discourse that says DRR, etc., should be the goal, and the ability, practically, to deliver on this goal. In parallel, the importance of organisational design is emerging within the reconstruction field (e.g. Lizarralde et al. 2009; Davidson 2009). Bridging this gap necessitates broadening our understandings of reconstruction towards incorporating an organisational and managerial body of knowledge and explore reconstruction programmes from the organisational dimension (Arefian 2010; Davidson2009). The most difficult task in post-disaster reconstruction is organising the necessary processes and procedures, particularly regarding the participants and their roles that makes organising reconstruction a design problem (Davidson 2009). Although the importance of organisational design is an emerging notion within reconstruction field by Davidson, the question of '*how*' to design post-disaster reconstruction programme organisations remains.

A brief example from Pakistan shows how interdependencies in high-level policies for reconstruction were the source of operational bottlenecks during implementation. Reconstruction in Pakistan, after the destructive earthquake on 05/10/2005 in the Kashmir region, incorporated people's participation and reducing future disaster risks. Reconstructed houses were required to comply with earthquake-resistant codes, which were to be introduced in the cities' urban development plans. Thus, each city's town plan was a blueprint for rebuilding the city. Everything involved in rebuilding the city first needed to be addressed in the city's long-term town plan. Whilst this town plan was under production in the aftermath, the financial help for people was attached to the stage of their housing reconstruction and their compliance with new codes (ERRA 2007). This created a locked triangulation since it tied housing reconstruction and introducing new building codes to the end of town planning studies; distributing part of financial assistance and tying the remaining financial assistance to the compliance of houses with the new building codes. During implementation, the existing bottlenecks were exposed. Significant delays and frustrations emerged (Integrated Regional Information Network—IRIN 2010; Quzai 2010; Integrated Regional Information Network—IRIN) n.d.). This led to a policy shift on paying the later financial instalments instead of completion of the roof to receiving approvals on drawings (ERRA 2007). Such delays and frustration indirectly encouraged informal housing and the policy shift, in turn, relaxed controls on the new houses' compliance with earthquake resistance codes (Quzai 2010).

At the operational level, complexities and chances of reoccurring problems in the field are even greater. Such a theoretical gap to help organising reconstruction activities means that organising reconstruction activities relies on developing know-how on a trial and error basis at local, national and international levels. Researchers and practitioners in various other reconstruction cases, such as reconstruction after the Haiti earthquake 2010, Indian Tsunami 2005, Gujarat earthquake 2001, the Marmara earthquake 1999, the Kobe earthquake 1995, and the Alto Mayo earthquake 1990, widely highlighted practical difficulties for organising operations and delivering reconstruction programmes in global practice. They relate such difficulties and problems to a variety of conclusions, namely, complex procedures of reconstruction programmes, which are multidisciplinary; multi-actor, and multi-objective (i.e. social, physical, financial); the large-scale nature of the programmes; post-traumatic uncertain and turbulent context in an aftermath; incapability of old relationships for exercising new approaches; lack of coordination between different actors; the reduction of mitigation paradigm to technical advice or implementing projects; lack of understanding between different disciplines; lack of strategy for housing, inadequate supply of material, labour and expertise to meet the reconstruction expectations; complex social, physical and political dilemmas in post-disaster situations (e.g. Global Facility for Disaster Reduction and Recovery 2016; Lassa 2012; Lloyd-Jones 2006; Ozerdem and Jacoby 2006; Davis 2007; Barakat 2003; Alexander 2002; Maskrey 1994; Johnson 2007; Comfort and Kapucu 2006; Le Masurier et al. 2006).

Such practical difficulties clearly indicate organisational aspects and '*how*' reconstruction activities are organised that are linked to the policy level—as shown in the case of Pakistan—or can be at an operational level for organising even well-intended programmes and implementing them towards their initial expectations. Even a structured approach and benefiting from experience does not necessarily promise the fluent delivery of reconstruction programmes towards its envisioned expectations in practice. This is especially a case in urban areas. The story of housing reconstruction in Bam, (the in-depth case of this research) shows that despite having reconstruction experience and attempts to approach the reconstruction systematically and purposefully, delivery of the reconstruction proved difficult as unforeseen problems emerged. After the destructive Bam earthquake on 26/12/2003, drawing on its previous experiences in reconstruction, Iran avoided the creation of the locked policy triangle that was an issue in Pakistan; and tried to address the reconstruction objectives systematically. They did this by introducing sets of practical considerations based on the objectives of the reconstruction programme: safeguarding the historical urban architectural identity; building earthquake-resistant buildings; and ensuring the mobilisation and participation of the beneficiaries. This delivery system was used for each of the 32,000 cases in Bam and its suburb Baravat for a participatory reconstruction. Thus employing the national capacity, the Bam case became an attempt to '*urbanise*' existing reconstruction know-how. Given the variety of expectations from reconstruction, it gathered a considerable number of participant organisations and reconstruction beneficiaries for administration, technical and financial assistance in the housing reconstruction programme. The Bam area was divided into operational and administrative zones with their allocated administrative and technical organisations to work with each other and the local people to undertake housing reconstruction. Despite this attempt to organise reconstruction systematically, in practice implementing the programme encountered unforeseen bottlenecks and problems, e.g. severe delays at one of the earlier stages of the implementation; reducing the initial architectural expectations; and emergent of abandoned structures and uncompleted houses at the end of the programme; and an imbalance in addressing the initial objectives. The programme evolved during implementation to address the emergent issues (Arefian 2016).

This book advances theoretical knowledge on this subject and contributes to bridging the previously identified gap between how reconstruction should be and how it is in practice. The evidence-based learning opportunities and the research propositions advance the practice theory knowledge contributes to bridging the aforementioned theoretical gap towards better organising reconstruction processes in practice. It also contributes to bridging the gap between theory and practice, given that the need of bridging the gap between theory and practice in the field of reconstruction is highlighted (Davidson 2009).

1.2 Systems and Urban Resilience to Disasters

The notion of resilience which is becoming increasingly used in policy, programming and thinking around DRR (alongside with climate change) is inherently linked with the notion of systems, rooted in the study of ecologic systems by Holling in 1973 (Bahadur et al. 2010; Petak 2002). Holling (1973) defined resilience as the capacity of any system to deal with external changes whilst maintaining its structure, functions and identity. In turn, the Oxford Dictionary (n.d.) defines a system as a complex whole, a set of things working together as parts of a mechanism or an interconnecting network. A system exists for a purpose and its whole is more than the sum of the system elements. Each part of the system is a system (subsystem) at a lower level, another whole, and based on the open systems theory, they interact with the system's contextual environment (Morgan 2006). As Bahadur et al. (2010) review, the widespread consensus that resilience involves the adaptation of multidisciplinary methods—as social and ecologic systems are highly integrated—acknowledges the need to employ instruments such as systems thinking. Thus, interactions between parts of the system are emphasised and the focus is on 'functioning' the whole system instead of 'fixed function' of each part (Martin-Breen and Anderies 2011).

According to UNISDR (2017), resilience to disasters is the 'ability of a system, community or society exposed to hazards to resist, absorb, accommodate, adapt to, transform and recover from the effects of a hazard in a timely and efficient manner, including through the preservation and restoration of its essential basic structures and functions through risk management'. As UNISDR (2009) clarifies the concept of resilience is relevant in both pre- and post-disaster situations. In recent years 'resilience' has been increasingly used to conceptualise the ideal characteristics of an urban system that can withstand natural hazard and the direct and indirect impacts of climate change (Johnson and Blackburn 2014). Bosher and Dainty (2006) explain that a resilient built environment is less exposed to disaster risks and recovers rapidly following a disaster event. They argue that 'the concept of "built-in resilience" is a key component of disaster risk reduction' (ibid., p. 4). In line with the proactive DRR approach, built-in resilience focuses on hazard mitigation and preparedness rather than post-disaster management (Coaffee cited in Bosher and Dainty 2006). Addressing the challenge involved in developing a more resilient built environment they highlight the need to develop 'multiple, mutually reinforcing strategies concurrently' for DRR and 'built-in' resilience if any real impact on resilience is to emerge. Thus, the built environment 'should be designed, located, built, operated and maintained in a way that maximises the ability of built assets, associated support systems (physical and institutional) and the people that reside or work within the built assets, to withstand, recover from, and mitigate for, the impacts of extreme natural and human-induced hazards' (Bosher 2008, p.13).

Resilient thinking in the context urban setting has advanced from engineering perspective towards socio-ecological perspective, recognising that the social systems—human settlements—are inextricably linked to each other and to the ecosystem they use and depend on (UN-Habitat 2017). To link the notion of disaster, resilience to the urban environment must take into account that an urban context means the existing concurrent urban systems and networks which their relations to each other might or might not be directly apparent. A city can be defined as a complex adaptive system within its region (Chelleri 2012). There are various elements in a city that when combined present qualities that may not be present individually. These urban systems are sociotechnical because they comprise technical elements, such as building and structures, and social elements, such as regulatory structures, rules and practices supporting interactions and relations and contain cultural codes and norms which also influence social behaviour. Changes in city are systemic that is change in one element of a system may induce changes in other elements and the sociotechnical system's behaviour can only be understood by looking at the entire system not its elements in isolation. Characteristics such as flexibility, redundancy, resourcefulness, safe failure, responsiveness, capacity to learn and dependency on local ecosystem are suggested as preliminary characteristics of urban systems those contribute to urban resilience (Silva et al. 2012). A resilient city requires DRR to known hazards and ultimately requires good city governance (Johnson and Blackburn 2014). The concept of resilience should be translated into the governance of urban systems in two directions: recovery (will be discussed in Chap. 3) that should act in the short term after a disaster, and the learning loop and subsystem transformations acting as milestones for any mid or long-term decision (Chelleri 2012).

Moreover, urban development is realised through projects and programmes which are introduced, formed and implemented for specific purposes. A city is a portfolio of various concurrent projects and programmes which has been formulated to meet their own purposes and that they may or may not overlap areas elements of the urban system. Projects and programmes are currently seen as temporary organisations (Hedeman and Heemst 2006) and in turn, organisations are sociotechnical systems. Thus, the delivery of those purposes temporary sociotechnical systems are formed within the bigger urban system. They bring various system elements or/and subsystems together in order to realise their envisioned purpose. Depending on the level of complexities managing the system might be difficult, (as it is the case in post-disaster reconstruction). Especially for urban built environment programmes as they involve many organisations and individuals, with increased levels of differentiation and interdependency that must be integrated towards achieving a strategy, a vision or defined objectives (Crawford et al. 2005; Davies and Mackenzie 2013), while taking note that we are talking about sociotechnical systems which bring technical systems that bring technical aspects of building and infrastructure together with regulatory processes and mechanisms, human relations, cultural norms and so on. Thus, systems approach must also take into account the relations between those system element/subsystems that corresponds with the 'business process' and its required alignment, For example, the process of urban housing development is, in fact, the system 'aligned process' that holds the whole system elements (e.g. people who want to develop their houses, architects and planners, municipalities, professional institutions and so on) together towards the delivery of the purpose of its related programme. As organisations are open sociotechnical systems, each part of their system is a system on its own that can be social or technical. The system dynamic includes relations/linkages between technical, social, managerial aspects of the system that here corresponds with the urban process of delivering urban housing. Any change in one element might induce a change in other elements or the way the entire system works.

1.3 Epistemological Positioning

In disaster studies historically there have been two dominant paradigms of defining the link between nature and society. They have emerged one after another, and provided epistemological perspectives for the field. The older behavioural paradigm perceived disasters as natural hazards, thus, scientific knowledge of the intrinsic characteristics of hazards was seen as the way forward of dealing with disaster causalities. Disasters were seen as 'acts of God'. Radical criticisms moved towards the structural paradigm, with advocates such as Kenneth Hewitt, Ben Wisner, Piers Blaikie, Terry Cannon and Ian Davis, suggested that while hazards are natural, disasters are man-made. The structural epistemological perspective linked disasters, society and development together and influenced further disaster studies in two distinct ways. First, it triggered the sphere of disaster risk reduction (Aragón-Durand 2009) and later resilience that are now at the centre of international attention by scholars, international agencies and NGOs. Second, it influenced disaster management by triggering a developmental approach towards recovery and reconstruction as a process of future disaster risk reduction. The twenty-first century started with the entitlement approach and saw attempts to refresh theoretical approaches to disaster-development studies, by interdisciplinary approaches by resorting to other advances in social science, e.g. feminism and management studies (Pelling 2003).

This research positions itself within the continuity of the recent interdisciplinary attempts in disaster-development studies and in particular reconstruction and recovery. At present interesting attempts are adopting an interdisciplinary approach to disaster from different angles. Research by Bosher (shadow organisations for resilience), or Pelling (adaptive capacity for climate change) are drawing on various aspects of management and organisation studies to try to explain the complexities of applying a structural approach in practice. In the case of post-disaster reconstruction Barakat (2003) highlights the need for organisational structure to bring all stakeholders together; and Lizarralde et al. (2009) stress the importance of organisation design for organising reconstruction activities and refers to reconstruction as a mega system that brings together sectoral layers of complexities. Davidson (2009) concludes that organising reconstruction is an organisational design problem. Here is where this research begins.

Additionally, the research should also be positioned within the organisation theory and management science. Organisations are open sociotechnical systems and management is an applied social science. A broad category in organisation theory, as Morgan (2006) reviews, is a mechanistic approach to organisations that see them as machines; this corresponds to the classical management approaches. He lists the pragmatic implications of the open systems approach to organisation theory. First, there is the ability to scan and sense changes in the task and contextual environment, able to bridge and manage critical boundaries and areas of interdependence, and capable of developing appropriate operational and strategic responses. The widespread interest in corporate strategy is the realisation that organisations must be sensitive to what is occurring in the world beyond. Second, systems are like Chinese boxes containing smaller size boxes: wholes within wholes. Organisations are systems that include subsystems, which are systems in their own right. The importance of managing such intra-and inter-organisational relations between these systems has been the interest of organisation theorists. The sociotechnical account of organisation extended to take account of relations between technical, social, managerial and strategic and environmental requirements. In this way ‘configuration of subsystems’ or ‘business processes’ as a way of managing the relationships between critical subsystems are important. In fact, everything depends on everything else. Thirdly, bringing in the concept of establishing ‘alignments’ between different system elements and subsystems helps to identify and eliminate potential dysfunctions because organisational subsystems should be aligned (Morgan 2006, pp. 34–39; Whitley 1984). The term ‘sociotechnical system’ was coined in the 1950s. This view is related to more recent explanations of human relations and recent advances in management. It is also expanded to address the relationships between technical, social, managerial and strategic aspects of an organisation (Morgan 2006).

The structural approach of linking disaster-development with society clearly manifests a constructionist ontological perspective. However, as Bryman (2008) explains there is a spectrum of approaches to constructionism. He states ‘The main concern of orientation here is the question of whether social entities can and should be considered objective entities that have a reality external to social actors, or whether they can and should be considered social constructions built up from the perceptions and actions of social actors…’.(Bryman 2008, p. 18). However, I acknowledge that a constructionist approach cannot be pushed to the extreme. Bryman (ibid.) explains that some scholars, e.g. Stauss et al. and Becker recognise that the constructionist position cannot be pushed to the extreme: it is necessary to appreciate that a social entity has a reality that 'persist and antedates the participation of particular people' and shapes their perspective, but it is not a static objective reality and it acts as a point of reference but it is always in the process of being formed.

1.4 Iran and the Strategic Case of the Bam Reconstruction

Iran is one of the most earthquake-prone countries in the world. Deadly earthquakes claimed more than 180,000 lives during the past century, and 35% of Asian earthquakes in the twentieth century occurred in Iran (Government of Iran 2005). For example, in 2003 alone 334 earthquakes were registered by the Iran Strong Motion Network (ISMN) at the Building and Housing Research Centre, BHRC (BHRC 2003). One of the most recent earthquakes that turned into a disaster was the 2003 Bam earthquake; the fourth in a row of disastrous earthquakes in recent decades that happened after Buyin Zahra in 1961 with 12,225 fatalities and 21,310 houses in need of reconstruction in 91 villages; Tabas in 1978 with 19,600 fatalities and 5,334 reconstructed houses in 90 villages and the small town of Tabas; and Manjil in 1990 with around 14,000 fatalities[1] and around 217,000 houses in 700 villages and 3 small towns required reconstruction (Akhoondi and Bahraini 2000; Ghafory-Ahtiany 1999).

Iran's experience in post-disaster reconstruction was enhanced by the lengthy large-scale post-war reconstruction experience of 3,891 villages and a number of cities in five provinces spanning 1352 kilometres from 1982 onwards (Amirahmadi 1987). Because of such extensive experience in reconstruction and organisational continuity, Iran was able to apply lessons learned from each disaster to the next one, leading to developing practical knowledge on managing the scale of the reconstruction, participating people and other aspects at the time of the Manjil earthquake in 1990. This resulted in the development of broad national policies for organising reconstruction that indicated the engagement of disaster-affected people in rebuilding their houses was vital for a successful reconstruction, as well as considering future disaster risk reduction when reconstructing houses. However, the

[1]In some research reports and papers, the number of fatalities is reported as 40,000; whereas in official organisations reports, e.g. Reports by the Housing Foundation of Islamic Republic (HFIR) and Building and Housing Research Centre (BHRC), the number is 14,000. Different interviewees at BHRC and HFIR explained that the number 40,000 was an error in early reports of the Manjil earthquake that has been repeated in a number of papers and documents produced by others (Akhoondi interview 2013; member of HFIR interview 2013; member of BHRC interview 2013). The issue of the overestimation of deaths often occurs and is reported in the Global Assessment Report (GAR) (UNISDR 2013).

Bam case—given its unprecedented characteristics—was different. The earthquake of 6.7 degrees in the mega Richter scale destroyed nearly 80% of the city's buildings, killed around 30,000 people and injured 25,000 in the city of Bam area of southeast Iran. The significant features of the earthquake included the shallowness of the earthquake epicentre and the severe destruction—the earthquake occurrence in a relatively large city with historical and cultural features, and there were heavy human losses and casualties in a concentrated zone (Joodi 2010). In fact, it was the first large-scale earthquake, after the Manjil one that had its epicentre underneath a relatively dense urban area, which was internationally famous for its ancient citadel (Arg-e-Bam), historical urban identity in the form of organic garden city characteristics and traditional urban architecture and elements. These led to debates on how reconstruction would address such characteristics and '*urbanise*' the existing know-how on rural reconstruction.

A combination of factors identifies the Bam case study as a strategic case for this research that is based on information-oriented and contemporary theoretical values (Flyvbjerg 2006; Yin 2009). They created an opportunity, during which, as Yin (2009) explains, an investigator has access to the situation for deeper analysis and observation, thus the single case study is worth to be conducted. First, objectives of the Bam housing reconstruction programme echo contemporary international reconstruction discussions about people's participation and reducing future disaster risk through reconstruction and in general in recovery. The Bam case, therefore, is a contemporary illustrative case and aligned with the current debates. Although it happened a year before the Indian Tsunami, the case sought to reduce future disaster risks, and acknowledged the importance of people's participation as one of its core principles. In addition to addressing the participation of beneficiaries and future disaster risk reduction, the Bam case also tried to address urban architectural identity in a historic city. The case presents great intrinsic complexity in the delivery of multidimensional integrated reconstruction. From this perspective, the case offers a laboratory with many learning opportunities, which might be useful for other international and national cases. Second, my professional involvement in the real-time phenomena of the case offered a rare opportunity with methodological value, helpful information and local networks, enabling to reach a deeper layer of understanding that is crucial for rich descriptions of qualitative research and theoretical generalisations. I was involved in post-disaster reconstruction of the Bam area, at the level of programme implementation from 2004 to 2007. This was within the capacity of a director of a local consultancy company for architectural and technical assistance for 2,100 locals to rebuild their houses, as well as project manager for a post-disaster streetscape urban design project and the reconstruction of the adjacent retail units. I visited Bam again in 2008 and 2010 to see how the reconstruction was progressing, and undertook the Ph.D. fieldworks in 2011 and 2013. The longevity of these experiences along with other involvements in the professional context in the country enabled me to achieve a deeper level of exploration and a deeper layer of understanding that is crucial for rich descriptions

Fig. 1.1 Location of the Bam area, source, United Nations, available from http://www.un.org/Depts/Cartographic/map/profile/iran.pdf, [Accessed 12/08/2015]

of qualitative research and theoretical generalisations. Third, a housing reconstruction programme is the first and most important platform for contemporary debates of people's participation in disaster studies and more than 50% of the reconstruction funds are spent on housing reconstruction programmes (Jha 2010). Fourth, building on previous experiences in reconstruction and developing know-how and practical knowledge, Iran attempted to apply a systemic approach to delivering its housing reconstruction programme. In my view, this systemic approach for undertaking urban housing reconstruction greatly adds to the instrumental value of the case. Understanding the case and its multidimensional complexities provide an illustrative example towards instrumental insights that can be theoretically generalised. It is helpful for other countries that experience more or less similarities, for example in having an organisation for reconstruction management. Figure 1.1 presents the geographical location of the Bam area.

1.5 The Overarching Approach and Analytical Pathway

This research is exploratory, theory building (inductive) and qualitative to allow deeper understanding, local contextualisation, causal inference and exposing different points of view. Given that available literature does not provide a conceptual framework, thus, the nature of this research was exploratory and through the provision of improved understanding it contributes to the generation of new knowledge and theories (Yin 2009 and 2003; Eisenhardt 1989). The context of selecting the research method is the existing theoretical in disaster literature about understanding the reality of organising reconstruction activities. The case study method is the appropriate research method if the research questions are 'how' or 'why' regarding a specific phenomenon (Yin 2009). The focus is on a holistic unit of analysis of the permanent housing reconstruction programme organisation in the Bam urban area that itself involved multiple participant organisations and beneficiaries. The housing reconstruction programme in the Bam area included 32,000 houses in the cities of Bam and Baravat (its suburban town), to be reconstructed individually. The unit of investigation is the multi-organisational system for delivery of the housing reconstruction programme, and my intention was to capture the wholeness of the system and all that belonged to it, including what influenced the system and what was influenced by it. The scope of this research is the sociotechnical multi-organisational system of the delivery of permanent housing reconstruction in the Bam urban area. I do not intend to provide a full survey of the case; and the research does not focus on any particular system element (i.e. participant organisations and beneficiaries).

Throughout this research, we see the overarching systems thinking approach within the light of strategic thinking. Morgan (2006) explains that open system epistemological advances led to advances in strategic thinking. Through the systems thinking approach, we can gain insight into the whole system through understanding linkages and interactions between system elements that are in place and related together towards the objectives of the system formation and implementation. The systems approach helps to handle the complexities and to support the decision-making process of many stakeholders. In a systems approach, each of these system elements work with each other towards the same goal, and their relationships and roles can be understood as part of the system (Lizarralde et al. 2009). Within the open system epistemology systems exist for a purpose; the system elements act and interact with each other towards a shared goal. I was aware that the system elements and their interactions are in place to contribute delivery of the system purpose, and the whole system is related and influenced by the broader local and national systems, defining the broader contexts. Lizarralde et al. (2009) in an attempt to link reconstruction and systems thinking suggest that since in reconstruction several sectoral layers of complexity work with each other, having some 'enriched system' approach allows them to be seen together.

This research, in fact, deals with this 'enriched system', within which all layers work with each other. This enriched system resembles the concept of 'mega system' by Davies et al. (2011) that mirrors the factual complexity in large-scale infrastructure construction activities, recently recognised as construction programmes. A mega system interrelates layers of complexity, and manifest itself in (multi) organisational design which defines the entities, their links, and process and embraces the cultural and behavioural aspects of people (or units or single organisations) who bring it to life. To deal with the 'enriched/mega' system of reconstruction programmes, I build on the notion of cascading which inherently linked with strategic thinking and systems approach (Armstrong 2009). The research conceptualises the (enriched) system and its layers of complexities according to strategic directions and objectives which must be achieved, therefore it explicitly connects sectoral layers of complexity, entities that are working together and understands the patterns. It is complemented with understanding interrelations between different existing patterns on different strategic directions and objectives at certain points of decision-making for the formation and implementation of the 'enriched system'. This supports the capturing 'wholeness' of a system; and it helps understanding the process through which the whole system is brought to life by people. The notion of cascading helps understanding the patterns within an 'enriched system'/'mega system', and supports decision-making for organisational design in a practical and effective way, respecting the strategic direction. It differentiates the systems needed for different strategic objectives, and therefore supports explicit mergers and combinations of such systems (creating a mega system) if there two or more strategic objectives are to be met. Akin to sectoral subsystems in these strategic objective-oriented systems the scope of decision-making and activities must be understood carefully. This creates a comprehensive matrix of understanding. Throughout the book 'system' refers to 'enriched/mega system'.

Given the aforementioned imbalance in disaster-development literature, the research starts with building a communication between organisation theory and disaster-development studies, offering a theoretical discussion and narrowing down to reconstruction. Theoretical discussions in Chap. 2 synthesise multidisciplinary literature to understand organisational characteristics of reconstruction activities that can be used as entry points to organisation theory and helpful for organising reconstruction. They offer insights towards understanding organisational aspects of reconstruction programmes. Discussions purely rely on multidisciplinary literature and various international empirical reports to propose a theoretical conceptual framework which is inductively related and positioned to the broader multidisciplinary literature on disaster-development and organisation theory. The framework integrates tentative insights to explore the case further. The case in turn verifies, polishes and develops the initial framework.

To understand the bigger picture both the initial formation of the delivery system and its adjustments during implementation have been analysed in this research. Organisations, as sociotechnical systems, are not frozen in time; and organisational charts and graphs do not necessarily project the way things are done in reality. They are dynamic not static. Such dynamism is reflected in formal or informal corrections and adjustments to the delivery system during the implementation. Adjustments to reconstruction programmes during implementation offer sharp entry points that help to explore deeper underlying dynamic contributors and their multidimensional interrelations during the stages of the delivery system formation and implementation. Those adjustments are corrective measures to deal with initial system formation issues because what is formed before the implementation requires further adjustments to deal with realities on the ground whether to be as a result of emerging problems, new requirements or changes in the broader sociopolitical and operational contexts. Cross-verification of information, qualitative coding and categorisation, and timeline were used to find entry points for the analysis.

The tentative theoretical and conceptual framework developed in Chap. 2 offered specific perspectives to analyse the case on both initial formation and adjustments in practice. To understand and analyse how the initial formation was conducted and implemented in practice, I tried to understand the complex situation of linkages that create a chain of potential influences that stipulates a complex chain of events over an extended period of time. Yin (2009) provides an example. The intervention in a programme could initially produce results and outcomes that these immediate outcomes themselves could produce intermediate results and in turn, these intermediate outcomes could influence the final results; these linkages can be qualitative and both linear and nonlinear. Given the interpretative nature of this research, I acknowledged that these logical chained interrelations explain the most likely contributors to phenomena rather than definite causal factors. I traced back and identified the potential linkages between how the system was formed with previous reconstruction (post-war and post-earthquake) experiences and existing policies in reconstruction, as well as the broader national context of urban development practice, socio-economic and specific characteristics of the case. I identified the adjustments that the system went through during the implementation of the programme. I identified potential drivers and triggers of that adjustment from one side and then analysed how those adjustments acted in turn as intermediary contributors to future implications and overall way the programme worked.

This approach linked back the case to the broader multidisciplinary literature. The fieldwork and multi-perspective analyses and examinations of the case showed emergent findings and revelations that, in turn, verified, polished, sharpened and focused the proposed conceptual model by taking it to a deeper level of exploration and understanding. Eisenhardt (1989) states that a prior specification of constructs is valuable to shape theory building research because if these construct are important as the case progress, then researchers have a firmer empirical grounding for developing theory. During the investigation, I compared the findings, which were unfamiliar to me prior to the fieldwork with the broader literature again. This inductive qualitative case study strategy not only helped to make sense of the phenomenon and to explore the focus subject towards theory development on organising reconstruction, it also paved the way to introduce new entry points to reconstruction studies that deserve further investigations per se.

A holistic approach to an investigation requires avoiding a one-dimensional picture. Therefore, the research brought out the plural knowledge, understanding and interpretations not only from organisations which formed the programme but also the ones which implemented them in the field and the beneficiaries who were impacted by them. In fact, this group included me too. Creating a collection of knowledge on the same reality from different perspectives, which are related to different roles and characteristics of various system elements, is a key strategy for understanding deeper influential interconnections and complexities. Those interactions do not immediately influence the programme performance but they do influence the performance through intermediary roles. Each interpretive perspective of participant organisations and locals (as system elements) sheds light on one or more dimensions. Such a strategy was not only effective for strengthening the credibility of the research, but it also helped me to understand the bigger picture instead of focusing on each individual system element. In total, 59 in-depth semi structure interviews with key and/or knowledgeable people covered the identified groups: (a) the case study sociotechnical system itself that encompassed organisations to formulate the programme; (b) the case study system elements (organisations and individuals) to implement the operational system for delivery of the reconstruction programme; and (c) other stakeholders related to the housing reconstruction programme in Bam, reconstruction and disaster management in the country and professionals active in housing development, including, participants in post-war reconstruction veteran professionals/academics.

Figure 1.2 schematically presents the analytical pathway of the case study research with regard to the stages which link with the previous literature review and theoretical discussions, a part of the iterative research design.

The outline of the text reflects the analytical pathway presented in Fig. 1.2 while keeping the storytelling order for readers. This book has eleven chapters, grouped into four parts based on their role in the research and storytelling. Part I consists of two chapters. It creates an entry point to the subject and the case, and provides a theoretical communication between organisation theory and disaster-development studies. Chapter 1 responds to the questions of *why*, *what* and *how,* positioning the research. Institutional and intrinsic values of the case are discussed, as well as the overall methodology and analytical pathway. Chapter 2 synthesises the multidisciplinary literature on reconstruction and recovery, as well as organisation theory. This part is concluded by introducing an inductive tentative and conceptual framework as a tool that integrates multiple lenses might be helpful for understanding the reality of organising reconstruction programmes. Part II consists of four chapters on the in-depth case study of the post-disaster housing reconstruction programme in Bam. It starts with Chapter 3 on understanding the institutional context at the national level, within which the reconstruction programme was conducted, and an overview of previous reconstruction experiences across the country. It is followed by Chap. 4 on contextualising Bam before, during and after the earthquake. It discusses the overall recovery approach and its links with development, and how the reconstruction programme introduced its objectives. Chapter 5 discusses how the reconstruction programme was planned and the practical considerations for the formation of the programme delivery system, as well as the introduction of the participant organisations. Chapter 6 identifies multiple adjustments made to the system. It discusses how the delivery system evolved and reached a level of maturity at a

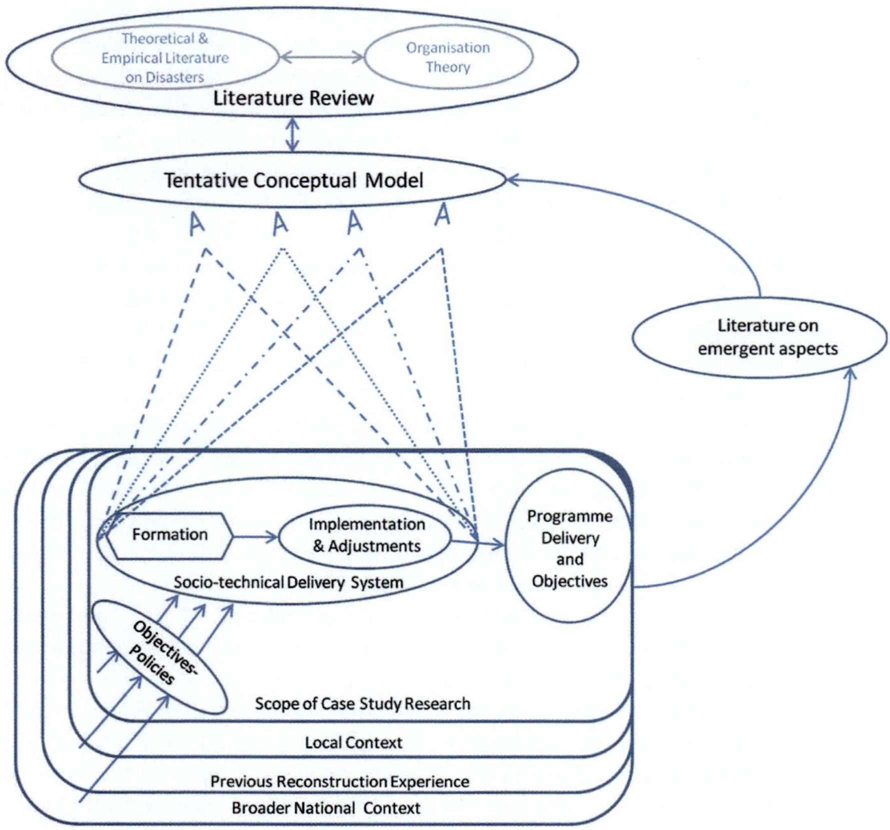

Fig. 1.2 Case study analytical pathway

certain point of the housing reconstruction implementation and its downsizing towards the end of the programme. Part III includes four exploratory analytical chapters. Each analytical chapter employs a perspective from the tentative model, developed in Part I, to explore, identify and analyse aspects and elements that influenced the system performance (through its formation and implementation) towards approaching the programme objectives. Chapter 7 analyses organisational configuration and hidden consistencies or inconsistencies related to the main identified organisational attributes. Chapter 8 analyses the strategic nature of the programme, starting with an overview of its objectives, which were approached through the programme delivery system. It then analyses the delivery system initially formed and its adjustments, including the introduction of objectives themselves and their relationships. Chapter 9 analyses the role of social constructs in the way the programme approached its objectives (through its delivery system). Chapter 9 analyses how the multi-organisational nature of the system influenced approaching strategic objectives through the system formation and adjustments. Part IV contains the final chapter. Chapter 11 offers concluding discussions which integrate the above-identified aspects and elements of the system formation and evolution. This chapter is the essence of the research and a synthesis of my findings. It marks the end of the book with recommendations on areas for further research that emerged from this research.

References

Akhoondi A, Bahraini H (2000) Reconstruction management of the disaster-affected areas; the experience of reconstruction of housing in the earthquake-affected areas of the Guilan and Zandjan provinces, Iran 1990–1992. The Institute of Publication and Print of the University of Tehran, Tehran (in Persian)

Alexander DE (2002) Principles of emergency planning and management, Illustrated edn. Oxford University Press, England

Amlrahmadl H (1987) Destuction and reconstruction: a strategy for the war-damaged areas of Iran. Disasters 11:134–147. https://doi.org/10.1111/j.1467-7717.1987.tb00628.x

Aragon-Durand FDJ (2009) Unpacking the social construction of 'Natural' disaster through policy discourses and institutional responses in Mexico: the case of Chalco Valley's Floods, state of Mexico. (Unpublished PhD thesis). Development Planning Unit, the Bartlett, University College London, London.

AON Benfield (2015) 2014 annual global climate and catastrophe report: impact forecasting. AON Benfield. http://thoughtleadership.aonbenfield.com/Documents/20150113_ab_if_annual_climate_catastrophe_report.pdf. Accessed 8 Sept 2017

AON Benfield (2017) 2016 Annual global climate and catastrophe report. AON Benfield. http://thoughtleadership.aonbenfield.com/Documents/20170117-ab-if-annual-climate-catastrophe-report.pdf. Accessed 8 Sept 2017

Arefian F (2010) Co-operative management knowledge and the inter-boundary collaboration in post-disaster reconstruction. In: TG63—Special Track 18th CIB World Building Congress May 2010 Salford, United Kingdom. Presented at the CIB World Cogress 2010, CIB TG63—Disasters and the Built Environment, Salford, UK, pp 145–159

Arefian FF (2016) Getting ready for urban reconstruction: organising housing reconstruction in Bam. In: Arefian FF, Moeini SHI (eds) Urban change in Iran: stories of rooted histories and ever-accelerating developments. The urban book series. Springer International Publishing, Cham, Heidelberg, New York, Dordrecht, London, pp 231–257

Armstrong M (2009) Armstrong's Handbook of management and leadership: a guide to managing for results, 2nd ed. Kogan.

Baas S, Ramasami S, Dey De Pryck J, Battista F (2008) Disaster risk management systems analysis: a guide book. Volume 13 of Environment and natural resources management. Food and Agriculture Organization of the United Nations, Rome

Bahadur AV, Ibrahim M, Tanner T (2010) Renaissance? Unpacking of resilience for tackling climate change and disasters. Institute of Development Studies, Brighton, UK. http://www.fsnnetwork.org/sites/default/files/ids_resilience-renaissance.pdf. Accessed 9 Feb 2015

Barakat S (2003) Housing reconstruction after conflict and disaster. Network paper, ODI, London. http://www.odihpn.org/documents%5Cnetworkpaper043.pdf. Accessed 23 Aug 2009

Bender S (1992) Disaster prevention and mitigation in latin America and the caribbean: notes on the decade of the 1990s. In: Aysan Y, Davis I (eds) Disasters and the Small Dwelling: perspectives for the UN IDNDR. James & James, pp 45–57

Benson C, Twigg J (2007) Tools for mainstreaming disaster risk reduction: guidance notes for development organisations. http://www.preventionweb.net/english/professional/publications/v.php?id=1066. Accessed 22 Nov 2010

BHRC, Catalogue of earthquake strong motion records (2003). http://www.bhrc.ac.ir/ismn/catalogue/pdf/2003.pdf. Accessed 17 Oct 2010

Blaikie P, Cannon T, Davis I, Wisner B (1994) At risk: natural hazards, people's vulnerability, and disasters. Routledge, London, New York

Bosher L (2008) Introduction: the need for built-in resilience. In: Bosher L (ed) Hazards and the built environment: attaining built-in resilience. Routledge, London, pp 3–20

Bosher L, Dainty A (2006) Disaster risk reduction and "built-in" resilience. Disasters: the Int J Disaster Stud Prac 35(1): 1–18. https://doi:10.1111/j.1467-7717.2010.01189.x

Bryman A (2008). Social research methods, 3rd ed. Oxford University Press, Oxford; New York.

Chelleri L (2012) From the "Resilient City" to urban resilience. A review essay on understanding and integrating the resilience perspective for urban systems. Documents d'Ànalisi Geogràfica 58(2):287–306. http://www.raco.cat/index.php/DocumentsAnalisi/article/viewFile/252964/3414

Christopolos I, ALNAP, Glemminge development research (2006) Provention consortium forum 2006. Strengthening global collaboration in disaster risk reduction. The elusive 'window of opportunity' for risk reduction in post-disaster recovery. ipcc-wg2.gov/njlite_download.php?id=5282. Accessed 20 Dec 2010

Comfort LK, Kapucu N (2006) Inter-organizational coordination in extreme events: the world trade center attacks, 11 September 2001. Nat Hazards 39(2):309–327. https://doi.org/10.1007/s11069-006-0030-x

Cuny F (1991) Review of twelve years' experience of disasters and small dwellings. In: Aysan Y, Davis I (eds) Disasters and the small dwelling: perspectives for the UN IDNDR. James & James, pp 23–29

Da Silva J, Kernaghan S, Luque A (2012). A Systems approach to meeting the challenges of urban climate change. Int J Urban Sustain Dev 4(2):125–145. http://www.tandfonline.com/doi/abs/10.1080/19463138.2012.718279. Accessed 2 Sep 2015

Davidson C (2009) Multi-actor arrangements and project management. In: Lizarralde G, Johnson C, Davidson C (eds) Rebuilding after disasters: from emergency to sustainability. Spon Press, London, New York

Davies A, Brady T, Prencipe A, Hobday M (2011) Innovation in complex products and systems: implications for project-based organizing. In: Cattani, G (ed) Project-based organizing and strategic management. Emerald, Bingley, UK pp 3–26

Davies A, Mackenzie I (2013) Project complexity and systems integration: constructing the London 2012 Olympics and Paralympics Games. Int J Project Manage 32(5):773–790. https://doi:10.1016/j.ijproman.2013.10.004

Davis I (2007) Learning from disaster recovery; guidance for decision makers. International recovery platform (IRP). http://www.undp.org.cu/crmi/docs/irp-decmakersleslearned-td-2006-en.pdf. Accessed 17 Nov 2010

Earthquake Reconstruction & Rehabilitation Authority, ERRA (2007). Urban development strategy, building back better planned cities. Government of Pakistan, Islam Abad

Eisenhardt KM (1989) Building theories from case study research. Acad Manage Rev 14(4): 532–550.

Flyvbjerg B (2006) Five misunderstandings about case-study research. Qual Inq 12, 219–245. https://doi:10.1177/1077800405284363

Gaillard JC (2017) Natural hazards and disasters. In: Richardson D, Castree N, Goodchild MF, Kobayashi A, Liu W, Marston RA (eds) International encyclopedia of geography: people, the earth, environment and technology. Wiley, Oxford, UK, pp 1–15. https://doi.org/10.1002/9781118786352.wbieg0774

Ghafory-Ashtiany M (1999) Rescue operation and reconstruction of recent earthquakes in Iran. Disaster Prev Manage 8(1):5–20

Global Facility for Disaster Reduction and Recovery, GFDRR (2016). What did we learn? the shelter response and housing recovery in the first two years after the 2010 haiti earthquake. Washington DC, USA. http://www.recoveryplatform.org/assets/publication/GFDRR/haiti-what-did-we-learn_GFDRR.pdf. Accessed 07 Aug 2017

Government of Iran (2005) National report of the islamic republic of iran on disaster reduction. Tehran http://www.unisdr.org/2005/mdgs-drr/national-reports/Iran-report.pdf. Accessed 22 Oct 2014

Guha-Sapir D, Hoyois Ph, Below R (2016) Annual disaster statistical review 2015: the numbers and trends. Brussels: CRED. http://cred.be/sites/default/files/ADSR_2015.pdf. Accessed 09 Aug 2017

Haigh R, Amaratunga D, Kerimanginaye K (2006) An exploration of the construction industry's role in disaster preparedness, response and recovery. In: The annual research conference of the Royal Institution of Chartered Surveyors. Presented at the COBRA 2006, The RICS, University College London, London

Hedeman B, van Heemst GV (2006) Programme management based on MSP an introduction. Van Haren Publication.

Holling CJ (1973) Resilience and stability of ecological systems. Annu Rev Ecol Syst 4:1–23. http://www.jstor.org/stable/2096802

Integrated Regional Information Network (IRIN) (2010) Pakistan: building back better in quake zone—News—Professional Resources—preventionweb.net (Website). http://www.preventionweb.net/english/professional/news/v.php?Id=13911. Accessed 03 Feb 2011

International Aid. ODI & GFDRR. London & Washington. http://www.odi.org/sites/odi.org.uk/files/odi-assets/publications-opinion-files/8574.pdf. Accessed 22 Jan 2015

International Federation of Red Cross and Red Crescent Societies (IFRC) (n.d.) About Disaster Management—IFRC (Website). URL http://www.ifrc.org/en/what-we-do/disaster-management/about-disaster-management/ Accessed 23 Jan 2015

International Federation of the Red Cross and Red Crescent Societies (IFRC) (2004) Chapter 4, Bam sends warning to reduce future earthquake risks. In: World disaster report 2004 focuses on community action. Report no. 7757. http://www.unisdr.org/eng/library/Literature/7757.pdf. Accessed 23 Nov 2010

Jha AK, Barenstein JD, Phelps PM, Pittat D, Sena S (2010) Chapter 13, Institutional options for reconstruction management, in: safer homes, stronger communities: a handbook for reconstructing after natural disasters. World Bank, New York, pp 197–207. http://www.housingreconstruction.org/housing/toc. Accessed 11 Jan 2011

Johnson C (2007) Strategic planning for post-disaster temporary housing. Disasters 31(4):435–458. https://doi:10.1111/j.1467-7717.2007.01018.x

Johnson C, Blackburn S (2014) Advocacy for urban resilience: UNISDR's making cities resilient campaign Env Urbanization 26(1):29–52. https://doi: 10.1177/0956247813518684.

Lassa JA (2012) Post-disaster governance, complexity and network theory: evidence from Aceh, indonesia after the Indian ocean tsunami 2004 (Working paper). Institute of Resource Governance and Social Change, IRGSC. http://www.irgsc.org/pubs/wp/IRGSCwp001complexity.pdf. Accessed 07 Aug 2017

Joodi M (2010) A brief review on bam reconstruction program. Presented in international recovery forum 2010, organised by International Recovery Platform (IRP), 16/01/2010, Kobe Japan. http://www.recoveryplatform.org/resources/meetings_and_trainings/514/international_recovery_forum_2010. Accessed 23 Dec 2010

Kellett J, Caravani A (2013) Financing disaster risk reduction; a 20 year story of international aid. ODI & GFDRR. London & Washington. http://www.odi.org/sites/odi.org.uk/files/odi-assets/publications-opinion-files/8574.pdf. Accessed 22 Jan 2015

Le Masurier J, Suzanne W, Shestakova Y (2006) An analysis of alliancing procurement method for reconstruction following an earthquake. In: Proceedings of the 8th national conference on earthquake engineering. Presented at the 8th national conference on earthquake engineering. San Francisco California, USA, paper 290. http://nisee.berkeley.edu/documents/elib/www/documents/200601/8NCEE/8NCEE-000290.pdf. Accessed 08 Dec 2014

Lizarralde G, Johnson C, Davidson C (2009) From complexity to strategic planning for sustainable reconstruction. In: Lizarralde G, Johnson C, Davidson C (eds) Rebuilding after disasters: from emergency to sustainability. Spon Press, London, New York

Lloyd-Jones T (2006) Mind the gap! post-disaster reconstruction and the transition from humanitarian relief. Royal Institution of Chartered Surveyors (RICS), London. http://www.rtpi.org.uk/media/3453/BEProfessionsGuide.pdf. Accessed 11 Nov 2010

Martin-Breen P, Anderies JM (2011) Resilience: a literature review. Bellagio Initiative. Brighton: Institute of Development Studies (IDS). http://opendocs.ids.ac.uk/opendocs/handle/123456789/3692. Accessed 9 Feb 2015

Maskrey A (1994) Disaster mitigation as a crisis of paradigms: reconstructing after the alto mayo earthquake, per. In: Varley A, Wiley J (ed) Disasters, development and environment. Chichester, New York

Morgan G (2006). Images of organization, updated edition (first edition 1997). Sage, Thousand Oaks.

Özerdem A, Jacoby T (2006) Disaster management and civil society: earthquake relief in Japan. I.B.Tauris, Turkey and India

Pearce L (2003) Disaster management and community planning, and public participation: how to achieve sustainable hazard mitigation. Nat Hazards 28(2–3):211–218. https://doi.org/10.1023/A:1022917721797

Pelling M (2003) Chapter 2, Cities as sites of disaster. In: Pelling M (ed) The vulnerability of cities: natural disasters and social resilience. Earthscan Publications, London, pp 19–45

Petak WJ (2002) Earthquake Resilience through Mitigation: a system approach. Proceedings of the Second Annual IIASA-DPRI Meeting 'INTEGRATED DISASTER RISK MANAGEMENT' Megacity Vulnerability and Resilience. July, 2002, Laxenburg, Austria, pp 29–31

Quzai U (2010) Pakistan: implementing people-centred reconstruction in urban and rural areas. In: Lyons M, Schilderman T, Boano C (eds) Building back better: delivering people-centred housing reconstruction at scale. Practical Action Pub, Warwickshire, UK, pp 113–134

Spence R (1992) Materials and construction techniques for disaster protection. In: Aysan Y, Davis I (eds) Disasters and the small dwelling: perspectives for the UN IDNDR. James & James, pp 36–44

UN (2010) Evaluation of UNDP contribution to disaster prevention and recovery; executive summary. www.undp.org/execbrd/word/dp2011-4.doc. Accessed 22 Jan 2011

UNDP (2004) Reducing disaster risk: a challenge for development. UNDP: bureau for crisis prevention and recovery, New York. http://www.undp.org/cpr/disred/rdr.htm. Accessed 24 Oct 2012

UN-Habitat (2017) Trends in urban resilience. Nairobi GPO Kenya: UN-Habitat. https://unhabitat.org/books/trends-in-urban-resilience-2017/. Accessed 8 July 2017

UNISDR (2005) World conference on disaster reduction, Kobe, Hyogo, Japan Hyogo framework for action 2005–2015: Building the resilience of nations and communities to disasters (No. A/CONF.206/6), World conference on disaster reduction. Kobe, Hyogo, Japan. http://www.unisdr.org/files/1037_hyogoframeworkforactionenglish.pdf. Accessed 22 Nov 2010

UNISDR (2009) UNISDR terminology on disaster risk reduction. UNISDR, Geneva, Switzerland. http://www.preventionweb.net/english/professional/terminology/v.php?Id=7823. Accessed 24 Sep 2011

UNISDR (2013) Global Assessment report on disaster risk reduction (GAR) 2013 (UN Report). http://www.preventionweb.net/english/hyogo/gar/2013/en/gar-pdf/GAR2013_EN.pdf. Accessed 20 May 2014

UNISDR (2015) Sendai framework for disaster risk reduction 2015–2030 (No. UNISDR/GE/2015-ICLUX EN5000 1st edition). Geneva, Switzerland. http://www.preventionweb.net/files/43291_sendaiframeworkfordrren.pdf. Accessed 20 Nov 2016

Whitley R (1984) The scientific status of management research as a practically-oriented social science. J Manage Stud 21:369–390. https://doi:10.1111/J.1467-6486.1984.Tb00234.X

Wisner B (2004) Chapter 9, Towards a safer environment. At risk: natural hazards, people's vulnerability and disasters. Routledge, London, pp 321–376

Yin RK (2009) Case study research: design and methods. Sage, London, New Delhi, Singapore.

Chapter 2
A Synthesise of Multidisciplinary Theoretical Discussions

Abstract In the absence of an existing theoretical model or point of reference on organising urban reconstruction processes, this chapter provides an exploratory and multidisciplinary theoretical discussion on the subject matter bringing together organisational and disaster-development discourses. It creates an iterative communication between '*what*' is expected and '*how*' to organise reconstruction towards delivery of such expectations. This communication between disaster-development studies and organisation theory also extends to construction management. Reconstruction and recovery are strategically and systemically linked. As the physical agent for multidimensional recovery and the process of the production of lost assets, reconstruction activities must integrate means and tools for the delivery of 'improvements' and expected objectives as part of recovery's big picture. This chapter identifies linkages and interactions among contemporary perceptions of disasters, post-disaster reconstruction, recovery and development, creating a baseline for understanding common contemporary expectations exist from reconstruction activities. The integration of disaster risk reduction measures and strategies as well as fit-to-purpose people's participation are the two most common expectations of reconstruction programmes. Reconstruction programme are positioned as sociotechnical systems within organisation theory. This enables further interdisciplinary analytical inductive discussions. This chapter extracts common reconstruction organisational characteristics that can be looked for in organisation theory. Respectively, multiple lenses to organisation theory that could be related and influential on organising reconstruction activities are explored. Each of them offers an insight that sheds light on reconstruction programme and its sociotechnical system and related process. The combination of those creates a tentative theoretical conceptual model as a point of reference for multiple insights that further will be applied to the case and will be verified and polished later.

© Springer International Publishing AG, part of Springer Nature 2018
F. F. Arefian, *Organising Post-Disaster Reconstruction Processes*,
The Urban Book Series, https://doi.org/10.1007/978-3-319-70911-6_2

2.1 Temporary Organisations and Management Knowledge Development

Current advances in interrelations of organisations, systems and strategic management theoretically connect reconstruction projects and programmes to the discourse of (temporary) organisations and sociotechnical systems. Therefore, the first thing on exploring '*how*' to '*organise*' reconstruction activities and processes is to understand those interrelations from the perspective of organisation theory. Organisations, which can be permanent or temporary, are open sociotechnical systems. As a result of open systems epistemological approaches to organisations, organisations are seen as sociotechnical systems. The term was coined in the 1950s and this view is related to more recent explanations of human relations and recent advances in management. It opposes a relatively rather old-fashion category in organisation theory that is a mechanistic approach to organisations that sees them as machines. Whilst the mechanistic approach to organisations corresponds to the classical management approaches, in further developments of a descriptive epistemological perspective of organisations, management is characterised as an applied social science (e.g. Morgan 2006; Hedeman and Heemst 2006; Whitley 1984). Pragmatic implications of the open systems approach to organisation theory are: *First*, the ability to scan and sense changes in the task and contextual environment, able to bridge and manage critical boundaries and areas of interdependence, and capable of developing appropriate operational and strategic responses. The widespread interest in corporate strategy is the realisation that organisations must be sensitive to what is occurring in the world beyond. *Second*, systems are like Chinese boxes containing smaller size boxes: wholes within wholes. Organisations are systems that include subsystems, which are systems in their own right. The importance of managing such intra- and inter-organisational relations between these systems has been the interest of organisation theorists. The sociotechnical account of organisation extended to take account of relations between technical, social, managerial and strategic and environmental requirements. In this way, 'configuration of subsystems' or 'business processes' as a way of managing the relationships between critical subsystems are important. In fact, everything depends on everything else. *Third,* bringing in the concept of establishing 'alignments' between different system elements and subsystems helps to identify and eliminate potential dysfunctions because organisational subsystems should be aligned (Morgan 2006, pp. 34–39).

The Oxford Dictionary (n.d.b) defines the word 'management' as *'The process of dealing with or controlling things or people'*. According to Wren (2005, p. 12), the word 'management' means 'the art of arranging physical and human resources towards purposeful ends'. Strategic management, as Armstrong (2009) explains, involves adopting a broader and/or longer-term view of what needs to be done and ensuring that the activities are carried out and contribute to achieving those strategic goals. The Oxford Dictionary defines the word 'strategic' '*Relating to the identification of long-term or overall aims and interests and the means of achieving them*'. It is derived from the word stratagem, originally denoting a military ploy in the fifteenth century and denotes '*a plan or scheme, especially one used to outwit an opponent or achieve an end*' (Oxford English Dictionary n.d.a). Management process therefore is a systematic way of doing things and achieving strategic goals and purposes requires strategic management, which in essence relies on strategic thinking (Armstrong 2009). Strategic management is now mostly accepted and employed in various organisations in public and private sectors and business administration because organisations must perform well in the present to succeed in the future. As he continues, organisations concurrently are concerned with the broader issues they are facing and the general directions in which they must go to deal with these issues and achieve longer-term objectives. As Hanaggan (2002) highlights, strategic management is about a sense of purpose, looking ahead, planning, positioning, strategic fit, leverage and stretching. It needs creativity and innovative thinking to make sense of organising, supervising and controlling. Strategic management is not strategic planning because it also deals with the implementation, although it may include planning. It consists of strategic analysis which is concerned with the expectation and purpose of an organisation, its resources and capabilities. It includes strategic choice, which is a question of considering options, evaluating and selecting. It also includes strategic implementation which is about organisational structure and design, resource allocation, control and managing strategic change.

The way management as a body of prescriptive knowledge supports strategic decision-making in practice has evolved in recent decades. Harpham (2009) reviews the background of this, and sets out that project management as an idea has been around for a long time but as the management science it can be traced back to 1950s and 1960s. From then, it evolved through cumulative development, for example, focusing on time span and project control by developing integrated, computer-based, management systems capable of integrating time, cost and quality (at least in theory). Following that the UK Association for Project Management (APM) and similar initiatives in Europe and the US listed competencies required of a good project manager. The concept of 'Management by Project' led to the recognition of the knowledge and skills needed for all general managers. During the late 1980s and early 1990s, the words 'Programme Management' entered the sphere of project management. The idea was to bridge the gap in organisations between the high-level strategy and projects. Programmes were seen as the layer that converted the strategy into a coherent set of projects to achieve the objectives, and to implement the strategy and the corporate vision. One of the first definitions of a programme was *'a set of related projects with a common strategic goal or aim'*. Simultaneously, the idea of benefits management came to a head when the focus of project (and programme) had become to deliver the benefits of the project (or programme) rather than the project (or programme) itself. Since the start of the twenty-first century, both programme management and benefit delivery are high in the management science discourse.

Modern project management as it has developed over the last 50–60 years is designed for complicated projects—rather than complex projects whose output is in the form of material artefacts, such as construction of physical infrastructure or the design and construction of an aircraft. However, emerging large-scale projects are complex and are usually described as 'megaprojects' in transportation, energy, water and telecommunication infrastructure, for example, the London Olympic and Paralympic 2012. Referred as construction programmes, such the built environment megaprojects are difficult to manage as they involve many organisations and individuals, with increased levels of differentiation and interdependency that must be integrated towards achieving a strategy, a vision or defined objectives. Interestingly, even early management projects initially considered as complicated, are becoming more complex and extend to new foci, a broader life cycle and new delivery mechanisms, such public–private partnerships, strategic alliances and others which replace traditional contractual arrangements (Crawford et al. 2006; Davies and Mackenzie 2013).

Three levels of implementation have been identified for the application of strategic management: the highest level is portfolio with long-term goals; the middle level is programmes with medium-term goals; and the lowest level is projects with short-term goals. The portfolio level deals with the selection, prioritisation and control of projects and programmes in line with its strategic objectives and capacity to deliver. Therefore, the purpose of programme management is to bridge the gap between corporate strategy (that should be reflected in portfolio) and projects, and to provide coordination and a common direction across projects when running complex organisations (e.g. Harpham 2009; Morris and Jamieson 2005; Pellegrinelli et al. 2011). Programme management does not operate in isolation though. According to Morris and Jamieson (2005), projects and programmes have often a two-way relationship with the corporate environment in which they evolve. Reporting an existing confusion in the literature (and variation in practice) about what is really involved in programme management, they state that most commentators define programme management as involving the management of a collection of interrelated projects to achieve strategic objectives and deal with change. According to Pellegrinelli et al. (2011), programmes are a relatively fresh idea in the management science discourse. They point out that programmes are the integral part of research in other organisational areas that require a further research agenda, including, organisation theory, organisational change, strategic management, leadership and competence. Moreover, such advances in this field of organisation and management studies have been informing sectoral areas of project/programme management, for example, in the field of development and construction management by researchers such as Morris and Jamieson (2005), and Davies and Mackenzie (2013).

In this context, projects and programmes are positioned as '*temporary organisations*', which connect series of activities and projects for the delivery of outcomes and objectives (Hedeman and Heemst 2006). Since organisations themselves are seen as sociotechnical systems, each project or programme requires a *temporary sociotechnical system* to connect actors and activities (and projects in case of a programme implementation) through which the project/programme will be delivered and the purpose of the project/programme implementation will be achieved. In this way, the notions of programmes and projects are related to the open system and sociotechnical account of organisations and organisation theory. The delivery system formation and configuration is related to the concept of organisation design.

Organisation design refers to the formation of an organisation, regardless of being a permanent or temporary one. It positions itself between planning and implementation (Goold 2002). Since reconstruction programmes bring various participant organisations together (Comfort and Kapucu 2006), the reconstruction programme organisation design refers to a sociotechnical system that includes a number of organisations, holds them together and integrates their activities towards the purpose of the programme. Davies et al. (2011) within construction management calls such systems for megaprojects (construction programme) as super systems that connect various different participant organisations, each one with its separate and different internal systems. Crucially, a complete organisation design must deal with both formal organisational structure and organisational connective process, and soft issues, including people and culture (Goold 2002). The formal structure includes the allocations of responsibilities and the reporting skeleton, creating the 'lines and boxes or organisational charts' of the typical organisation design. An organisational connective process is the mechanism through which the units relate to each other. It links relationships and collaboration between sister-units (lateral relationships). Through this connective process, the components of all organisations work with each other within a certain workflow (Goold 2002, p. 4). The formal organisational structure—skeleton and process—is brought to life by people in the organisation and is influenced by their behaviour, culture and values (Goold 2002). Advances in organisation theory and perceiving organisations as sociotechnical systems rest on the recognition of the influences the organisational society has on the way an organisation works (Morgan 2006). The Chartered Management Institute, CMI (n.d.b) defines organisational culture as the way things are done in an organisation, the unwritten rules that influence individual and group behaviour and attitudes. Organisational culture is defined and influenced by the structure of an organisation, the behaviour and attitudes of its personnel, its management and leadership style.

An organisation design approach to projects (in its broad meaning) characterises them as ad hoc (adhocracy) temporary organisations (Smyth and Morris 2007). As Hanisch and Wald (2012) review, for a long time and similar to permanent organisations in organisations theory, projects were studied as entities detached from their environment. Perceiving project and programmes as temporary organisations changed the approach towards emphasising the influence of context and contingencies in effectiveness of a temporary organisation. At present, there are interesting attempts to understand the formation of those temporary organisations within contemporary complex contexts of mega construction projects (construction programmes). For example, recently, Davies and Mackenzie (2013) pose a research question on 'How organisational structure and process be established to cope with a high degree of project complexity', for which they chose a single case study, the construction programme for the London 2012 Olympic and Paralympic Games. Acknowledging that there is no single best managerial strategy or magic managerial solution for every complex project, they found different levels of systems integration between and within individual component systems as a way to decompose a large level of complexities. They relate mega systems to the programme level that integrates various project systems together at lower level. They argue that the structure and process (organisation design indeed) is the systems integration.

Notably, organisation theory is plural in nature, so that we can find different entry points to understanding organisations and analysing them. Given the pluralism of organisation theory, Morgan (2006) suggests the application of various entry points for understanding organisations, so that multiple perspectives, chosen by the researcher based upon purpose, leads to a better understanding of organisations. Reflecting this on projects and programmes whilst addressing the contextual complexities, there are calls for the application of multiple perspectives of such temporary organisations (Hanisch and Wald 2012). The identification of required perspectives to understand projects and programme comes from an in-depth understanding of the overall field and subject of those programmes and projects towards extracting organisational characteristics which in turn will specify required entry points and perspectives to understand those organisations. For example, large-scale post-disaster reconstruction programmes, are to some extent different in nature than aforementioned high-profile construction programmes. In reality, such high-profile complex construction programmes take place in non-disastrous and non-traumatic situations and they are not participatory in form of engaging ordinary people in the delivery of the programme. Even if such construction programmes (megaprojects) are defined as public–private partnerships, the private sector side of those partnerships are usually large-scale construction companies. But a participatory reconstruction deals with beneficiaries as a component of the mega system of delivery within a complex post-traumatic disastrous context, which as mentioned by Comfort and Kapucu (2006), is an extreme event. Here is where we need the communication between organisation theory and disaster-development studies narrowing down to reconstruction.

2.2 The Contemporary Mission of Reconstruction

2.2.1 Shifting the Entry Point to Disasters

With growing international concern about the rising frequency and severity of natural hazards and disasters, there has been an increased drive internationally to reduce the destructive effects on the lives and livelihoods of individuals and communities. Thus, the discourse in the 1990s and early 2000s led to moving the trend of tackling disastrous incidents from the reactive act of managing disasters after their occurrence (disaster management, DM) to the proactive act of reducing the risk of their happening (disaster risk reduction, DRR) with increasing the resilience of the community, and shifting the entry point to tackle disasters from after to before (e.g. Gaillard 2017; UN 2010; Bosher 2008; Pearce 2003; UNDP 2004; IFRC 2004). In the pre-disaster stages, HFA 2005–2015 pursued the following expected outcome:

The substantial reduction of disaster losses, in lives and in the social, economic and environmental assets of communities and countries (UNISDR 2005, p. 3). The successor instrument of HFA is the Sendai Framework 2015–2030, an intergovernmental agreement which aims for the more inclusive following outcome: *The substantial reduction of disaster risk and losses in lives, livelihoods and health and in the economic, physical, social, cultural and environmental assets of persons, businesses, communities and countries* (UNISDR n.d.d; WCDRR n.d.).

The discourse of DRR is interrelated with overlapping notions of disaster, vulnerability, resilience and development. In principle, DRR should be an integrative process involving a large array of stakeholders, comprising both top-down and bottom-up actions (Gaillard 2017). Within the terminology prepared by the United Nations International Strategy for Disaster Risk Reduction (UNISDR 2009a, b), DRR refers to:

> The concept and practice of reducing disaster risks through systematic efforts to analyse and manage the causal factors of disasters, including through reduced exposure to hazards, lessened vulnerability of people and property, wise management of land and the environment, and improved preparedness for adverse events.

Within the 2017 updates on the terminology (UNISDR 2017), DRR refers to:

> Disaster risk reduction is aimed at preventing new and reducing existing disaster risk and managing residual risk, all of which contribute to strengthening resilience and therefore to the achievement of sustainable development.
>
> Annotation: Disaster risk reduction is the policy objective of disaster risk management, and its goals and objectives are defined in disaster risk reduction strategies and plans

Thus, the current concept of DRR in international agreements appears to advocate analysing and managing causal factors of disasters. For example, the Sendai Framework elaborates on a number of issues given the experience of HFA implementation and portrays Four Priorities for Action:

- Priority 1. Understanding disaster risk
- Priority 2. Strengthening disaster risk governance to manage disaster risk
- Priority 3. Investing in disaster risk reduction for resilience
- Priority 4. Enhancing disaster preparedness for effective response and to 'Build Back Better' in recovery, rehabilitation and reconstruction (UNISDR 2015, n.d.a).

In 1970s, distinction was posed between natural hazard and disaster, broadening the perspective from physical science to incorporate human systems and its relations with the natural and the built environment (e.g. Mileti, Drabek and Haas, cited in Wenger 1977). This shifted the focus from risk to vulnerability. Blaikie et al. (1994) declare that 'a disaster is an intersection of two opposing forces: those processes generating vulnerability on one side, and physical exposure to a hazard from the other', thus simply put, Risk = Hazards + Vulnerability. Disasters as both technical and social process/events are linked to society, environment and culture, and the definition of disaster, or the 'what' question, is highly related to the 'why' question and notion of vulnerability (Oliver-Smith 1998). With the emergence of the notion of vulnerability in relation to natural hazards, Blaikie et al. reintroduced human relation to disaster discourse and defined vulnerability as the following:

> By vulnerability we mean the characteristics of a person or group and their situation that influence their capacity to anticipate, cope with, resist and recover from the impact of a natural hazard (an extreme natural event or process).
>
> (Blaikie et al. 1994, p. 11)

The degree of vulnerability is determined by a combination of several factors, including awareness of natural hazards, the condition of human settlements and infrastructure, public policy and administration, and risk management, social capital and other related factors. The specific dimensions of social, economic and political vulnerability are also related to inequalities, to gender relations, economic patterns and ethnic or racial divisions (Wisner 2004). To UNDP 'Vulnerability is the concept that explains why, with a given level of physical exposure, people are more or less at risk' (UNDP 2004, pp. 30–31). The chain of explanations helps to understand progression of vulnerabilities as causes of disasters. It is reflected in the Pressure and Release model, PAR, introduced by Blaikie et al. (1994). They suggest that understanding pressures and releasing them proactively is needed to tackle effects of disasters. In other words, reducing disaster risk goes through reducing vulnerability (Blaikie et al. 1994). Figure 2.1 presents the PAR model.

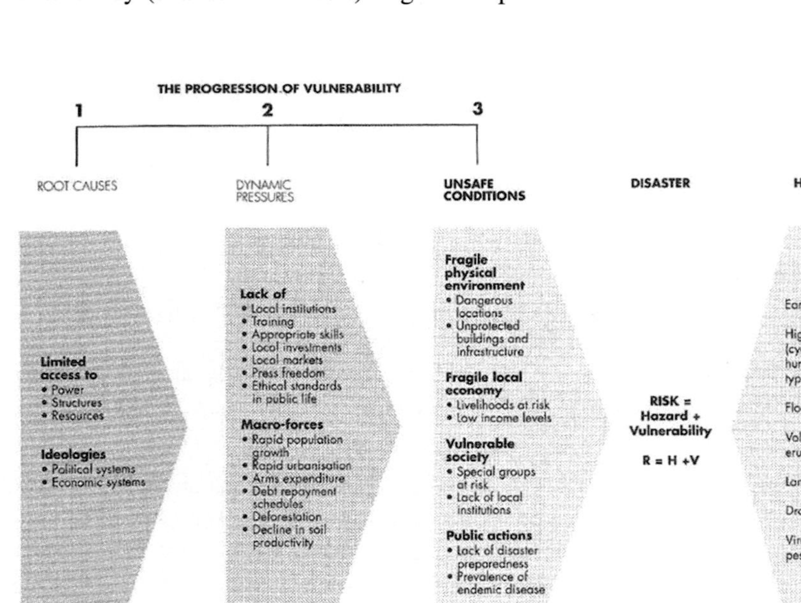

Fig. 2.1 PAR model: pressures that result in disasters: the progression of vulnerability (Wisner et al. 2004, p. 47) (Reprinted with permission)

There are strong interrelations of disasters and development efforts from the DRR perspective because disasters expose underlying vulnerabilities and are manifestations of failure of development programmes and activities. Development initiatives do not necessarily reduce vulnerability to natural hazards. Instead, they can unwittingly create new forms of vulnerability or exacerbate existing ones, sometimes with tragic consequences. From the perspective of vulnerability, urban areas regardless of their size and population are at risk (e.g. Bosher 2008; Twigg 2004; Benson and Twigg 2007; Wisner 2004; Pelling 2003a; Blaikie et al. 1994). Benson and Twigg (2007) highlight the increasing recognition of the need for mainstreaming disaster risk reduction into development from the late 1990s. To mainstream DRR, its strategies and measures should be integrated into the overall development framework and process. Therefore, 'disaster risk reduction as an integral component of the development process is a means rather than rather than an end in its own right'. (Gaillard 2017; Benson and Twigg 2007, pp. 5–6).

It is assumed that the strategic integration of disaster risk management within development planning can make a significant contribution to meeting the Millennium Development Goals—MDGs—and later Sustainable Development Goals—SDGs—(UN-Habitat 2017; UNDP Bureau for Crisis Prevention and Recovery 2004, p. 57). Hyogo Framework for Action (HFA) emphasises that disaster risk reduction, including reducing vulnerability to natural disasters, is an important cross-cutting element that contributes to the achievement of sustainable development (UNISRD 2005). At present, there are attempts to bring the notion of disaster risk reduction into development activities in urban areas at more local urban governance levels. For example, a campaigning initiative on urban risks has a focus area of 'invest wiser, build safer "The Making Cities Resilient:" My City is getting ready!' launched in May 2010 by UNISDR (UNISDR n.d.b).

Resonating Benson and Twigg's (2007) discussion on integrating DRR strategies and measures into the development process, Ofori (cited in Bosher and Dainty 2006) details components of the development process within which resilience as a key component of DRR must be built-in, including building regulation and development control, procurement practices, design processes, construction and the operation of the built facility. Bosher and Dainty (2006) stress the importance of developed or developing country contexts, for example, the existing operational frameworks, in this discourse. Broad principles can be useful as a point of departure for the development of context-sensitive resilience frameworks (ibid, p. 18).

However, disasters do not wait for the built environment and cities to get ready! The reactive approach to disasters, so-called disaster management, historically is a cycle that consists of four interrelated stages: response, recovery, mitigation and preparedness. In this approach, reconstruction is positioned from the later stages of recovery to the early stages of mitigation and can even take 10–25 years (e.g. Alexander 2002). A recent development of this is that these phases are not separated, they are undertaken concurrently, earlier activities influence later activities; there is a continuum of relief, rehabilitation and reconstruction (e.g. Davis 2007; Zetter and Boano 2010).

2.2.2 Recovery: The Linkage Between Post-Disaster and Development

Recovery (consequently reconstruction) is the meeting point of these proactive and reactive entry points to disasters. The current epistemological developments in disaster studies characterise recovery with a multiplicity of tasks and objectives, along with the discourse that linked disasters to vulnerability and development from the late 1970s, the perception of recovery changed from replacing lost assets into an opportunity for improvement: seeing recovery as a window of opportunity (Alexander 2002). Thus, if recovery aims to be anything more than the re-establishment of the previous conditions that made people, schools, hospital vulnerable in the first place, there must be a broader development vision. Continuing the previous business as usual will only reproduce the preconditions for yet more disasters (e.g. Cuny 1991; Alexander 2002; Susman et al.; Blaikie et al., Hewitt cited in Wisner 2004, p. 51). Following the same argument, in the 2017, updated version of the 2009 UNSDR terminology recovery is defined as

> The restoring or improving of livelihoods and health, as well as economic, physical, social, cultural and environmental assets, systems and activities, of a disaster-affected community or society, aligning with the principles of sustainable development and "build back better", to avoid or reduce future disaster risk.
>
> (UNSDR 2017)

This approach first provides longer-term prospects for recovery, putting strategic emphasis on recovery for linking the post-disaster situations to development because disaster-affected communities and the structures that supports them must be back to 'normal' but with less vulnerability and more resilience than they had before the occurrence of the disaster. Second, it broadens the dimensions of a desired recovery, making it multidimensional that goes beyond the physicality of the lost assets and integrates qualitative elements as well, e.g. psychological, social and economic, etc. Five aspects of a fully integrated recovery process: psycho-social, physical, economic and environmental recovery, as well as rebuilding the local administration and damaged local institutions are emphasised (e.g. Wisner 2004; Davis 2007; International Recovery Platform 2007; UNISDR n.d.c). Disasters expose failures in development but simultaneously open a window of opportunity, and although the realism of the full delivery of such idealistic expectations has been questioned there is acknowledgment that some improvements systems are possible (e.g. Alexander 2002; Christopolos et al. 2006; Wisner 2004). Some observations also suggest that the recovery operation occurs within a highly politicised environment (e.g. Quzai 2010; Duffiled cited in Brun and Lund 2008; Davis 2007). The multidisciplinary nature of desired recovery also connects with the UN Cluster Approach in disaster management that is sectoral division of responsibilities and accountability for international humanitarian organisations. The Cluster Approach addresses a broad spectrum of timing after a disaster, from emergency relief onwards, and it has defined sectoral themes which nevertheless are overlapping. The themes driven from the Cluster Approach themes, defined by Inter-Agency Standing Committee (IASC), the International Recovery Platform (n.d.) identifies the following themes for recovery: gender, governance, livelihoods, environment, shelter, infrastructure, health, psycho-social issues and climate change adaptation in recovery.

Thus, recovery by creating opportunities for improvement must create the linkage between post-disaster and development. It must provide necessary changes and improvements to improve *community resilience* for future hazards (Wisner 2004).

It is noteworthy that the word development, as used in development studies is represented as 'an economic, social, and political process, which results in a cumulative rise in the perceived standard of living for an increasing proportion of a population' (Hodder quoted in Pelling 2003c, p. 4). Thus, the question is how recovery activities, with limited resources, to raise the 'living standards' in a catastrophic post-disaster in the long-term? Some researchers argue that the word 'development' does not have a unique and specific meaning but has an ambiguous meaning instead (e.g. Wisner 2004; Vincent 2004). As Cowen and Shenton (cited in Vincent 2004, p. 112) explain the history of word development can be traced from its original English meaning as 'autonomous process of immanent change', that led to the newer meaning of a detectable process intended to address problems though immanent changes. It is, in fact, an improvement process that can be directed by the state, multilateral agencies and non-governmental organisations, communities, individuals, external agents or development facilitators (Vincent 2004). It appears that within disaster recovery studies, the word 'development' is associated with progressive process towards improvement by addressing the problems as Vincent emphasises. He makes the case that whilst in development studies the tendency gradually leaned towards far-reaching expectations from development, but the word development in essence means incremental improvement (ibid).

However, the multidimensional characteristics of recovery makes it full of complex social, physical and political dilemmas, e.g. speed versus safety, speed versus quality or speed versus participation in decision-making (Davis 2007). Despite such recognised challenges, a forward-looking contemporary approach puts strategic emphasis on recovery (at least in theory), and embracing multidimensional expectations. For example, attempts to improve social relations that define access to land, credit, information and power, with the endorsement and support of political consciousness and public awareness were influenced by the new-Marxist approach (Wisner 2004; Pelling 2003b). This brought utopian (if not revolutionary) approach to recovery that resembles the far-reaching expectations in development discussed previously. But idealistic expectations are challenged by a number of researchers, advocating a more practical and realistic approach and expectations from recovery. For example, Wisner (2004) explains that whilst in the 1990s, the phrase 'developmental recovery' was fashionable and emphasised but the meaning of development, it is associated with gradual improvements, and within each society, there is room for improvement. Christopolos et al. (2006) also argue that the assumed window of opportunity opened by a disaster is illusive and disaster is not a good entry point for the actual achievement of mainstreamed disaster risk reduction, although it can provide space for improvements. Wisner (2004) acknowledges that full delivery of various expectations is idealistic, but improvements are possible within societies by working within the existing systems and the failures in their delivery do not invalidate the intentions. Therefore, this is the realm of shared responsibility and marginal reforms.

2.2.3 Reconstruction of the Built Environment: Physical and Strategic Agent for Recovery

The built environment is the manifestation of permanent physical assets that are destroyed and lost as a result of the natural hazard. Reconstruction of the built environment is a constituent of recovery and as Johnson (2007) states it is a system that exists within a bigger system of recovery which has various dimensions, e.g. psycho-social, physical, economic, environmental and administrative. Although at first glance it appears to be the physical agent for recovery to replace the lost assets, but as Davis (2007) states, it is also perceived as facilitative of other aspects of recovery. Since1990s, the focus in reconstruction has been shifted from replacing lost products into facilitating the process of producing products, with focus on housing reconstruction. It is in parallel to the shifts in perception of disasters, development and recovery to incorporate areas for improvements (e.g. Cuny 1991; Spence 1991; Wisner 2004).

In recent decades, perceptions and definitions of reconstruction add non-physical expectations to reconstruction that blends the notions of reconstruction and recovery in a way that they are sometimes used interchangeably. For example, Kondo and Karatani (2016) use the term 'housing recovery' to imply the meaning that is not reconstruction as action but rebuilding which includes reconstruction process and decision-making. Dynes and Quarantelli (1989) highlighted the intertwined relations between reconstruction and recovery, by stating that reconstruction can only be understood in a larger context of recovery and that reconstruction affects recovery in turn; in fact, the physical rebuilding in reconstruction is a social process. Thus, reconstruction is expected to reflect all qualitative principles of successful disaster mitigation, such as people's participation (Maskrey 1994). Focusing on permanency and sustainability, reconstruction becomes a rebuilding measure which involves not only constructing physical structures (housing provision, utility and infrastructure) but also building the confidence, self-esteem, self-dependency, mutual support and mutual trust and, the rebuilding of communities (Delaney and Shrader cited in Thurairajah et al. 2008; Ozerden and Jacoby 2006). OCHA describes reconstruction as 'rebuilding of entire communities, including livelihoods, such that they are able to support themselves and have reduced vulnerability to future natural hazards'. There are reported empirical examples of reconstructions that contribute to recovery, not only physical recovery but also psycho-social recovery, strengthening local economy, people's livelihoods and environmental recovery (Davis 2007).

In parallel to the above, the concept of involving people in reconstruction of their houses and community emerged as a win–win scenario for all stakeholders whilst promoting non-physical dimensions of recovery. Called people's engagement or people's participation, it is advocated by NGOs, international agencies, researchers and practitioners. Such emphasis can be traced back to researchers such as Maskrey. Following examinations of his professional involvement with programmes developed in Peru to lessen the impact of natural disasters in 1989, he suggested that community participation is a key component in the success of such programmes. The self-reliant people in hazard-prone geographic areas are able to use resources and organisation to survive the worst effects of natural disasters. Thus, people's participation has a built-in empowering element, instead of creating dependency under relief programme approaches (Maskrey 1989). This will be discussed further.

Optimistic expectations from reconstruction are reflected in political promises made after a disastrous incident, of building better than before. For example, the former USA president called for 'bigger and better New Orleans' after the Hurricane Katrina in 2005 (Wilford 2008); the former president of Iran, Khatami, promised to build a 'stronger Bam' after the Bam earthquake in 2003 or Clinton, the former USA president, called for 'building back better' after the South Asian Tsunami in 2005. More recently, the Reconstruction Promotion Committee under a newly elected government after the Tohoku earthquake in 2011, called for a 'New Tohoku', stressing the need for a robust, resilient social infrastructure, able to mitigate future disasters and for self-reliant local societies, utilising regional and endogenous resources (Reconstruction Agency 2013, quoted in Dimmer 2014). Simultaneously, international agencies echo such a developmental approach. For example, UNDP with its presence in 166 countries declares that it promotes the development dimensions of post-disaster recovery, and the need to 'build back better' to reduce risk and vulnerability to future natural hazards (UN 2011, p. 7; UNDP Evaluation Office 2010). Such slogans for optimistic expectations from reconstruction, fit well in disaster struck places, as disaster-affected people has seen the disaster and have desires for improvement (Wilford 2008).

However, despite advances on 'what' is expected from reconstruction, organising the delivery of such expectations in practice has proved problematic. As International Recovery Platform, IRP (2007) reports, rapid and poorly considered reconstruction recreates the very conditions of vulnerability that expose people to the possibility of further losses in the future. Reconstruction is a complex process, often dogged by unforeseen setbacks (Alexander 2002). The existing gap in managerial and organisational aspects of reconstruction is also identified by some researches. For example, Lloyd-Jones (2006) by examining empirical cases of Indonesia (Aceh) and Sri Lanka after the Indian Ocean Tsunami in 2004 and the Northern Pakistan earthquake in 2005 identifies the organisational and managerial gap in dealing with reconstruction activities. As he observes, despite the improvements in the emergency response to disasters, permanent reconstruction is often inefficiently managed, uncoordinated and slow to get off the ground. Acknowledging that many issues of reconstruction and re-housing originate in factors outside the area of construction planning and management, he suggests shaping a more coherent and flexible framework for planning and managing reconstruction, supporting and making the best uses of local efforts and resources and targeted external inputs could achieve both speed and effectiveness (ibid). Another example is the retrospective analysis of reconstruction of Alto Mayo after the earthquake in 1990, in which Maskrey (1994) concludes that the idea of improvement in reconstruction cannot be reduced to provision of technical advice or implanting projects. He suggests it must also refer to the reconstruction of an institutional and sociopolitical framework that brings together the different actors, such as the state, people and their organisations, NGOs and funding agencies discover and establish new relationships in order to converge.

Turner (1972b) provides an interesting example of his professional experience in post-disaster housing reconstruction that reflected the difficulties of organising self-help housing methods. After the earthquake in Peru in 1957, and as an architect in the Office of Technical Assistance to the Popular Urbanizations of Arequipa (OAIA), he was engaged in the housing reconstruction programme of 150 houses in the Tiabaya village. In collaboration with local institutions, the programme was based on aided self-help people's participation to reduce reconstruction costs. The housing was to be structurally identical whilst many individual architectural varieties were possible. The owners were only charged with the cost of construction material. However, the complicated administrative process was conditioned by the geographic dispersal of the individual lots, the variety of lot shapes, sizes and slopes, as well as lack of control over participation and getting the participants to work in addition to a large number of interest groups. All that contributed to the emergence of problems and unnecessary sacrifices required by the system.

Urban reconstruction is even more challenging than rural reconstruction. To explain challenges in urban reconstruction, Shilderman (2010) states that regulatory frameworks in cities are more complicated and varied in their bureaucracy or efficiency. Quzai (2010) also concerned with the case of the Kashmir earthquake reconstruction in 2005 identifies dependency of the housing reconstruction on the development of municipal infrastructure. The latter would invariably be constructed by engaging contractors and lead to lack of efficiency and transparency, and issues regarding land ownership for urban reconstruction. Such intrinsic challenges make urban reconstruction potentially less attractive for humanitarian organisations and international agencies than rural reconstruction, as it was the case in Pakistan after the Kashmir earthquake in 2005. Cities are dynamic systems with multiple interconnected factors and population density must be taken into account for the purpose of reconstruction, which in itself entails complexities (Groupe URD 2011).

Recovery and reconstruction are strategically interconnected (Arefian 2016; GFDRR 2016). Spencer reports that until about 1970 the assumption was that reconstruction programmes should replace houses amounting to the lost ones. But, as a result of consensus on the reconstruction role to avoid recreating the pre-existing vulnerable situation of building (including houses) stock without changing the essential nature of buildings, the focus of reconstruction was shifted from replacing lost products into facilitating the process of producing products (Spence 1991). Parallel to shifts in the perception of recovery and its links to development, the perception of reconstruction broadened from replacing destroyed physical assets, to reflecting all qualitative principles of successful disaster mitigation (Maskrey 1994).

Within the disaster response community, the relation between recovery and reconstruction is very close, that, as Dynes and Quarantelli (1989) highlighted, they sometimes used interchangeably or always with together. The reason perhaps is that reconstruction is a big milestone in the recovery process through time. Reconstruction, within the disaster management cycle, starts within the stage of recovery onward (Alexander 2002). Although recovery is traditionally split into different phases: '*relief*', '*rehabilitation*' and '*reconstruction*', it gains greatly from being regarded as a fully integrated process. Thus, initial decisions made in the heat of a post-disaster activity may well have long-term consequences. The implications are for emergency service staff to become aware of their role and interdependence with the overall recovery system and for urban planning staff to recognise their links to initial decision making by other participants (Davis 2007; Wisner 2004, Vicent 2004). This implies the need for a strategic approach in the sense of convergence. Such a continuum shows that recovery starts from day one, includes reconstruction progressively over the time, and finally joins the usual development of a community.

The above is also linked with LRRD, which is based on the linkages between disasters and development, the important role of recovery for making things better than they were before, and the facilitative role of reconstruction The concept of a continuum in the disaster response community: LRRD (linking relief, rehabilitation and development) has become an increasingly important element of international development policy (Davis 2007; Lloyd-Jones 2006). This emphasises keeping the '*red thread*' in mind during a variety of short-term and long-term activities. Reconstruction manifests the strategic characteristics and importance of recovery in contributing towards a longer-term vision. There is a connectedness between short-term interventions (even at the stage of emergency relief) and longer-term goals. In practice, disaster recovery in areas prone to natural hazards (e.g. earthquake, floods or volcanic eruptions) must incorporate preventive measures to reduce the risk of other disasters in future. This means the need for 'reconstruction-plus' as the integral part and physical agent of disaster recovery. However, based on a number of international cases (e.g. Sri Lanka and Thailand after the Indian Ocean Tsunami 2004, Nahrin earthquake, Afghanistan 2002, Kashmir earthquake, Pakistan 2005), he argues that in practice, the effectiveness of reconstruction, within this idealised 'linking of relief, rehabilitation and development' is constrained by the lack of planning and targeted funding, and ineffective management (ibid).

This strategic and systematic interconnection, in fact, means that the path towards a fully integrated recovery embracing qualitative expectations, such as improvements, passes through a reconstruction. In fact, recovery lends its DNA. Reconstruction, as the process of delivering products is a strategic physical agent of an integrated recovery within which technological, administrative, social and financial incentives must work with one another to have a longer-term vision insight. Reconstruction, therefore, is a transformative stage that connects post-disaster to development.

2.3 Reconstruction Organisational Identity: Programmes Versus Projects

Reconstruction is the manifestation of physical recovery. Post-disaster reconstruction has been the subject of a significant body of research with particular emphasis on developing countries that may be less able to deal with the causes and impacts of disasters (Haigh et al. 2006). Contemporary discussions on reconstruction, including the definition by OCHA of 'rebuilding entire communities, including livelihoods, such that they are able to support themselves and reduce vulnerability to future natural hazards', indicate the functional nature of reconstruction. The 2017 update on the UNISDR terminology, refers to reconstruction as:

> The medium- and long-term rebuilding and sustainable restoration of resilient critical infrastructures, services, housing, facilities and livelihoods required for the full functioning of a community or a society affected by a disaster, aligning with the principles of sustainable development and "build back better", to avoid or reduce future disaster risk.
>
> (UNISDR 2 February 2017)

All such discussions deal with the mission of reconstruction which links it to a high level of recovery programme and the formulation of strategic objectives (the what). However, they do not help to understand the organisational nature of reconstruction. Hereby, I intend to unpack this discourse and look at the built environment reconstruction from another angle in order to position it in a more unambiguous way to be in relation to organisation theory.

Reconstruction must deliver its mission and duties and successfully implement its related strategies through groups of activities. Examples of the mission will be discussed later in this chapter. The first thing one must know of organising activities for such reconstruction is to understand the organisational identity of those activities, in other words, to position the activity—or the group of activities—within a range of organisational characteristics. Why is this an important question? Because understanding the position of activities helps to understand those activities from the perspective of organisational characteristics, behaviour and management needs that are useful for forming a fit-to-purpose organisation. How can we successfully organise and manage something without knowing its organisational identity and behaviour? Is it a single stand-alone activity or is it a group of activities? Is a group of activities in reconstruction a project? Is it a big project, a programme or anything else? It is important to understand what they are from a management point of view and to positioning the reconstruction endeavours accordingly is the basic prerequisite for unfolding its complexities strategically. The first step towards understanding reconstruction organisational complexity is to return to basics semantically.

At present, there exists some semantic confusion regarding organisational identity of activities in the reconstruction field. Reflecting the confusion previously mentioned by Morris and Jamieson (2005), the terms 'project' and 'programme' especially are subject to semantic confusion, and are sometimes used interchangeably. The everyday terminology for recognising the totality of reconstruction activities includes 'reconstruction project', 'reconstruction programme'. Examples of addressing housing reconstruction include 'housing project', 'the whole housing projects' and 'housing programme'. As Quarantelli (1995) points out, the sociology of science for developing the names of concepts moves from everyday terms to detailed distinctions to explain certain phenomena. Zetter and Boano (2010) also identified the semantic confusion between shelter and house, linking it to a greater level of inherent complexities in reconstruction.

Addressing this current semantic confusion on project and programme is important because it is not merely about names or labels. They directly conceptualise the behaviour of organisations from the perspective of system's theory to understand them. Based on supporting strategic thinking Hedeman and Heemst (2006, p. 29) provide basic definitions. A project is 'a series of connected activities within a temporary organisation for delivering a pre-defined outcome under set conditions'. Therefore, project management can be defined as: 'a co-ordinated organisation and management of a series of connected activities for delivering a pre-defined outcome'. A programme is 'a series of connected projects and activities within a temporary organisation for delivering one or more strategically important pre-defined objectives'. Thus, programme management is defined as 'the co-ordinated organisation, management and implementation of a series of connected projects and activities for realising one or more strategically important pre-defined objectives' (ibid, p. 34).

The built environment reconstruction as a whole corresponds with multi-sectorial physical recovery within the notion of disaster management, for example, reconstruction of hospitals and health-related facilities, educational amenities, roads and infrastructure, retail units and so on. It deals with a certain level of implementation of construction-related activities. In a city, it cross-cuts the broader existing urban system on a specific sector, for example, health, education, or housing at the very operational physical delivery of amenities and facilities. Similarly, it cross-cuts the sectorial division-based UN cluster system that has been in place for coordinating responsibilities and accountability for international humanitarian organisations. The delivery of those hospitals, schools, infrastructure, houses and so on, can include single projects or a programmes that are a number of connected projects and activities. Thus, city reconstruction as a whole and as a certain component of disaster management is a portfolio and collection of groups of activities (which can be projects or programmes) that are a collection of various construction projects and urban development programmes. This corresponds with the higher level of implementation of strategic management. Reconstruction programmes link various projects and bottom line tasks and incentives to achieve the related objectives and approach to the strategic goals. They correspond with the middle level of implementation in strategic management. For example, housing reconstruction for which one of the objectives may be people's participation is a collection of numerous small construction projects which may be tied to social incentives and financial support. They are directed by broader policies by means of specific mechanisms. All this happens at a certain time in a specific locality with its own social, political and economic, administrative and technological characteristics which form part of the broader regional or national characteristics of those matters in the broader context. Those characteristics either within or outside and around the projects and programmes influence them and are a source of uncertainty (Davies 2013; Morris and Geraldi 2011).

We previously discussed that three implementation levels and their strategic relations are linked with the success of strategic management and delivery of the end-goal objectives: portfolio (higher level with long-term goals), programmes (middle level with medium-term goals) and projects (lower level with short-term goals) Association for Project Management 2006). Given the above, the application of those implementation levels can also be applied on contemporary reconstruction. Figure 2.2 shows reconstruction activities and the relationship between project, programme and portfolio in reconstruction. The focus of this research is on reconstruction programmes that tie reconstruction projects and social, financial other tasks and incentives together in order to achieve strategic objectives.

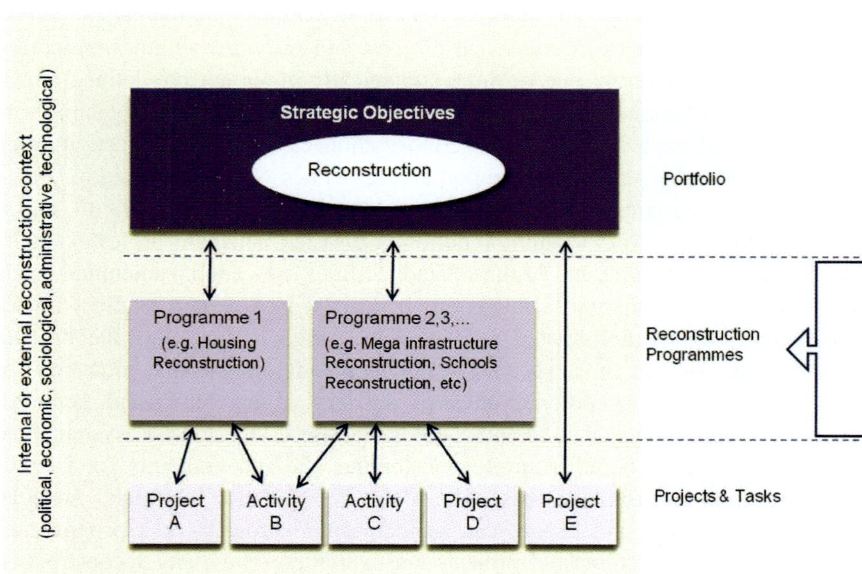

Fig. 2.2 Levels of activities in post-disaster reconstruction and the focus of this research

2.4 Objectives of Reconstruction: How is Progress Understood?

Objectives of a reconstruction programme set the expectations that the construction activities should fulfil within the overall recovery agenda. For example, housing reconstruction programmes, as a measure by which all qualitative expectations can be persuaded are purposeful, and seek contributions towards the bigger picture and an improved future. They aim to tackle specific challenges, most commonly to reduce risks of future disasters and to benefit from people's participation in order to contribute the psycho-social, physical and economic recovery (or other perceived objectives). The objectives of a reconstruction programme gain importance by outlining the expectations from the reconstruction programme as the added value to the technicality of the construction activities.

According to the Chartered Management Institute (CMI), which is a chartered professional body for managers in the UK, an *objective* is a statement which describes what an individual, team or organisation is hoping to achieve (CMI n.d.a). Objectives of a reconstruction programme act as the reference points for generating and directing policies, projects and tasks, and should meet some criteria. As CMI explains, objectives should be SMART, namely, Specific, Measurable, Achievable, Realistic. The acronym 'SMART' was coined by Drucker in 1954 whilst discussing 'objective-based management' that implies focusing on achieving the quantitative results by engaging the organisation team members into setting the objectives towards getting the results (CMI n.d.a; Management Study HQ n.d.). Since then, however, organisational theory and management studies advanced towards incorporating the human side of an organisation, seeing an organisation equally as a social system, that includes notions of motivation, behaviour and needs within organisation theory (e.g. McGregor 1966; Morgan 2006). However, as CMI (ibid) states, the importance of objectives and also the term SMART has moved beyond the boundaries of management for getting quantitative results and is now in common usage among managers who agree on objectives to pursue many different purposes.

In reading the literature on disaster, recovery and reconstruction, it appears that integrating measures and strategies for reducing future disaster risks and people's participation are the most common expectations in the contemporary discourse on post-disaster reconstruction. They seem to be must-have desirable strategies for the bigger system of recovery in all cases and therefore become the reconstruction objectives (e.g. Lizarralde et al. 2009; Lyons et al. 2010; Maskrey 1994; Özerdem and Jacoby 2006). Therefore, I explore these two kinds of objectives further to understand what expectations reconstruction must fulfil through organised activities, and how reconstruction process performance contributes to approaching and achieving the objectives, and the overall recovery vision?

2.4.1 *Fit-for-Purpose People's Participation*

Having local people as active participants in disaster recovery discourse, including reconstruction, grew quickly from early 1980s onward, with growing numbers of professionals in NGOs, local and central government and international agencies advocating this (Maskrey 2011). In 1994, the first World Conference on Disaster Risk Reduction in Yokohama officially recognised participatory approaches as an integral element in recovery. HFA identified community participation as a cross-cutting issue (UNISDR 2005). And the third World Conference on disaster risk reduction in Sendai, 2015, outlined participation of society as a whole, meaning non-discriminatory participation as a guiding principle within the suggested elements for the Post-2015 framework (United Nations Intergovernmental Preparatory Committee 2014). The potential advantages of people's participation in reconstruction cover a wide spectrum, such as follows.

- Enhancing disaster-affected people's psychological healing and overcoming their trauma in an aftermath (Jha et al. 2010a; Barakat 2003)
- The possibility of addressing the meaning of home for individuals and families (Zetter and Boano 2010)
- Addressing people's household needs and culture (assuming that communities know better of their requirements) and reducing the risk of refusal by people, prevent standardisation of houses that is a result of one-design-for-all without taking different needs into account (Jha et al. 2010b; Lizarralde et al. 2009)
- Sharing the risks of implementing the whole reconstruction process, e.g. delay, failure (Jha et al. 2010b)
- Potential cost reduction, and potentially maximising the variety of solutions for individual problems and minimising the dangers of centralised decision-making in the construction industry in an aftermath, which is largely characterised by high levels of uncertainty (Lizarralde et al. 2009)
- Communities to become agents of change for empowering themselves within the local system (Archer and Boonyabancha (2011).

The discourse on people's participation in reconstruction was fundamentally linked to the rationale behind the participatory development in other fields, such as housing (Maskrey 2011). In development studies, Turner in 1972 highlights the necessity of seeing housing as a process rather than a product and understanding people as the principal actor of the housing process. He argues that perceiving housing as a product to provide people's needs turns people into consumers and passive beneficiaries. In contrast, if housing is treated as a process, it contributes to personal fulfilment, assuming that fulfilment and maturity depend on personal responsibility for decision-making that shapes one's own life. However, perceiving housing as a process and means to human ends requires transferring decision-making power so that they remain in the hands of the users themselves. Turner advocates housing as an activity in which users are the principal actors as a matter of economic, social and psychological common sense. This does not mean that every family should build its own house, but rather that households should have greater choices and be free to choose their own housing and to build or direct its construction if they wish and manage it in their own way; and local decisions should be made locally. He argues that the best results are obtained by the user who is in control of the design, construction and management of his own home. He provides examples of lower income owner-builder cases in the USA that often achieved cost savings of 50% or more, whilst also addressing their non-financial needs and lifestyle (Turner 1972a). People's participation in development is also linked with the agenda for social inclusion in development, aiming to the inclusion of marginalised or excluded groups, e.g. women, poor, ethnic minorities and others (Isin and Wood cited in Hickey and Mohan 2004). Participation aims to provide the opportunity of practicing formulation and claims of new rights or to expand and maintain existing rights (Hickey and Mohan 2004).

There is, however, a criticism of the official recognition of participatory development that is applicable to reconstruction field as well. As Williams (2004) reviews, in participatory development, communities are treated as fixed, unproblematic and idealised; communities are homogenised and structural issues, such as micro-level problematic and repression is overlooked (e.g. gender, class and ethnicity). Kothari (cited in Williams 2004) argues that participation actively depoliticises development by incorporating marginalised individuals in development projects without giving them the possibility of questioning the development. Such official people's participation, therefore, denies the role of development experts in shaping the process of participation (Cooke cited in Williams 2004). Consequently, people's participation obscures the role of development agencies and the motivations of development workers; therefore, it bypasses important questions about the nature of management and excludes key aspects of the development process from public scrutiny. The recent efforts to address such criticisms led to linking participation to citizenship that broadens the participatory vision from specific rural/disadvantaged groups into a rights-and-obligations-based approach for every single household. Therefore, through this participatory approach, citizens become part of the machinery of the governance and exercise their right and obligations in a progressive process (Willliams 2004).

These general critiques of participatory development apply to a reconstruction context as well, since the responsibility of the reconstruction programme performance falls onto the shoulders of disaster-affected peoples who are beneficiaries of the reconstruction programme (e.g. housing). This obscures the procedural mechanisms that were employed to exercise specific kinds of beneficiaries' participation. Williams (2004) points out incorporating people and shifting them from 'beneficiaries' and 'objects' to 'empowered subjects' paves the way to displace the blame of the 'failure' of projects from macro-level on to 'the people'. In participatory development, people engage in a process which has already been established by others, who can be the state, multilateral agencies or NGOs (Vincent 2004). People are not usually involved in early decision-making processes for the formulation and formation of a programme; rather they are participants in implementing the programme. In the post-disaster reconstruction discourse, Williams' argument is echoed by Lizarralde et al. (2009) who argue that the ‘potential advantages of participation’ in reconstruction are realised through the broader project organisational framework within which disaster-affected people’s roles, responsibilities and the mechanisms of their engagement are defined in a rapidly changing environment.

Furthermore, for the purpose of organising reconstruction and the formation of reconstruction programmes, there is an ambiguity about ‘community participation’. They highlight the lack of specific use of the term ‘community participation’. They identify the lack of definition in ‘project environment’. The consequence is that even if organisations are willing to accomplish the goal of community participation for a specific purpose, only very few know how to do so within the specific structures of project-by-project interventions (Davidson et al. 2007).

The above arguments highlight the practical challenges in organising participatory reconstruction projects and programmes. Whilst excluding people (in general or a group of people) from the reconstruction process proved problematic, including them through participatory approach raises many questions on effective implementation. For example, what exactly is meant to be achieved through participation in reconstruction in a given context? At what point should participation be embedded into the reconstruction process? Is it at the strategic decisions for the formation of the programme or is it about their individual houses only? How much authority can disaster-affected people have within the process of (re)construction of their individual houses? Can they decide on material, design of their homes or timing of rebuilding their properties? And so on.

The above questions can be linked to different types of people’s participation and the level of control, powers and choice people have in a reconstruction programme formation and implementation. Arnstein (1969) offered a typology of people’s participation ranging from non-participation to full citizen control. At the bottom is manipulation which in fact is non-participation in either planning or implementing the programmes. In the middle is how people are informed about their rights, responsibilities and options, an important step towards legitimate people’s participation. Nevertheless, on the whole, people have little opportunity to influence the programme that is ‘designed for their benefits’. At the level of consultation, people hear and are heard but they lack the power to insure that their views influence the programme. At the higher levels of partnership and citizen control, citizens have decision-making and negotiation power over a particular programme. The highest level consists of citizen control for which people govern a programme and have full managerial power over the programme. Building on Arnstein’s typology, Davidson et al. (2007) categorise people’s level of participation based on the level of control they have in the reconstruction programme. As they state, people participation in a reconstruction programme covers a wide spectrum and requires various approaches within organisation design to involve the community within the whole housing project (programme). At the top of the ladder are empowered people, who have important decision-making roles that resemble Archer and Boonyabancha’s (2011) suggestion on facilitating, whereby people gather around their similarities in groups, make decisions on initiatives, plan and implement them. At the bottom is manipulated people within housing reconstruction programme formation and implementation, in which people do not have any decision-making power. In the middle of these two extremes, there are other levels of participation with various levels of decision-making power in a housing reconstructing programme, such as collaboration, consultation and information (Davidson et al. 2007).

Such diverse perceptions of participation, combined with the cultural, social, traditional and political conditions before a disaster, create enormous possibilities for their application. This makes it almost impossible to propose a single theoretical model for the best level of participation. Davidson et al. (2007) challenged the usefulness of both terms of 'community' and 'participation' for the purpose of organising housing reconstruction programmes because of the following:

> The idea of community participation has been so widely expressed that it does not seem to mean anything clear anymore. The term "Community" has been—often arbitrarily—used to refer to a neighbourhood, a slum, a group of local NGOs, a group of militant leaders, the residents of a small town, a workers' union, a group of women, etc. In this sense, the term neither denotes what this group of people really have in common nor their differences. The term "Participation" is also randomly used to denote civil debate and communication, consultation, delegation of activities, partnership, self-help construction, communal meetings, political decentralization, etc. Probably the main difficulty in the application of this concept is that community participation has not been defined in terms of what it means in a project environment. This is crucial because housing in general and housing reconstruction in particular are carried out in a project-by-project mode. The consequence is that organizations would like to accomplish the goal of community participation, but very few know how to do so within the specific strictures of project-by-project interventions.
>
> (Davidson et al. 2007, p. 102)

Drawing on empirical evidence from four housing reconstruction case studies (i.e. the rural area of Eje cafetero in Colombia after the earthquake 1999; the case of La Hermandad after the earthquake in El Salvador after the earthquake 2001 for 3,000 houses; the Marmara reconstruction in Turkey after earthquake in 1999; and the rural area of Cankiri reconstruction in Turkey after the earthquake in 2000 for 1,892 houses), Davidson et al. (2007) conclude that there is no single universal 'best' approach because reconstruction is rooted in communities' socio-politico-economic contexts and the project-related dimension of both 'community' and 'participation' is neglected in the current literature. They identify a big gap between practice and theory, and it conclude that it is impossible to propose a single theoretical model for participation and to imply that it results in better projects because there are as many notions of what is *better* as there are participants and there are contexts.

Given the above discussions, we can conclude that although the exercise of people's participation has its own merit in development in general, in the reconstruction field it will be a confusing wishful approach if it is not supported by (a) a clear understanding of the purposes of their participation, and (b) facilitating it within organising the housing reconstruction programme. These are required characteristics of people's participation as an objective that contributes to an integrated recovery/longer-term vision. People's participation in reconstruction must contribute to the realisation of the potential advantages sought for the participation at the first place, e.g. speeding psycho-social recovery, cultural acceptance and avoiding 'one-size-fits-all' solutions. People's participation has a purpose that is stated or meant by the objective of people's participation. Therefore, 'fit-for-purpose' people's participation is what reconstruction activities are seeking, not broad misled participation merely for the sake of participation itself. Crucially, the broader organisational framework and the operational mechanisms of engaging people is a key aspect of exercising fit-for-purpose participation. It is the organisational courier for the delivery of such potential advantages meant by the objective. The above arguments are fundamental foundations for any discussion of this apparent win–win situation in a given context. They are interlocked with their political, social and traditional context.

- **Population, Culture and Lifestyle**

Addressing cultural and non-physical losses in a disaster, Boano (2007) and Wilford (2008) argue that physical assets of place are not merely physical assets because they do have meaning and identity. Thus, the collapse of the physical world, houses, shops, offices, infrastructure, is not merely the collapse of their materiality and economic assets since it entails cultural loss and wiping out the meaning and identity. Given home as an example, they are 'nodal points' of human lives; significant types of place that have socialcultural and emotional importance. As Nine (2017) explains, the home plays a role in the development of a person's memories, beliefs and values through the extended cognitive functions and the physicality of the home is influenced by those and is also influential. Homemaking against housing (re)construction, where housing reconstruction refers to the materiality of houses, the sense of belonging, non-material and symbolic dimensions make it a homemaking process (Boano 2007). In the theory of materiality, according to Wilford (2008), the material world is integral and intimately consistent with the human world. Therefore, meaning and material bring each other into existence. The physical world and the human world cooperate in the construction of meaning. A core concept in materiality is that meaning requires participation in the material world, individually and socially.

Wilford (2008) also argues that the study of the cultural dimensions of natural disasters must address the human world, the physical world and culture. He employs the concept of material culture from materiality theory to link the human world and the physical world. The physical world plays an integral role in constructing and creating meaning out of natural disasters. Disaster-affected people's participation is a core concept on achieving recreation of meaning, places and spaces. It is a way of integrating the human world with the physical world that leads to the recreation of meaning that was destroyed or damaged by the natural hazard (Boano 2007; Zetter and Boano 2010; Wilford 2008).

Whether to be intentional or not the lack of adequate attention to the importance of cultural factors severely reduces the effectiveness of post-disaster reconstruction activities (Lin and Lin 2016). There is a strong technocratic bias in reconstruction and an emphasis on technically 'safe' housing without certainty that such housing is culturally acceptable or affordable. Such technical bias is especially the case in large-scale programmes where external technical consultants may influence technology choices by people. It can be also a reflection of a pre-disaster situation where, local building knowledge is often devalued by outsiders—and indeed by local people, who prefer 'modern' building styles as symbols of development, and believe that they are more secure against natural hazards (Twigg et al. 2006).

Therefore, the built environment reconstruction is not just about reconstructing the materiality of houses or shops; it is enabling the recreation of space and place with meaning and identity within which considerations such as lifestyle, social norms and values are intertwined. We could see people's participation as a means to enable such a recreation of places and spaces (from technology choice and architectural design to creating meaning and identity) since it is integrating the human world with the physical world to recreate meaning that was lost or damaged (Wilford 2008; Boano 2007; Zetter and Boano 2010). People's participation is important in such aspects: reproduction of meaning, responsiveness of the new houses to the physical and sociocultural landscape, needs of their lifestyles, acceptance of the reconstructed houses, as well as supporting the healing process of the aftershock trauma (Boano 2007; Vincent 2004; Wisner 2004). Such conceptual importance of the sociocultural implications of people's participation in reconstruction activities interrelates with organising reconstruction activities that shape the procedural/operational mechanisms of realising people's participation.

- **Material Culture and Reconstruction in Historic Cities**

The notion of material culture directly addresses cultural heritage in the context of the built environment and historic values that can be in form of large-scale destruction of urban centres, severe destruction of significant or monumental buildings, landscapes and archaeological sites, creates a sudden deep interruption in the historic continuity of in a city's life cycle. This influences the sense of identity, belonging and the overall well-being of the city's population. It has raised the question of how to address cultural heritage during recovery and reconstruction. In parallel, there is an emerging international agenda with regard to the protection of cultural heritage. Recent disasters of both natural and human origins highlight the escalating risks and destruction of historic cities, for example: the 2015 Kathmandu earthquake in Nepal, the 2009 L'Aquila earthquake in Italy and the 2003 Bam earthquake in Iran in addition to damages to historic cities during conflicts, for example, damages to cultural heritage in Afghanistan and Syria.

At present, there is an emerging attention to the urban-related cultural heritage during reconstruction. With the growing number of historic cities in the Middle East and Central Asia that require urban reconstruction, UNESCO organised a multidisciplinary seminar on Post-Conflict Reconstruction in the Middle East Context 18–19/06/2015). The issue of post-crisis reconstruction has become an urgent and pressing issue for ICOMOS, an international NGO dedicated towards the protection and management of cultural heritage, with an initial remit of archaeological monuments and sites (ICOMOS 2014, 2016). ICOMOS' international scientific committee on risk preparedness, ICORP, is active and at present, the world heritage system already confronted post-crisis reconstruction for architectural and archaeological monuments (Kono and Misako 2016; ICORP n.d.). However, as stated by the president of ICORP beyond monuments, the challenge exists on traditional buildings and historic city centres which had specific characteristics, from morphology to architectural typology (Jigyasu 2016). Recently, ICOMOS organised a colloquium on Post-trauma Reconstruction (4/03/2016), stating there is a need for a comprehensive overview of the current state of events and discussions. The need remains for theoretical guidance, methodologies and operational frameworks through research and multidisciplinary cooperation between many different actors in urban contexts (Jigyasu 2016; Rössler 2016).

Theoretically, reconstruction activities must take into account how to deal with historic built environment, whilst dealing with urgent needs of disaster-affected population and complexities of reconstruction that is widely acknowledged. However, in practice, this adds further expectations from reconstruction activities on '*what*' to achieve whilst the main issue if '*how*' to achieve that still remains problematic. Different cases show that organising reconstruction in historic contexts in practice to operationalise strategies for safeguarding historic urban identity remains problematic, such as international and national coordination during the recovery phase, short timeframes for decision-making, involvement of the private sector and development pressures, lack of documentation to guide restoration and reconstruction choices (Jigsayu 2016; Arefian 2016; UNESCO 2015; Kuceravcova and Dzurdzenik 2016). Urban reconstruction—regardless of whether it happens in a historic or modern city—is more complex and uncertain than dealing with the same crisis in rural areas (Zargar 2004; Quzai 2010; Shilderman 2010; Johnson and Blackburn 2014). It overlaps with urban development planning, regulations and processes. An urban context means the existing concurrent urban systems and networks such as social, economic, environmental, climatic, physical cultural and governance/administration that their relations to each other might or might not be directly apparent (Chelleri 2012). This research enables a better understanding of the underlying dynamics of paying attention to urban cultural identity and continuity during city reconstruction process in general beyond repair of monumental archaeological and architectural buildings.

2.4.2 Facilitating Future Disaster Risk Reduction During Reconstruction

Whilst mainstreaming DRR is expected to be the way forward in a proactive approach, its practicality/realism is challenged when it is tried in post-disaster reconstruction. There is little available evidence of post-disaster reconstructions which in practice significantly reduced the risk of future disasters (Gaillard 2017).

Such expectations are linked with disaster discourses in 1990s, as Wisner (2004) reports the words 'holistic', 'comprehensive' or 'developmental' recovery were prominent in the discourse, which in principle had to address economic, political and social needs, in addition to rebuilding infrastructure and housing and opening a way to more resilient livelihoods. Thus, in practice, it should reverse the dynamic pressure and root causes that contributed to the disaster in the first place. Christopolos et al. (2006) also argue post-disaster is not a good entry point for mainstreamed DRR because of the pressing multiple demands during reconstruction. Despite deep complexities in addressing dynamic pressures and root causes of disasters in reconstruction researchers agree that slower and smaller reforms and changes can be effective (Oslon and Gawronski cited in Wisner 2004) and potentials for improvement towards mainstreamed full DRR exist (Christopolos et al. 2006). This pragmatic approach echoes the progressive version of development as a long-term vision for local people. Reconstruction, therefore, can be a starting point for progressive positive changes, such as introducing new regulations, techniques or construction processes, even if such positive changes are small. It is a physical opportunity supported by a collective mindset for introducing changes in structural and non-structural risk reduction elements that need to be mainstreamed into the central flow of government policies and planning, and disaster recovery may provide a catalyst for such changes. Careful reconstruction planning, balancing change and continuity, and integrating bottom-up and top-down actions towards DRR (Gaillard 2017; Davis 2007).

In-depth vulnerability analyses to reduce disaster risk and increase resilience, alongside exploring a diverse range of causal factors and dynamic pressures from an organisational perspective entail examining them against the existing capacity of change. The question arises whether the introduced changes are achievable. As Benson and Twigg (2007) point out, DRR is not a product by itself; it is a process through which DRR measures are integrated with development activities. Thus, what happens in reconstruction in this regard is to integrate newly introduced changes in construction processes, regulations or techniques into the process of reconstruction activities at all levels. But how can any change (optimistically improvement) be understood for integrated DRR measures and strategies as a result of the reconstruction activities? According to UN-Habitat (2017) global progress on the implementation of Sendai Framework, including on 'building back better', is measured through a set of indicators related to disaster risk reduction. However, the Framework is a long-term implementation until 2030, if at all. The importance of measuring vulnerability and developing indicators for disaster risk reduction was also mentioned in 2005 Hyogo Conference, focusing on pre-disaster initiatives (Birkmann 2006). Usually, in a chaotic after disaster scene, there is no a priori risk assessment to compare it with. However, there are flaws in the system, not least the absence of indicators to understand the performance of reconstruction activities in regard to DRR.

Given the above discussions, we can suggest the following possible measures of DRR to reflect reconstruction performance in regard to DRR: First, whether the introduced changes influence the built environment development activities and general everyday developmental initiatives (for example, towards safe construction) after the reconstruction period. Second, whether disaster-affected people as the subjects of disaster risks, perceive themselves less at risk as a result of reconstruction activities.

2.5 Influencing the Agenda for Recovery and Reconstruction

Recovery and reconstruction are situated in their larger local and national context which, in turn, is influenced by the broader societal, political and economic context. In an aftermath of a disaster (upon declaring a disaster internationally), they are also open to international aid and agencies involved. The international trend and perception on disaster-development discourse, as well as previous experience in dealing with disasters, recovery and reconstruction are influential. The humanitarian nature of the post-disaster situation, especially at the stages of emergency and relief brings in international aid; some extends to recovery and reconstruction, which influence recovery strategy directly or indirectly. Such influence is not isolated from the 'state of the art' of the disaster-development discourse at global level that is usually undertaken by international and development agencies at the time.

Looking at the Sendai framework 2015–2030, HFA 2005–2015, Developing Recovery Framework (by GFDRR) and campaigning initiatives run by international agencies, such as 'Making Cities Resilient: My City is Getting Ready', by UNISDR, it appears that a 'global mechanism' has emerged on disaster risk reduction and resilience. For example, although the Sendai Framework is a 'voluntary and non-binding agreement', it is part of the UN system that recognises that the State has the primary role to reduce disaster risk but that responsibility should be shared with other stakeholders including local government, the private sector and other stakeholders (UNISDR n.d.d). Well-funded expectations exist and the targets of the Sendai Framework were endorsed by the United Nations General Assembly and cohere with Sustainable Development Goals—SDGs (UN-Habitat 2017). Such initiatives reinforce contemporary advances and censuses on perceptions on disasters, recovery and reconstruction, whilst holding national governments, subnational local governance bodies and all stakeholders responsible for realising the proposed framework (UN Intergovernmental Preparatory Committee 2014; GFDRR 2014). For example, the campaign 'The Making Cities Resilient: My City is getting ready!' launched in 2010, addresses issues of local governance and urban risk. It links disaster risk reduction into development activities and has a growing number of local governance bodies and municipalities as its self-reporting members from around the globe. It has been carrying on beyond its original timeline of 2015, and the number of cities, which were registered for the campaign increased from 1995 cities on 7 August 2014 to 3621 cities on 3 August 2017 (UNISDR n.d.b). The publication on recovery prepared by the World Bank's GFDRR in 2014, emphasises the importance of a programmed approach to recovery and reconstruction, whilst focusing on decentralisation and engaging people (GFDRR 2014). As we know from the practice, such discussions will be backed by international aid if a disaster occurs. However, researchers such as Brun and Lund (2008) argue that post-disaster recovery creates a situation for pursuing developmental goals advocated by international agencies, for example, sustainable development goals. Berg and De Majo (2017) also argue UNISDR is an articulation of global norm and whilst the science has come to influence policies on the global strategy for DRR, in return, the policies without taking global economic and power relations can shape the future of science and the world. Duffield (cited in Brun and Lund 2008) reports of the generic criticism of international aid as an emerging liberal system of global governance.

Given the existing global agenda, a highly political post-disaster situation might create a contradictory environment (Davis 2007). For example; the government might regard the capitalist economic model as the root cause of the disasters that affected the country (ibid). Or some agenda is to be shaped that stretches (or even contradict) traditional 'business as usual' sociocultural and political norms in a country. Strategies such as gender equality and people's participation might be perceived as attempts to 'westernise' the community, although they are part of the Millennium Development Goals. This means that an agenda can be shaped without the national government being deeply committed. A similar contradiction might be observed in other post-disaster aspects, for example, disaster diplomacy, which means countries may collaborate on disaster-related activities which are not normally prone to collaboration for diplomatic or political reasons (Kelman 2006).

Additionally, an existing dynamic societal, economic and political context, and the developmental pathway at national and local level play role in setting the agenda and the recovery strategy. Reconstruction system and its bigger recovery system are embedded and happening within a bigger system, which has sociocultural, political, economic, administrative and also environmental dimensions. (Johnson 2007). Those dimensions potentially contribute to shaping the recovery strategies. Post-disaster recovery is built on a combination of certain societal, political, economic and diplomatic background that creates a complex dynamism during which the agenda shaped and strategies for recovery and reconstruction are devised. Understanding the above complex dynamism might not be easy, especially for the outsiders. For example, just taking the political factors alone Quarantelli (1990) stresses that the issue is not about proving that political factors influence the recovery process, they do without a need for documentation. The issue is that we really do not have even a minimum understanding of how these political factors interact in the field. It is not always true that the more powerful always win over less powerful groups or that there is always a consensus on disaster recovery (ibid).

Adding to the above is previous experience in dealing with disasters, which contributes to shaping the agenda and organising reconstruction activities. It also influences the perception of success or failure and might generate the know-how. If there is prior experience, it means the state or other national or local organisations have dealt with a number of disasters in the same or a larger geographic area. Their prior experiences affect the perception of the recovery process (Quarantelli 1990). Disasters disrupt the development path of a locality that is a component of the national development path (Baas et al. 2008). This logically plays a role whilst setting the agenda for recovery to overcome such disruption (and hopefully to achieve improvements at the same time).

However, shaping the agenda and establishing strategies for recovery and reconstruction per se (even if well-intentioned and committed) will not necessarily lead to a satisfactory practice of organising reconstruction. For example, even where measures for disaster risk reduction had been taken, they were uncoordinated (Davis 2007).

2.5.1 Who Develops Reconstruction Programmes?

The combination of international concerns and consensus manifested in the HFA and Sendai Framework, for example, as well as societal, political economic and developmental contexts at local and national levels in the aftermath of disasters influence shaping the agenda soon after a disaster. It was detailed in this Chapter. But who develops the strategy and organisation designs and who implements the programmes at lower level to translate the strategy into lower and operational levels?

Davis (2007) poses some key questions on who should develop and implement reconstruction programmes. Should recovery and reconstruction management be handled by a special task force or undertaken by normal line ministries of the government? Or should reconstruction be placed in the hands of a new organisation? He suggests that recovery management and effective recovery requires a single point of overall responsibility in government and this may be best achieved by having a dedicated organisation established at the apex of political power and decision-making with a clear mandate supported by appropriate legislation, adequate resources, direct links to all line ministries and knowledge of the dynamics of the disaster recovery process.

According to a UNDP (2008) report, 28 countries in 2008 were in need of building basic capacity for recovery and disaster risk reduction in correspondence with the HFA priorities two and five. The question is what happens to those 28 countries if they are hit by a disaster, as was the case in Pakistan, Haiti and Indonesia? It appears that in such cases negotiations and arrangements in aftermath between national governments, international agencies and humanitarian communities are crucial and international organisations undertake this task directly for the major reconstruction programmes (subject to government capacity), e.g. housing reconstruction. Also, those high-profile international negotiations and national momentum can lead to the establishment of basic national institutions to deal with reconstruction. For example, as Quzai (2010) reports of the Pakistan case, the

Earthquake Reconstruction and Rehabilitation Authority (ERRA) was established soon after the disastrous Kashmir earthquake in 2005. UN-Habitat actively supported the reconstruction for designing and managing the implementation of the rural housing programme, whose results are praised by the government and international agencies. However, complexities of urban reconstruction tend to result in lack of such support for urban housing reconstruction. This represents the other side of the coin, whereby the urbareconstruction—regardless of whethern development strategy has been unable to overcome the complexities of coordinating housing reconstruction with services and infrastructure and, in some cases, resettlement. Thus, he reports, urban dwellers did not receive the same level of efficient reconstruction response as rural counterparts.

The same UNDP (2008) report identified a second group of 33 countries that had already established mechanisms but need improvements to deal with comprehensive disaster risk reduction and recovery in accordance with the all Hyogo framework's priorities. Therefore, in countries with established institutions and mechanisms for recovery, such as Iran and Turkey those institutions are ultimately responsible for developing such programmes, although they might be supported by international agencies. In the guide for Developing Disaster Recovery Framework by GFDRR emphasises that national governments are responsible for developing recovery and reconstruction programmes, although they might be helped by international organisations (GFDRR 2014). It must, however, be noted that even countries with an established national organisations for handling early recovery and reconstruction (according to UNDP's categorisation) do not necessarily achieve satisfactory results for reconstruction programmes. An example is the response to the Hurricane Katrina in 2005 in the USA. It exposed conflicting policy goals between the reconstruction plan of rapid recovery, safety, betterment, as opposed to the actual decisions made for reconstruction. The reconstruction of neighbourhoods after Katrina clearly demonstrated a rush to rebuild the familiar (Kates et al. 2006).

2.6 Reconstruction Is not Re-Construction; It Is 'Re-Construction + X'

Previous discussions on the contemporary expectations from reconstruction showed that 'the way' these construction activities are conducted towards the delivery of expectations is important. In the word 'reconstruction', the prefix 're' indicates constructing again, indicating that it is like construction project-based, whether it is a big project or a collection of projects. The logic behind this is that the prefix of 're' in reconstruction addresses redoing construction projects (Davidson 2009), therefore, what happens in a construction programme happens here. This logic highlights the technicality of reconstruction. Consequently, tackling the chaotic situation of reconstruction is assumed to become the responsibility of the construction industry. For example, Ofori (2008; cited in Haigh et al. 2006) focuses on the importance of improving the construction industries of developing nations; he highlights the need to equip them to manage the post-disaster scenario (projects and programmes). Traditionally, the construction industry deals with technicalities. As Parkins (1996) argues, engineers take for granted decision-making for design, construction and production as they are guided by codes, specifications and drawing, but the management of people and organisations is radically unlike the engineering process because people are not predictable. Although recently with the emergence of more complex projects, e.g. megaprojects, the need for embracing complexities of non-technical aspects of such activities are being highlighted by a number of researchers in this field (e.g. Morris and Jamieson 2005; Morris and Geraldi 2011; Davies and Mackenzie 2013). According to Morris (2011), various existing guiding standards and handbooks for project management do not offer understanding of the broader project system. It is necessary to see what needs have to be managed for the successful delivery of a project or programme, instead of executing technicalities (Morris and Geraldi 2011).

As a fundamental part of the physical recovery, the facilitating agent and '*process of the production*' of the lost assets, reconstruction activities (projects and programmes) must integrate means and tools for the delivery of '*improvement*' and expectations from a reconstruction as part of the recovery's big picture. These values are in fact the objectives, which reconstruction is expected to facilitate achieving them whilst constructing the lost physical assets. Organising reconstruction programmes and processes goes through the formation of the delivery sociotechnical system for reconstruction programmes. Organisation design for a reconstruction programme, therefore, is the formation of the sociotechnical delivery system that deals with the formal structure and connective process as regards organisational society and culture. It uses available resources and has to be brought to life effectively and efficiently to deliver defined strategic objectives of reconstruction programmes overarched by a developmental recovery vision.

Be it explicitly or implicitly, researchers and practitioners in disaster discourse include the importance of designing organisations in the search of effectiveness in achieving reconstruction objectives. They do this, even if they do not refer to it as the *organisation design* or even acknowledging organisational aspects of reconstruction programmes. One can identify implicit or explicit concluding remarks towards organisation design at the broader scope of programme level within post-disaster recovery and reconstruction literature. Many of those remarks are stated within the case of housing reconstruction programmes. A conclusion of some examples has been extracted from earlier discussions on reconstruction:

- Davidson (2009) refers to organisation design as a system that connects networks of projects.
- Lizarradle et al. (2009) refer to an 'enriched system' that connects layers of complexities, economic, political, social, technical, etc., for exercising the general complexity. They refer to it as an enabler for the contribution of stakeholders to reconstruction tasks.

- Lyons et al. (2010), IRP (2007) and Lankatileke (2010) refer to a need for a ‘coherent model’ for housing reconstruction.
- Quzai (2010) and Zetter and Boano (2010) highlight the importance of housing design and delivery process in reconstruction. Zetter and Boano indicate this process is a production of negotiations and policies in an aftermath.
- Pelling (2003b) refers to a platform for integrating social and physical approach, and Barakat (2003) for housing reconstruction
- Johnson et al. (cited by Davidson et al. 2007, p. 114) refer to a framework resulted from meta-procurement activity, within which participant organisations (including community) are mobilised and relations between them established, ‘that is to say the project organization is *designed*’.

Davidson (2009) explicitly points out the challenges of ‘designing’ relationships between participants in the best interest of recovery effort. He highlights the most difficult tasks in post-disaster reconstruction is organising the necessary processes and procedures, particularly regarding the participants and their roles. Practical decisions have to be made, in a context of competing interests. The potential participants include affected people, community-based organisations, local and central government, NGOs and international agencies; and the success of post-disaster reconstruction largely depends on the complex relationships between, them, which have different organisational sizes, capabilities and internal interests. He defines organisation design as identifying the actors and their relationships and concludes that organising reconstruction is an organisation design problem. In the reconstruction field, Davidson et al. (2007) examined the formal structure of four case studies in Colombia, 1999; El Salvador, 2001; Turkey, 1999; and Turkey, 2000, and highlighted the need for balance between various participants in reconstruction. In order to tackle theoretical problems of ‘community participation’, he suggests that instead of trying to determine the appropriate roles of community participation (only one of the objectives) in isolation, we need to relate it to organisation design of a project (programme) as a whole because they (community, users) are only one of the many actors in a project. The solution, as Davidson suggests is to tackle the organisation design as a whole, and to see reconstruction from the perspective of a project management point of view. The advantages of this wider view of organisation design are that users’ participation is—in a system approach—seen in terms of risk reduction, overall performance, results obtained versus objectives, resource management, etc. This organisation design includes the role of beneficiaries in terms of interrelationships and interactions with and between the other members of the project team. Organisation design is at the centre of the search for better project performance (ibid). His statement here clearly shows that his use of the term ‘project management’ is generated from daily use terminology and what he addresses is the sphere of management of activities rather than a technical project.

The best interest of recovery programmes are the ones that were set as the reconstruction objectives, for example, disaster risk reduction, benefits of people’s participation and so on. Adding such values (objectives) to the reconstruction is something more than re-construction (re-technicality); it is 're-construction +X'. This X is the medium between technical science and social science, between homemaking and constructing the materiality of houses. But what is this X? In my view, in participatory reconstruction, this X is the organisational mechanism that allows reconstruction to be a platform for local people to experience something new, which they had not experienced before and to be champion of change in reducing disaster risks and setting an example for other parts of a city or country. Simultaneously, it provides space for people to reproduce their culture and lifestyle within the reconstruction period.

From the perspective of organisation theory, this X is *the organisational integration process* that brings technical, social and financial ingredients of reconstruction together in order to achieve envisioned developmental objectives. For example, 'Reconstruction +X' must create fit-for-purpose people's participation and integrates practical DRR measures into the rebuilding process, in order to reduce disaster risks that caused a specific natural hazard to become a disaster. Given the systemic and strategic continuity of reconstruction, recovery and longer-term vision, this X is the organisational carrier to facilitate disaster-affected people's participation (beneficiaries) in a way that the operational process contributes to the delivery of the improvements, which supposedly could be delivered. Throughout this research, whenever I use the word reconstruction, it refers to 'reconstruction + *organisational integration process*'. From disaster-development studies perspective, it is crucial to recognise the organisational identity of this X. As an organisational element, the X is determined whilst organising reconstruction through the delivery, such as setting objectives, formation of delivery system and introducing practical operational considerations and putting them together for the actual implementation of a programme. This will be discussed further.

2.7 Multiple Organisational Perspectives for Understanding Reconstruction Organisations

Given the above discussions, we are able to identify entry points to organisation theory for further understanding reconstruction organisations by offering multiple perspectives which are inherently relevant to the subject of programmes and projects as suggested by organisational theorists (e.g. Morgan 2006). They build on common subject-oriented characteristics of reconstruction activities. Unsurprisingly, these perspectives are overlapping and will lead to a better understanding and analysing them. These organisational natures of reconstruction programmes influence organising reconstruction activities and are mirrored in both disaster-development studies and organisation theory. Each of these perspectives is discussed further towards criteria and considerations for the initial organisation design and for managing it effectively during the implementation.

2.7.1 Organisational Form of Reconstruction Programmes

Organising for 'reconstruction + X' is an organisation design problem, not for the sake of construction itself. This design problem is at the programme organisation level of delivering the benefits of reconstruction to contribute to multidimensional recovery as part of the big picture. This directly addresses the 'X' and is inherent in contemporary perception on reconstruction. The question, therefore, is that if the organisation for a reconstruction programme is formed responsively, it contributes to approaching strategic objectives of reconstruction (for example, as a pathway towards envisioned recovery integrated with DRR). Within the open systems epistemological approach to organisations, there are variable combinations that have been studied of their influence on organisation, including size, age, tasks performed environment and the overall strategy (Griener cited in Badir 2006). In order to find an entry point into organisational studies on forming an organisation here below is extracts of some common organisational characteristics from previous discussions on reconstruction and recovery that indirectly indicate those.

- First, reconstruction and recovery are multi-actor and multi-organisational systems with diverse capabilities, structures and cultures (e.g. GFDRR 2016; Lassa 2012; Davis 2007; Christopolos cited in Pelling 2007). This may range from public to private sector, from international to national and local institutions, and from organisations to individuals.
- Second, the context and environment in post-disaster aftermath are chaotic, complex, dynamic and turbulent (e.g. Lassa 2012; Johnson 2007; Le Masurier et al. 2006; Coppola 2006). This provides a great deal of uncertainty and unexpectedness compared to a normal situation.
- Third, reconstruction is project and programme based, even if the institution managing total reconstruction is not changing, as is the case generally for built environment formation. Reconstruction programmes are formed for each reconstruction case, based on the size of the programme, available resources and other practical considerations. They use extra resources, and bring together various stakeholders that might not have worked with each other before and which need to devise together a shared vision and objectives. Resembling 'megaprojects', such construction programmes constitute a mega system for delivery which integrates a number of participant organisations—a variety of organisational types each with their own system (Davies and Mackenzi 2013).
- Fourth, each reconstruction programme organisation is, to some extent, new and resorts to degrees of innovations (or changes) to address previous failures (at least in theory). Qualitative measures and expectations of reconstruction cannot be delivered if the exact same old mechanisms and/or ways of doing things stay in place after the failure of old process of the built environment formation and maintenance exposed by natural hazard (Maskrey 1994). For example, to integrate measures for disaster risk reduction to an existing process means that the result is something new and inexperienced, otherwise, the natural hazard

would not have turned into a disaster. Reconstruction programmes, therefore, are experiencing new ways of forming the built environment in a disaster-affected area. Even if the same reconstruction management applies a previously experienced approach (e.g. in a neighbouring country or city), the programme specifications vary, e.g. the scale of a disastrous destruction, size of the disaster-affected population and the geographic and social context are different. Thus, project/programme organisations should be formed accordingly and there is (has to be) an element of changing old processes and measures at least to some degree.

- Fifth, such reconstruction programme organisations are temporary and time restricted (Coppola 2006). Reconstruction activities are temporary within the time frame of project and programme specifications. They should connect to development activities and business as usual (Hamdi 2010). Since the reconstruction process is not the only job for the national authorities, international agencies and NGOs, the whole process must be returned to the locals at some point.

Even the above organisational examination (with only having organisational form in mind) shows that the combination of such organisational characteristics increases likelihood of encountering difficulties in organising reconstruction programmes and the unexpectedness and setbacks of forming the organisational system and implementing it from an organisational management perspective.

The extensive debate on forms of organisations initiated in the 1970s moved towards a contingency view that can be seen as marshalling and 'operationalising' the systems view in organisation discourses (e.g. Hanisch and Wald 2012; Kay 2008; Morgan 2006). In brief, the contingency view argues that there is no single 'best' form of organisations, the way of managing and organising. Successful organisations are most likely those that develop the most suitable 'fit' between organisation structure and contingencies (Hanisch and Wald 2012; Kay 2008). According to Davies et al. (2011), Mintzberg was among the first organisational theorists to articulate the strengths of the project form in a rapidly changing environment. Mintzberg (1980) linked the contingency view to the abstract organisational 'configurations' based on a variety of different conditions and changing variables. This led to introducing the typology of 'ideal', abstract or 'pure' types of organisations so that each one is more likely to suit different conditions or contingency factors. Starting with a simple structure, machine bureaucracy, professional bureaucracy and adhocracy (which is an abstract form of matrix and network organisations), it extended to include missionary (religious) organisations, and later virtual organisations (e.g. online campaigning organisations) and other new typologies. According to him, adhocracy structural configurations are potentially suitable for young and temporary organisations working in complex environments, as well as projects and programmes. Mintzberg explains that adhocracies demand an organic structure, coupled with innovation in a highly dynamic context. Adhocracies later were referred to as innovative organisations (Mintzberg 1980; Mintzberg et al. 2003). Despite entering a new era, the contingency view is considered as a basis of forming and analysing organisations (e.g. Morgan 2006; Buelens 2006; Mintzberg 1980). According to Kay (2008), it is still a dominant strand of thought in organisational behaviour. Nevertheless, the variety of classic and new potential forms of organisations offers starting points for thinking about forming organisations that are uniquely tailored to meet specific challenges up front. There is no need for an organisation to fall into a single abstract typology but it is necessary for it to be contingent and fit (Herber et al. 2000).

Given the common organisational characteristics of reconstruction programmes, one can identify how such common characteristics broadly correspond to specific conditions of the independent variables in the contingency view. 'Adhocracy organisations' can be seen as the basis of typical configurations for reconstruction programme organisations which are temporary, new and within complex dynamic environments in the aftermath of a disaster. Whilst adhocracies have the least association with the classical principles of management, they are also vulnerable because they demand a good deal of decentralisation, and liaison devices to encourage mutual adjustment as the key coordinating mechanism. The operational-level strategy is not top-down and everyone within the organisation takes part in it (Mintzberg et al. 2003).

2.7.2 *Strategic Nature of Reconstruction Organisations*

As theoretically discussed, reconstruction is a strategic agent for facilitating the goals and the big picture envisioned in an aftermath for redevelopment of the built environment, mitigation and multidimensional recovery. Reconstruction programmes do so if they achieve the strategic objectives or at least are progressing on this. This clearly has an association with organisation theory that is about strategic thinking and management. Disaster risk reduction and people's participation are inherited in the contemporary overall recovery strategies. As Armstrong (2009) remarks on organisational performance, strategies are developed for the realisation of a vision. Strategies at broader and higher levels are objectives for the lower level system (Ansoff 1990), such as integrating measures for disaster risk reduction, or meaningful people's participation in reconstruction in general, and reconstruction programmes in particular. It means that the strategic nature of recovery and reconstruction transfers to reconstruction programmes and projects. Thus, reconstruction programmes have to be strategic, in their introduction, design and implementation.

Such strategic and long-term purposeful nature of reconstruction programmes and their related organisations indicate the need for employing strategic thinking in managing them. According to Ansoff (1990), who is called the father of corporate strategy (CMI n.d.c), strategic management is the systemic approach in general management of purposeful activities. It can be traced back to the 1950s and 1960s. This is now widely employed in human organisations and business administration as they are ambitious for survival and improvements in rapidly changing environments (Hannagan 2002). As Armstrong (2009) states achieving strategic goals requires managing strategically. This involves adopting a broader and longer-term view of what needs to be done and ensuring that the activities are carried out and contribute to achieving those strategic goals. Hanaggan explains that strategic management consists of decision and actions used to formulate and implement methods for providing the best fit between the organisation and its environment, to enable it to achieve organisational objectives. It is about a sense of purpose, looking ahead, planning, positioning, strategic fit, leverage and stretching. To be effective, strategic thinking should be creative (rather than rational) and be supported (but not substituted) by analysis (Markides 2008).

Progressively, the domain of the strategic management application transcended business-related professional boundaries and entered the areas of project management in the construction industry too. At present, attempts to incorporate strategic thinking in the construction industry and the development field is moving beyond pure technicality towards incorporating social aspects and seeing projects as a whole. According to Morris and Geraldi (2011), the focus of studying construction and development projects broadened from delivery execution management towards an enlarged perspective of what had to be managed in order to develop and deliver the project, as an organisational entity, successfully within its context. Morris (2011) emphasises that all projects are unique, calling for recognition of multiple agendas, the complexities and contingencies of context for projects and programmes that requires moving beyond a traditional technical approach to project management. Identifying areas for the further development of project management, he stresses the realisation of projects and programmes requires social skills as well as technical skills for moving beyond execution management towards managing projects and programmes as a whole entity including strategy (ibid). In parallel, some concepts that are used by the thinker and practitioner in the development field, Hamdi (2010) are rooted in strategic management, for example, calling for strategic action plans and applying corrective adjustments to reflect emerging issues in the field. Strategic thinking, reflected in designing and managing strategically to achieving objectives and approaching a vision greatly relies on taking a systems approach (e.g. Morgan 2006; Ghoshal et al. 2003; Ansoff 1990). Despite enormous prescriptive discussions on strategic management and its various capabilities for professional managers, strategic management as a theory focuses on cascading as a core element of strategic decision-making (Armstrong 2009).

Cascading, which is also called 'horizontal and vertical coordination' (FAO 2009) refers to keeping the core direction at various levels of activities and methods. Horizontal and vertical interplay do often occur at the same time leading to an increasingly complex system of interactions (Tesliar et al. 2016). Cascading is seen as a core element in strategic decision-making (Armstrong 2009). On vertical cascading, Armstrong (2009) explains keeping the core direction between various levels of strategy and cascading them is a core element in strategic decision-making. Strategy needs to be translated correctly and accordingly into lower level objectives. Also, practical considerations and policies that are at the same level must be put in place by organisational operational means effectively in order to keep direction of implementing strategy and achieving strategic objectives.

This refers to horizontal cascading. Cascading, therefore, can be seen as forming and directing 'alignments' in and between different subsystems within a whole system which was a principal systems theory (Morgan 2006). Figure 2.3 presents this schematically. The point here is that policies as 'plans for action adopted by organisation' and organisation design are not legitimate on their own. They are justified only if their results and benefits are in line with the higher level strategy, including strategic objectives and corporate strategy. They exist to serve the strategic objectives, not themselves per se. Cascading objectives are the key element for approaching and achieving objectives (Armstrong 2009).

Morris and Jamieson (2005) highlight the inherent dynamism in cascading and alignment from higher level strategies to project and programme operational levels. Although strategies are to be aligned and cascaded from higher levels to programme and project levels in a systematic and hierarchical level that provides cohesion, visibility and an effective means of communication, not all is deliberate, and there is emergence and iteration. In parallel, they highlight the importance of process as a means of translating higher level strategies into objectives in operational considerations of projects and programmes, ensuring the continuity of strategy in a systematic and structured way.

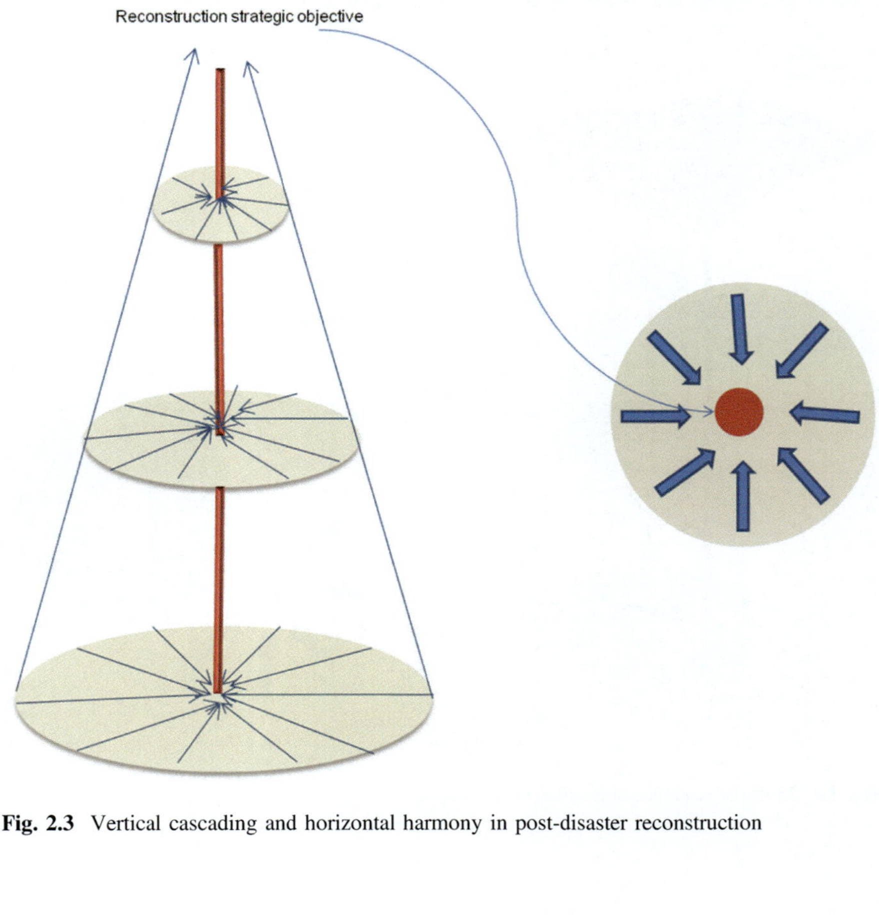

Fig. 2.3 Vertical cascading and horizontal harmony in post-disaster reconstruction

Acknowledging the above, logically, upon targeting more than one objective (for example, on disaster risk reduction, people's participation, cultural historic landscape and poverty reduction), each one must be addressed accordingly. Clearly each strategic objective needs its specific targeted policies, tasks and practical considerations, as well as probably system elements and subsystems for organisation design. All these multiply operational complexity of, and policy making scene towards achieving all objectives. The more multiple strategic objectives, the greater are the chances of contradictions at operational policies and frameworks. Approaching a single element strategy or one strategic objective in practice is easier than two elements of strategy and two strategic objectives. As Bonabeau (2007) states with a growing complexity, the risk of failure of a system increases. The more complex a system is, the more the system itself becomes vulnerable to possible internal risks (ibid). This is simply because the system incorporates an increasing number of interrelated activities, entities and elements. It also means that the alignment of the internal policies of literal actors with the overall strategy and strategic objectives is more difficult as the system created is highly complex, increasing the chance of bottlenecks and unforeseen situations in combination with the chaotic context in an aftermath. Therefore, the necessity of commitment to the strategic concepts and objectives has vertical scale and horizontal dimension, vertical alignment and horizontal cascading. Figure 2.4 schematically present this.

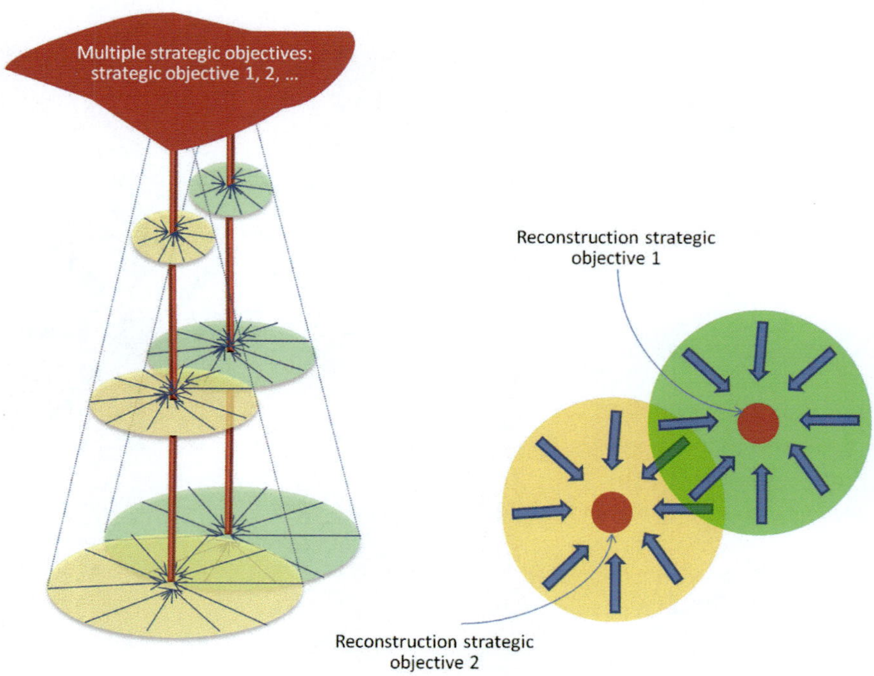

Fig. 2.4 Multiple strategic objectives in reconstruction

2.7.3 Social Nature of Reconstruction Programme Organisations

Drawing on inspirations from biological systems and psychological advances, organisations accommodate human relations rather than machine-like precision (e.g. Perrow 1973; Holbeche 2007; Morgan 2006). As Perrow (1973) reviewed, all varied schools of organisation theory seemed to agree that organisations are open systems. Thus, the mechanical school of organisation theory that treated organisations as machines (engineering systems) moved towards the growth of understanding organisations as social systems. Such advances in the 1950s created a dual focus in understanding organisations that, as he (ibid, p. 11) explains, 'bewilderingly' increases the complexities of understanding organisations by 'forcing ever more variables into our consciousness'. Organisations are best understood as sociotechnical systems, cooperative systems. Such understanding expanded to take account of relations between technical, social, managerial and strategic and environmental requirements, as in systems view everything is related to everything else, although in different degrees of tension and reciprocity. Every organisation, organisational unit, department and task group interacts through its activities with its larger systems, whilst sociological, psychological and cultural aspects of units also interact. Also, notions of power and conflict enter the systems. In fact, everything depends on everything else in sociotechnical systems (e.g. Perrow 1973; Morgan 2006).

The advocated importance of people's participation in reconstruction activities by researchers such as Davis (2007), adds more weight on the social nature of reconstruction programme sociotechnical systems. When disaster-affected people, who are the beneficiaries of reconstruction activities, are participating in reconstruction programmes they are part of the related reconstruction programme organisational society. They are part of machinery for doing things; thus, their sociocultural values and beliefs will influence organising reconstruction programmes. Thus, local people' sociocultural characteristics are not the external context of reconstruction programmes and activities anymore. In contrast, they are at the heart of the internal organisational environment of organising reconstruction. Whilst organisations as systems are open to their environment, the participation of local people blurs the boundaries between the organisational system and its outer environment because the outer environment is penetrating the inner system of the organisation, making it an extremely open system. In this regard the social underlying characteristics and dynamics, e.g. culture, power and political economy environment, influence organising reconstruction immensely, at both stages of the formation and implementation. Such blurred boundaries between the internal and external environment, in my view, potentially adds to the complexities of organising reconstruction programme, which even in non-participatory practices have to attend the sociocultural characteristics of a disaster-affected area.

According to Zetter and Boano (2010, pp. 225–226), many of the current challenges for integrating cultural considerations into reconstruction activities, e.g.

of creating place and space, are symptoms of wider challenges of creating coordinated organisational and strategic framework. They conclude that in order to achieve a wide range of objectives and to have better places and spaces, there is a need to coordinated planning at all levels; and the reconciliation of conflicting mandates. Their conclusion directly addresses the strategic nature of the reconstruction and the black box of the organisation design for reconstruction programme, as one of those aforementioned levels. Moreover, they link the cultural integration of the built environment reconstruction to its sociocultural context with the strategic nature of reconstruction programmes, indicating that they come together through an organisational framework (of participants and their connective process). In this way, organisation design and management of the reconstruction programmes are influenced by the social nature of the participatory reconstruction programme organisation. Simultaneously, it influences the sociocultural relevance of the outcomes of the reconstruction activities. However, at the heart of the problem sits organisation design, and organisation design and management of the built environment reconstruction is the subject of this research.

Reminding Davidson's (2009) statement that organising for reconstruction is a design problem, this design problem is about linking various projects and bottom line tasks, incentives and policies, to achieve strategic objectives. Within a participatory reconstruction programme, people and those participant organisations are all system elements of the multi-organisational sociotechnical system. People who participate in reconstruction programme as a whole are one element within the organisational arrangements for that reconstruction programme (Davidson et al. 2007), even if the size of this specific element is to be very large. If the size of this system element is too large this, in fact, increase the overall size of the programme. Figure 2.5 shows this. Even if disaster-affected people are categorised into different groups and it is to be reflected in specific organisational arrangements, that group again is one element within its related organisational arrangements.

Derived from the attention to the social nature of organisations the notion of organisational learning emerged in the 1980s with an upsurge in the 1990s (Curado 2006; Miner and Mezias 1996). According to Curado (2006), organisational learning is a social phenomenon, and many researchers believe it is the product of the involvement of members in an organisation in interaction and sharing of experiences and knowledge. As she reviews, the ontological dimension of organisational learning (the subject who learns) exists at both individual and collective (organisation) level. They coexist and organisational knowledge is leveraged through individuals in different ways (such as instruction). Simultaneously, the organisational code adapts itself to the beliefs of its members (system elements). In this way, mutual learning produces results on both individual and organisational levels. The task performance during a process over time is converted to knowledge which in turn changes the organisation's context and affects future experience (Argate and Miron-Spektor 2011). The learning process might be linked to knowledge creation directly based on a unit's (system element) experience or based on knowledge transfer from one unit (organisational system element) to another.

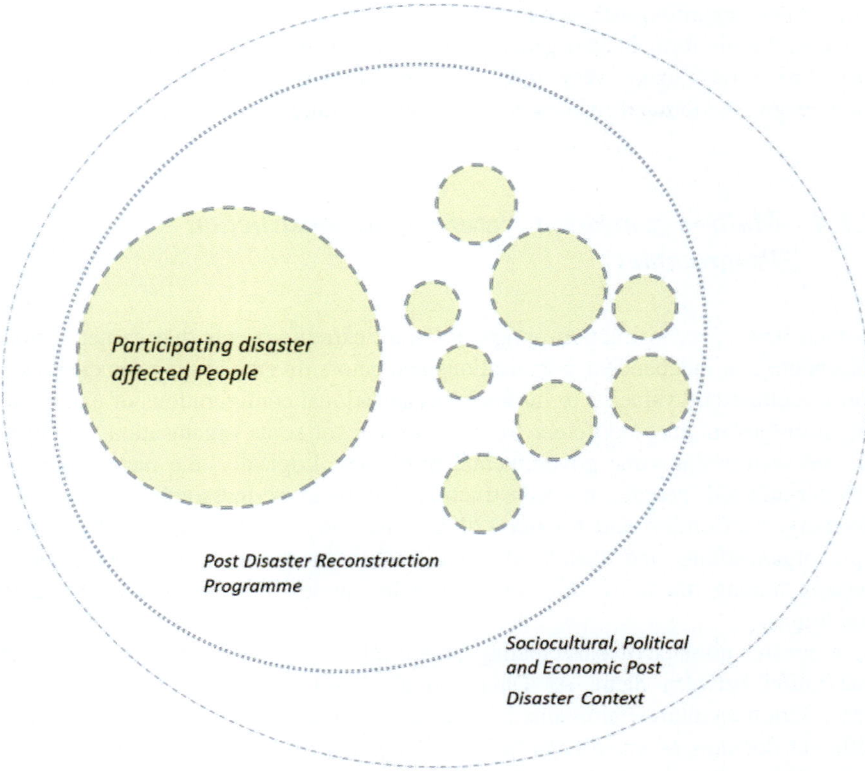

Fig. 2.5 Schematic system elements of a participatory reconstruction programme

As Argate and Miron-Spektor (2011) argue, an important dimension in the organisational learning process is whether it should be top-down or bottom-up. They also identify further research areas of organisational structure and its social network, power balance, co-location or remoteness of organisational members to advance organisational learning. However, the scientific conception of knowledge in organisations is still at an early stage of development (Curado 2006). Interestingly, researchers such as Curado (2006) and Beeby and Booth (2000), state the notion of organisational learning is also connected to strategic management. Knowledge is seen as a key of organisational success, capability development, strategic flexibility and faster reaction to environment changes. As Argate and Miron-Spektor (2011) highlight knowledge is a challenging concept to define and measure and assessing knowledge by understanding changes in practices and performance is advantageous as it captures both tacit and explicit knowledge. Beeby and Booth (2000) also state that organisational learning is consistent with a contingency view to organisations which sees them as systems continuously adapting to their environments. Echoing this, Curado (2006) concludes that organic organisation designs with less formalisations and more decentralisations are more likely to facilitate organisational learning. However, Van Den Eede et al. (2004) argue that even highly formalised organisations which any deviation from regulations and processes might make very risky are able to foster organisational learning, knowledge creation and management, through making sense and projection.

2.7.4 Multi-organisation Nature of Reconstruction Programmes

In their nature, reconstruction programmes are extreme events that garner various dependent and independent organisations and actors diverse in their organisational forms, culture and values. For instance, organisational configurations of participant organisations might be different from each other, or some organisations might be private sectors but some governmental ministries. Logically, the participation of disaster-affected people in reconstruction programmes massively expands this diversity. As Comfort and Kapucu (2006) state, the task of integrating these multiple organisations and system elements into a functioning inter-organisational system among them is difficult, especially under the chaotic post-disaster conditions.

A greater number of participating organisations intensifies the need for better integration between them. As Davies et al. (2011, p. 14) point out 'different organisational cultures, problems in contractual relations and the need to integrate different domains of knowledge make collaborative projects very difficult to execute'. This collaborative system of various previously independent organisations, also called inter-organisational sociotechnical system, creates a great level of interdependency and interconnections. The more their interdependency increases, the more problems arise of achieving integration between them (Comfort and Kapucu 2006; Badir 2006). Comfort and Kapucu (2006) count at least three basic sets of conditions that influence the interactions among participating organisations in such extreme events: organisational procedure and process that shapes actions both within and among the participating organisations; technical structure that supports information search and exchange; and cultural openness to new information, new strategies for addressing unforeseen problems, also a willingness to adapt to extraordinarily difficult conditions. According to them, these conditions shape the evolution of inter-organisational systems in fundamental ways, since the interaction among organisations within the system shapes the next round of actions within the system. The result is the emergence of a complex and adaptive system that responds both to the demands from the environment and the degree of pressures or support from other organisations within the system as it evolves.

From the purely organisational perspective, such multi-organisational complex systems benefit from adaptive flexibility of their system, which relies on lateral communication instead of vertical power to achieve coordination (Herber et al. 2000). In the case of extreme events, such as post-disaster reconstruction, Comfort and Kapucu (2006) uses the concept of 'complex adaptive systems' and 'learning organisations' to state that benefiting from the advantages of such collaboration requires coordinated actions among participating organisations, and as she concludes, and to do so an auto-adaptive system is needed. She sees the auto adaptation system as a form of mutual adjustment, which itself is one method for a coordination mechanism (Mintzberg 1980) that is an element of integration between organisations within inter-organisational systems (Badir 2006).

- **Organisational Integration Process Maintains the Whole System of a Reconstruction Programme**

Organisation design should address the organisational process as well. Within inter-organisational sociotechnical systems, such as the ones for reconstruction programmes, this organisational process connects various organisations and becomes an integration process which finds its critical importance because it is the actual basis of bringing the coordinating mechanism between various activities to life, from the start of a programme to the finish. This process cross-cuts hierarchies and diverse organisational structures (Galbraith 1983). As Morgan (2006) reviews, the systems view is linked to the integration process that maintains the whole system with its multidimensional aspects because systems are like Chinese boxes in that they always contain wholes within wholes. The importance of managing such intra-and inter-organisational relations between these systems has been the interest of organisation theorists. The sociotechnical account of organisation expanded to take account of relations between technical, social, managerial and strategic and environmental requirements. In this way, 'configuration of subsystems' or 'business processes' as a way of managing the relationships between critical subsystems are important. This organisational connective process refers to links between various organisational elements, and integrates the system elements to achieve unity of efforts within the organisation system to accomplish the organisation's task (Herber et al. 2000 reproduced in Mintzberg et al. 2003, pp. 234–241; Lawrence and Lorsch cited in Badir 2006). There are theoretical debates on organisation design on whether this process is part of the organisational structure (Goold 2002), or it is an independent attribute that contributes to the formation of an organisation (Galbraith 1983). Similarly, various labels such as the connective tissues (Goold 2002), or a way of coordination mechanism (Mintzberg 1980) have been used for that. The connective process is the manifestation of the functional coordination mechanism that determines the way the necessary activities and tasks are interconnected, their order and their interdependencies (Mintzberg 1980). In case of multi-organisational systems that are trying to develop new products, a well-integrated process leads to systemic transition so that the output of one function becomes the input of another function at the right time. This may result in seamless transferral between activities (Badir 2006).

For each reconstruction case which requires multi-organisational socio-technical systems this integration process finds its immense and critical importance. This is because it is the actual basis for bringing the programme to life during implementation. This process cuts across organisational hierarchies and the formal structures of all organisations engaged in a reconstruction programme. Despite the above theoretical debates, this connective integration process presents the outline of how a reconstruction sociotechnical system will be in practice implemented, the workflow. A well-integrated process in a multi-organisational system leads to systemic transition so that one function's output becomes another function's input at the right time; this may result in seamless transferral between activities. As Badir (2006) points out, the organisational integration process that creates unity and harmony between actors and subsystems strongly is interrelated with the notions of coordination and communication. Figure 2.6 schematically shows this connective process in participatory reconstruction programmes.

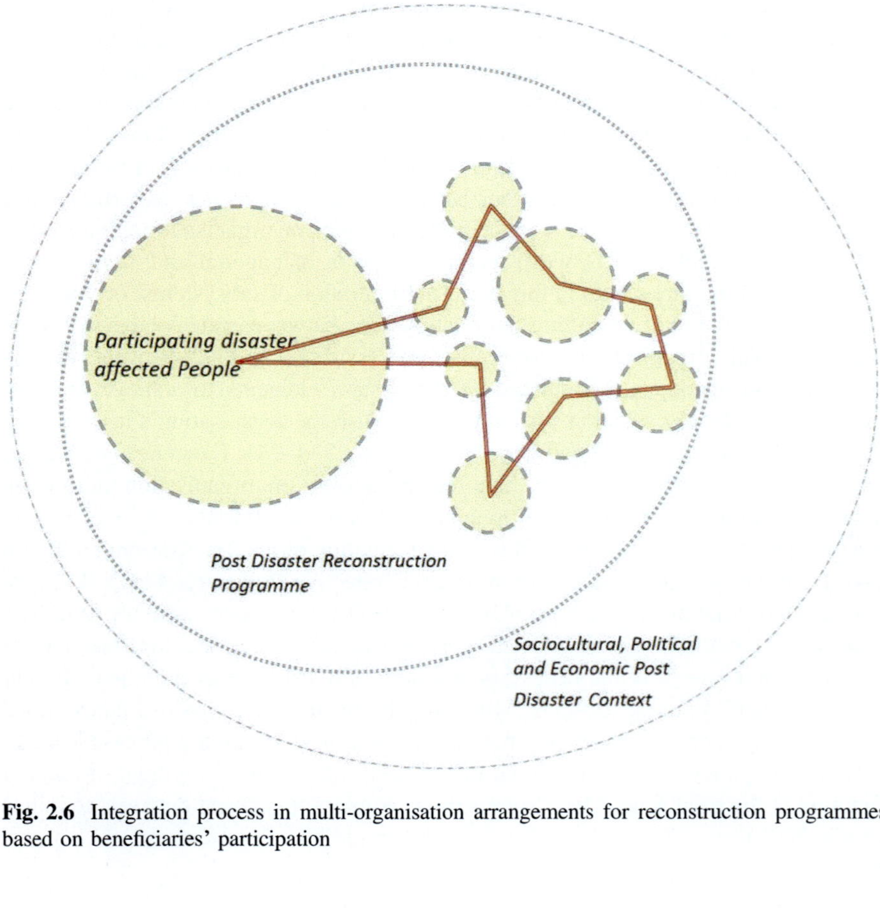

Fig. 2.6 Integration process in multi-organisation arrangements for reconstruction programmes based on beneficiaries' participation

2.8 Conclusion: A Tentative Conceptual Framework of Understanding Reconstruction Organisations

Multidisciplinary theoretical discussions in this chapter synthesised disaster, reconstruction and organisation theory in order to bridge the gap of understanding how to achieve those contemporary expectations from reconstruction programmes. Current linkages between disasters and development, recovery and reconstruction indicated what the mission of reconstruction programmes is, conceptualise reconstruction as 'reconstruction + X', which not only reconstructs lost assets but also facilitates the achievement of strategic objectives and added values. This chapter, in turn, extracted common organisational characteristics of reconstruction programmes, which were also positioned within organisation theory. 'Adhocracy organisations' were identified as the basis of typical configurations for reconstruction programme organisations which are temporary, new and within complex dynamic environments in the aftermath of a disaster. The main overlapping aspects of organisational nature of reconstruction programmes influence organising reconstruction activities.

Integrating the above multiple organisational perspectives on understanding reconstruction organisations, we can develop a tentative conceptual framework that offers an integrated tool of insights for understanding and examining the case study, which in turn contributes to further development of this model. Figure 2.7 presents a tentative conceptual model as a point of reference that integrates multiple lenses for understanding organising for reconstruction programmes through organisation design and management.

Each of the insights provides an entry point to organisation theory and sheds light on reconstruction organisation as a sociotechnical system, for understanding and examining the way reconstruction programmes are organised:

The first insight addresses organisational form of reconstruction programmes that are linked to organisation design and contingency theory, fit for organisational attributes, their consistency and harmony with each other, regardless of their form (e.g. Mintzberg 1980, Galbraith 1983; Mintzberg et al. 2003; Morgan 2006).

The second insight is the strategic nature of reconstruction programmes (e.g. Hrebiniak and Joyce 1984; Armstrong 2009; CMI n.d.d) towards the reconstruction programme objectives, including, the integration of DRR measures and the advantages of people's participation in addition to any other objective.

The third insight is the social construct of reconstruction programme organisations (Williams 2004; Vincent 2004; Lizarralde et al. 2009) that in addition to sociocultural aspects addresses relations within the system and between the system and its context.

The fourth insight is the multi-organisation nature of reconstruction programmes (Comfort and Kapucu 2006; Badir 2006), for which the sociotechnical system of

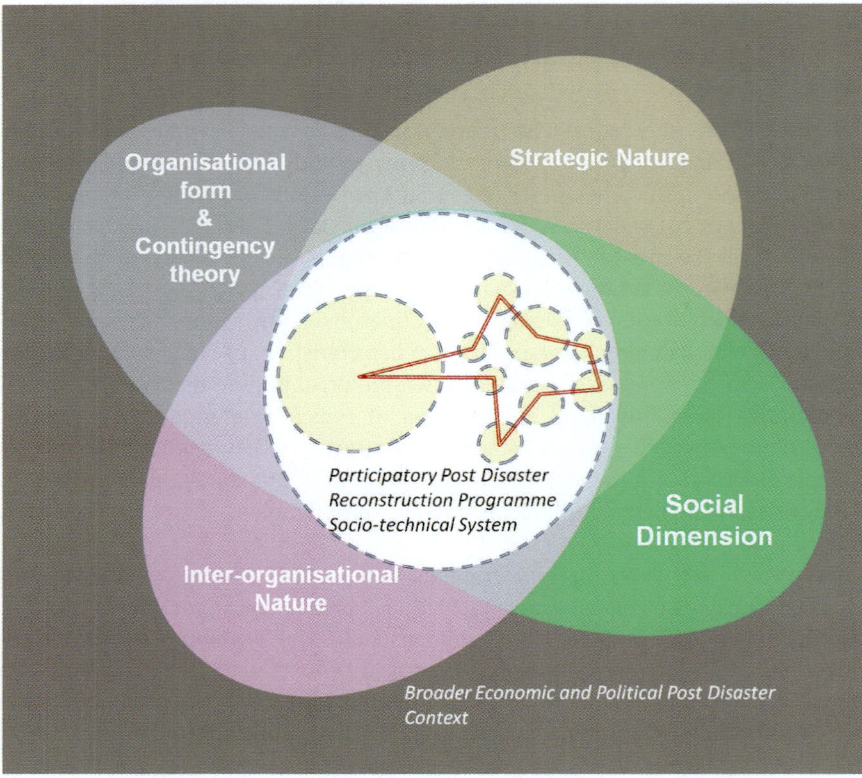

Fig. 2.7 Initilal tentative, conceptual and multi-perspective framework for understanding design and management of reconstruction programme organisations

the reconstruction programme becomes in fact a mega system towards the integration of those organisations.

Later in Part III, I will employ the multiple perspectives of this framework to analyse the case study and understand it further. Due to the exploratory nature of this research, on the one hand, this preliminary tentative framework sets directions to examine the case, and on the other hand, within an iterative process, the case helps to develop the model further. The case, therefore, will be a laboratory to be examined and to refine this framework, taking it to a detailed level. Notably, developing this framework purely relied on multidisciplinary literature review (both theoretical and empirical) and inductive discussions. Therefore, in addition to its use in the analyses of the case, this model also links the case study to the broader existing literature. Any emergent notions within or related to the analyses will be addressed through looking at the literature and the development of the model.

References

Alexander DE (2002) Principles of emergency planning and management, Illustrated edn. Oxford University Press, England

Ansoff H (1990) Implanting strategic management, 2nd ed. Prentice Hall, New York

Archer D, Boonyabancha S (2011) Seeing a disaster as an opportunity—harnessing the energy of disaster survivors for change. Environ Urbanization 23(2):351–364. https://doi.org/10.1177/0956247811410011

Arefian FF (2016) Getting ready for urban reconstruction: organising housing reconstruction in Bam. In: Arefian FF, Moeini SHI (eds) Urban Change in Iran: stories of rooted histories and ever-accelerating developments. The urban book series. Springer International Publishing, Cham Heidelberg, New York, Dordrecht, London, pp 231–257

Argate L, Miron-Spektor E (2011) Organisational learning: from experience to knowledge. https://server1.tepper.cmu.edu/Seminars/docs/ARGOTE_SPEKTOR_OS_%20inpresspdf. Accessed 3 Aug 2012

Armstrong M (2009) Armstrong's handbook of management and leadership: a guide to managing for results, 2nd edn. Kogan

Arnstein SR (1969) A ladder of citizen participation. J Am Plann Assoc-JAIP 35(4):216–224

Association for Project Management (2006) APM body of knowledge. The Association, High Wycombe, Buckinghamshire

Baas S, Ramasami S, Dey De Pryck J, Battista F (2008) Disaster risk management systems analysis: a guide book. Volume 13 of environment and natural resources management. Food and Agriculture Organization of the United Nations, Rome

Badir Y (2006) The role of network lead company in integrating new product development processes across strategic partners. PhD thesis, Ecole PolytechniqueFederale De Lausanne, Lausanne

Barakat S (2003) Housing reconstruction after conflict and disaster. Network Paper, ODI, London. http://www.odihpn.org/documents%5Cnetworkpaper043.pdf. Accessed 23 Aug 2009

Beeby M, Booth C (2000) Networks and inter-organizational learning: a critical review. Learn Organ 7(2):75–88

Benson C, Twigg J (2007) Tools for mainstreaming disaster risk reduction: guidance notes for development organisations. http://www.preventionweb.net/english/professional/publications/v.php?id=1066. Accessed 22 Nov 2010

Berg M, De Majo V (2017) Understanding the global strategy for disaster risk reduction: Understanding the global strategy for DRR. Risk Hazards Crisis Public Policy 8:147–167

Birkmann J (2006) Indicators and criteria for measuring vulnerability: theoretical bases and requirements. In: Press TERI (ed) Measuring vulnerability to natural hazards. United Nations University Press, Tokyo, New York, Paris, pp 55–78

Blaikie P, Cannon T, Davis I, Wisner B (1994) At risk: natural hazards, people's vulnerability, and disasters. Routledge, London, New York

Boano C (2007) Dynamics of linking reconstructionand development in housingand settlements for forced migrants in post-disaster. Oxford Brooks University

Bonabeau E (2007) Understanding and managing complexity risk. MIT. Sloan Manage Rev 48. http://sloanreview.mit.edu/the-magazine/articles/2007/summer/48416/understanding-and-managing-complexity-risk/. Accessed 25 Nov 2010

Bosher L (2008) Introduction: the need for built-in resilience, In: Bosher L (ed) Hazards and the built environment: attaining built-in resilience. Routledge, London, pp 3–20

Bosher L, Dainty A (2006) Disaster risk reduction and "built-in" resilience. Disasters: Int J Disaster Stud Prac 35(1):1–18. doi:https://doi.org/10.1111/j.1467-7717.2010.01189.x

Brun C, Lund R (2008) Making a home during crisis: post-tsunami recovery in a context of war, Sri Lanka. Singap J Trop Geogr 29(3):274–287. https://doi.org/10.1111/j.1467-9493.2008.00334.x

Buelens M (2006) Organisational behaviour, 3rd edn. Mcgraw-Hill, London

Centre for Research on the Epidemiology of Disasters—CRED (n.d.) Disaster list | EM-DAT EM-DAT, the international disaster database. http://www.emdat.be/disaster_list/index.html. Accessed 2 Dec 15

Centre for Research on the Epidemiology of Disasters, CRED (2016) The Human cost of natural disasters, 2015, a global perspective. Centre for Reserach on the Epidomology of Disasters, DRED. http://www.preventionweb.net/files/42895_cerdthehumancostofdisastersglobalpe.pdf. Accessed 09 Aug 2017

Chartered Management Institute (CMI) (n.d.a) Setting SMART objectives. Checklist 231. http://www.managers.org.uk for CMI members. Accessed 14 Feb 0211

Chartered Management Institute (CMI) (n.d.b) Organisational culture. Checlist 232. http://www.managers.org.uk for CMI members. Accessed 13 Mar 2011

Chartered Management Institute (CMI) (n.d.c) Igor Ansoff father of corporate strategy. (Thinker 052). www.managers.org.uk. Accessed 14 Feb 2011

Chartered Management Institute (CMI) (n.d.d.) Strategic planning. Checklist 064. http://www.managers.org.uk for CMI members. Accessed 15 Feb 2011

Chelleri L (2012) From the "Resilient City" to urban resilience. A review essay on understanding and integrating the resilience perspective for urban systems. Documents d'AnàlisiGeogràfica, 58(2), pp 287–306. Available from: http://www.raco.cat/index.php/DocumentsAnalisi/article/viewFile/252964/341459

Christopolos I, ALNAP, Glemminge Development Research (2006) Provention consortium forum 2006. Strengthening global collaboration in disaster risk reduction. The Elusive 'Window of Opportunity' for Risk Reduction in Post-Disaster Recovery. ipccwg2.gov/njlite_download.php?id=5282. Accessed 20 Dec 2010

Comfort LK, Kapucu N (2006) Inter-organizational coordination in extreme events: the World Trade Center attacks, 11 Sept 2001. Nat Hazards 39:309–327. https://doi.org/10.1007/s11069-006-0030-x

Coppola DP (2006) Introduction to international disaster management, 1st edn. Elsvier, Burlington, Oxford

Crawford L, Morris P, Thomas J, Winter M (2006) Practitioner development: from trained technicians to reflective practitioners. Int J Project Manage 24:722–733. https://doi.org/10.1016/j.ijproman.2006.09.010

Cuny F (1991) Review of twelve years' experience of disasters and small dwellings. In: Aysan Y, Davis I (eds) Disasters and the small dwelling: perspectives for the UN IDNDR. James & James, pp 23–29

Curado C (2006) Organisational learning and organisation design. Learn Organ 13(1):25–48

Davidson C (2009) Multi-actor arrangements and project management. In: Lizarralde G, Johnson C, Davidson C (eds) Rebuilding after disasters: from emergency to sustainability. Spon Press, London, New York

Davidson C, Johnson C, Lizarralde G, Dikmen N, Sliwinski A (2007) Truths and Myths about community participation in post-disaster housing projects. Habitat Int 31(1):100–115

Davies A, Brady T, Prencipe A, Hobday M (2011) Innovation in complex products and systems: implications for project-based organizing. In: Cattani G (ed) Project-based organizing and strategic management. Emerald, Bingley, UK, pp 3–26

Davies A, Mackenzie I (2013) Project complexity and systems integration: constructing the London 2012 Olympics and Paralympics games. Int J Project Manag 32(5):773–790. https://doi.org/10.1016/j.ijproman.2013.10.004

Davis I (2007) Learning from disaster recovery; guidance for decision makers. International Recovery Platform (IRP). http://www.undp.org.cu/crmi/docs/irp-decmakersleslearned-td-2006-en.pdf. Accessed 17 Nov 2010

Dimmer C (2014) Evolving place governance innovations and pluralising reconstruction practices in post-disaster Japan. Plann Theory Pract 15(2):260–265

Dynes RR, Quarantelli, EL (1989) Reconstruction in the context of recovery: thoughts on the Alaskan earthquake. http://udspace.udel.edu/handle/19716/513. Accessed 08 April 2014

EM-DAT, the OFDA/CRED International Disaster Database (n.d.) (Website) www.emdat.be, UniversitéCatholique de Louvain, Brussels, Belgium. Accessed 05 Jan 2015

Gaillard JC (2017) Natural hazards and disasters. In: Richardson D, Castree N, Goodchild MF, Kobayashi A, Liu W, Marston RA (eds) International encyclopedia of geography: people, the earth, environment and technology. Wiley, Oxford, UK, pp 1–15

Galbraith JR (1983) Strategy and organization planning. Hum. Resour Manage 22(1–2):63–77. https://doi.org/10.1002/hrm.3930220110

Ghoshal S, Lampel J, Quinn JB (2003) The strategy process: concepts, contexts, cases, 4th edn. Pearson Education, pp 234–240

Global Facility for Disaster Reduction and Recovery, (GFDRR) (2016) What did we learn? the shelter response and housing recovery in the first two years after the 2010 haiti earthquake. Washington DC, USA. http://www.recoveryplatform.org/assets/publication/GFDRR/haiti-what-did-we-learn_GFDRR.pdf. Accessed 07 Aug 2017

Global Facility for Disaster Reduction and Facility (GFDRR) (2014) Guide to developing disaster recovery framework (World Reconstruction Conference Version), GFDRR, Washington DC

Goold M (2002) Designing effective organizations: how to create structured networks. Jossey-Bass; Wiley (distributor), San Francisco Calif

Groupe URD (2011) Cities and crises; urban fragility, risk and prevention; humanitarian action in cities; and the challenges of reconstruction (Introductory report for a conference on this title to be held on 26 April 2011 in France). http://www.urd.org/IMG/pdf/Cities_and_crises_-_topics_overview.pdf. Accessed 24 Oct 2014

Haigh R, Amaratunga D, Kerimanginaye K (2006) An exploration of the construction industry's role in disaster preparedness, response and recovery. In: The annual research conference of the royal institution of chartered surveyors. Presented at the COBRA 2006, The RICS, University College London, London

Hamdi N (2010) The placemaker's guide to building community. Earthscan Publications, London, Washington DC

Hanisch B, Wald A (2012) A bibliometric view on the use of contingency theory in project management research. Project Manag J 43:4–23. https://doi.org/10.1002/pmj.21267

Hannagan T (2002) Mastering strategic management. Palgrave, Basingstoke Hampshire [u.a.]

Harpham A (2009) Bridging the gap between corporate strategy and project. Introduction to programme management. www.pm4success.com. Accessed 22 Aug 2009

Hedeman B, van Heemst GV (2006) Programme management based on MSP an introduction. Van Haren Publication

Herber J, Singh JV, Useem M (2000) The design of new organisational forms. In: Wharton and managing technologies, pp 376–392 (Reprinted in Mintzberg H, Ghoshal S, Lampel J, Quinn JB (2003) The strategy process: concepts, contexts, ases, 4th edn. Pearson Education, pp 234–240)

Hickey S, Mohan G (2004) Relocating participation within a radical politics of development: critical modernism and citizenship. In: Mohan G, Hickey S (eds) Participation: from tyranny to transformation?: exploring new approaches to participation in development. Zed Books, London, New York, pp 59–74

Holbeche L (2007) Chapter 5, The evolution of organization theory. In: Understanding change. Routledge, London

Hrebiniak L, Joyce WF (1984) Implementing strategy. Macmillan; Collier Macmillan, New York, London.

ICOMOS (2014) Tangible risks, intangible opportunities: long-term risk preparedness and responses for threats to cultural heritage proceedings of the ICOMOS scientific symposium, 31 October 2012, Beijing, China. http://openarchive.icomos.org/1509/1/ICOMOS_2012_Proceedings_final.pdf. Accessed 09 Mar 2016

ICOMOS (2016) Post -Trauma Reconstruction Colloquium proceedings, 4 March 2016, Charenton-le-Pont—France. http://openarchive.icomos.org/1707/1/ICOMOS%2DPost%2DTrauma_Reconstruction_Proceedings%2DVOL1%2DENGok.pdf. Accessed 09 April 2016

International Federation of the Red Cross and Red Crescent Societies (IFRC) (2004) Chapter 4, Bam sends warning to reduce future earthquake risks. In: World disaster report 2004 focuses on community action, Report no. 7757. http://www.unisdr.org/eng/library/Literature/7757.pdf. Accessed 23 Nov 2010

International Recovery Programme (IRP) (2007) Learning from disaster recovery; Guidance for decision makers. http://www.unisdr.org/preventionweb/files/3619_LearningFromDisaster Recovery.pdf. Accessed 17 Nov 2011

Jha AK, Barenstein JD, Phelps PM, Pittat D, Sena S (2010a) Chapter 13, Institutional options for reconstruction management. In: Safer homes, stronger communities: a handbook for reconstructing after natural disasters. World Bank, New York, pp 197–207. http://www.housingreconstruction.org/housing/toc. Accessed 11 Jan 2011

Jha AK, Barenstein JD, Phelps PM, Pittat D, Sena S (2010b) Chapter 6, Reconstruction approaches. In: Safer homes, stronger communities: a handbook for reconstructing after natural disasters. World Bank, New York, pp 93–107. http://www.housingreconstruction.org/housing/toc. Accessed 11 Jan 2011

Jigyasu R (2016) cause culture cannot wait! Post-earthquake recovery of Nepalese cultural heritages. In: Post-Trauma Reconstruction. Proceedings of the 1-Day Colloquium at ICOMOS headquarters, 4 March 2016. Presented at the ost-Trauma reconstruction, France, pp 7–8

Johnson C (2007) Strategic planning for post-disaster temporary housing. Disasters 31(4):435–458. https://doi.org/10.1111/j.1467-7717.2007.01018.x

Johnson C and Blackburn S (2014) Advocacy for urban resilience: UNISDR's making cities resilient campaign. Environ Urbanization, 26(1):29–52. https://doi.org/10.1177/0956247813518684

Kates RW, Colten CE, Laska S, Leatherman SP (2006) Reconstruction of New Orleans after Hurricane Katrina: a research perspective. PNAS 103(40):14653–14660

Kay J (2008) Structure of strategy. Business strategy review (Special report: Strategy classics) 19, 74–85. https://doi.org/10.1111/j.1467-8616.2008.00553.x

Kelman I (2006) Acting on disaster diplomacy. J Int Aff 59(2):215–240

Kent A (n.d.) In the shadow of the storm: Getting recovery right one year after Typhoon Haiyan (Oxfam Briefing Paper). Oxfam, Oxford, UK. http://oxfamilibrary.openrepository.com/oxfam/bitstream/10546/333586/1/bp-shadow-storm-haiyan-recovery-philippines-061114-en.pdf. Accessed 07 Aug 2017

Kondo T, Karatani Y (2016). Chapter 1Ab, Spatial planning for housing recovery after the great east Japan earthquake. In: Stefan G, Michio U (eds) Spatial planning and resilience following disasters: international and comparative perspectives. Policy Press, pp 41–54

Kono T, Misako O (2016) Nara + 20 and reconstruction community involvement in the task of reconstruction. In: Post-Trauma reconstruction. Proceedings of the 1-Day Colloquium at Icomos Headquarters, 4 March 2016. Presented at the Post-Trauma Reconstruction, Charenton-le-Pont, France, pp 11, 12

Kuceravcova A, Dzurdzenik J (2016). Chapter B3, Role of coordination in building spatial resilience after disasters. In: Stefan G, Michio U (eds) Spatial planning and resilience following disasters: international and comparative perspectives. Policy Press, pp 277–282

Lankatilleke M (2010) The people's process: the viability of an international approach. In: Lyons M, Schilderman T, Boano C (eds) Building back better: delivering people-centred housing reconstruction at scale. Practical Action Pub, Warwickshire UK, pp 63–73

Lassa JA (2012) Post disaster governance, complexity and network theory: evidence from Aceh, indonesia after the Indian ocean tsunami 2004 (Working paper). Institute of Resource Governance and Social Change, IRGSC. http://www.irgsc.org/pubs/wp/IRGSCwp001 complexity.pdf. Accessed 07 Aug 2017

Le Masurier J, Suzanne W, Shestakova Y (2006) An analysis of alliancing procurement method for reconstruction following an earthquake. In: Proceedings of the 8th national conference on earthquake engineering. http://nisee.berkeley.edu/documents/elib/www/documents/200601/8NCEE/8NCEE-000290.pdf. Accessed 08 Dec 2014

Lin J-J, Lin W-I (2016) Cultural issues in post-disaster reconstruction: the case of Typhoon Morakot in Taiwan. Disasters 40:668–692. https://doi.org/10.1111/disa.12172

Lizarralde G, Johnson C, Davidson C (2009) From complexity to strategic planning for sustainable reconstruction. In: Lizarralde G, Johnson C, Davidson C (eds) Rebuilding after disasters: from emergency to sustainability. Spon Press, London, New York

Lloyd-Jones T (2006) Mind the gap! Post-disaster reconstruction and the transition from humanitarian relief. Royal Institution of Chartered Surveyors (RICS), London. http://www.rtpi.org.uk/media/3453/BEProfessionsGuide.pdf. Accessed on 11 Nov 2010

Lyons M, Schilderman T, Boano C, D'Urzo S (2010) Conclusions. In: Lyons M, Schilderman T, Boano C (eds) Building back better: delivering people-centred housing reconstruction at scale. Practical Action Pub, Warwickshire UK, pp 345–359

Management Study HQ (n.d.) Advantages and disadvantages of management by objectives (MBO). Management Study HQ; Library for managerial guide and training. http://www.managementstudyhq.com/advantages-and-disadvantages-of-MBO.html. Accessed 27 Aug 2014

Markides C (2008) Strategy as balance. business strategy review (Special report: strategy classics) 19(3):50–91. https://doi.org/10.1111/j.1467-8616.2008.00553.x

Maskrey A (1989) Disaster mitigation: a community based approach. Oxfam, Oxford

Maskrey A (1994) Disaster mitigation as a crisis of paradigms: reconstructing after the Alto Mayo Earthquake, Per. In: Varley A, Wiley J (eds) Disasters, development and environment. Chichester, New York

Maskrey A (2011) Revisiting community-based disaster risk management. Environ Hazards 10 (1):42–52. https://doi.org/10.3763/ehaz.2011.0005

McGregor D (1966) The human side of enterprise. In: Bennis WG, Schein EH (eds) leadership and motivation, essays of Douglas McGrego. MIT Press, Cambridge, MA, pp 3–20

Miner AS, Mezias SJ (1996) Ugly duckling no more: pasts and futures of organizational. Learning 7(1):88–99

Mintzberg H (1980) Structure in 5's: a synthesis of the research on organization design. Manag Sci 26(3):322–341

Mintzberg H, Ghoshal S, Lampel J, Quinn JB (2003) The strategy process: concepts, contexts, cases, 4th edn. Pearson Education

Morgan G (2006) Images of organization, updated edition (first edition 1997). Sage, Thousand Oaks

Morris PWG, Jamieson A (2005) Moving from corporate strategy to project strategy. Project Manag J 36(4):5–18

Morris PWG (2011) A brief history of project management. In: Morris PWG, Pinto JK, Soderlund J (eds) The Oxford handbook of project management. Oxford University Press, New York, pp 253–272

Morris PWG, Geraldi J (2011) Managing the institutional context for projects. Proj Manage J 42 (6):20–32. https://doi.org/10.1002/pmj.20271

Nine C (2017) The wrong of displacement: the home as extended mind: the wrong of displacement. J Polit Philos. https://doi.org/10.1111/jopp.12133

Oliver-Smith A (1998) Global changes and the definition of disaster. In: Quarantelli E (ed) What is a disaster? perspective on the question. Routledge, London, pp 177–194

Oxford Dictionary (n.d.a) Oxford Dictionary. http://www.oxforddictionaries.com/definition/english/strategic. Accessed 17 Nov 2014

Oxford Dictionary (n.d.b) Oxford Dictionary. http://www.oxforddictionaries.com/definition/english/stratagem. Accessed 17 Nov 2014

Özerdem A, Jacoby T (2006) Disaster management and civil society: earthquake relief in Japan. I. B.Tauris, Turkey and India

Parkins J (1996) Preface. In: Management decisions for engineers, ICE Publishing. https://doi.org/10.1680/mdfe.25011

Pearce L (2003) Disaster management and community planning, and public participation: how to achieve sustainable hazard mitigation. Nat Hazards 28(2–3):211–218. https://doi.org/10.1023/A:1022917721797

Pellegrinelli S, Partington D, Geraldi JG (2011) Programme management: an emerging opportunity for research and scholarship. In: Morris PWG, Pinto JK, Soderlund J (eds) The Oxford handbook of project management. OUP Oxford, pp 253–272

Pelling M (2003a) Chapter 2, Cities as sites of disaster. In: Pelling M (ed) The vulnerability of cities: natural disasters and social resilience. Earthscan Publications, London, pp 19–45

Pelling M (2003b) Conclusion. In: Pelling M (ed) Natural disasters and development in a globalizing world. Routledge, London, New York

Pelling M (2003c) Paradigms of risk. In: Pelling M (ed) Natural disasters and development in a globalizing world. Routledge, USA & Canada

Pelling M (2007) Learning from others: the scope and challenges for participatory disaster risk assessment. Disasters 31(4):373–385. https://doi.org/10.1111/j.1467-7717.2007.01014.x

Perrow C (1973) The short and glorious history of organizational theory. Org Dyn 2(1):3–15. https://doi.org/10.1016/0090-2616(73)90007-7

Quarantelli EL (1995) Patterns of sheltering and housing in US disasters. Disaster Prev Manag 4:43–53. https://doi.org/10.1108/09653569510088069

Quarantelli EL (1990) Similarities and differences in institutional responses to natural and tecnological disasters. University of Delaware Disaster Research Center. http://udspace.udel.edu/bitstream/handle/19716/519/PP147.pdf?sequence=3. Accessed 09 April 2014

Quzai U (2010) Pakistan: implementing people-centred reconstruction in urban and rural areas. In: Lyons M, Schilderman T, Boano C (eds) Building back better: delivering people-centred housing reconstruction at scale. Practical Action Pub, Warwickshire UK, pp 113–134

Rössler M (2016) Preface. In: Post-Trauma reconstruction. Proceedings of the 1-Day Colloquium at Icomos Headquarters, 4 March 2016. ICOMOS, Charenton- le-Pont, France, pp 3–4

Shilderman T (2010) Putting people at the centre of reconstruction. In: Lyons M, Schilderman T, Boano C (eds) Building back better: delivering people-centred housing reconstruction at scale. Practical Action Publications, Warwickshire UK, pp 7–33

Smyth HJ, Morris PWG (2007) An epistemological evaluation of research into projects and their management: methodological issues. Int J Project Manag 25:423–436

Spence R (1991) Materials and construction techniques for disaster protection. In: Aysan Y, Davis I (eds) Disasters and the small dwelling: perspectives for the UN IDNDR. James & James, pp 36–44

Sudmeier-Rieux K, Ash N, FAO (2009) Environmental guidance note for disaster risk reduction: healthy ecosystems for human security, Revised Edition. Ecosystem management series. IUCN, Gland, Switzerland. http://data.iucn.org/dbtw-wpd/edocs/CEM-008.pdf. Accessed 16 Sept 2017

Jaroslav T. Alena K, and Jan D (2016) Chapter B3, Role of coordination in building spatial resilience after disasters. In: Stefan G, Michio U (eds) Spatial planning and resilience following disasters: international and comparative perspectives. Policy Press, pp 277–282

Thinker 052 (n.d.) http://www.managers.org.uk for CMI members. Accessed 19 Jan 2011

Thurairajah N, Amaratunga D, Haigh R (2008) Enabling women's empowerment in post-disaster reconstruction. In: RICS Construction and Building Research Conference, Royal Institution of Chartered Surveyors (eds) COBRA 2008: the construction and building research conference of the royal institution of chartered surveyors held at Dublin Institute of Technology, 4–5 Sept 2008. The Royal Institution of Chartered Surveyors, London

Turner JFC (1972a). Chapter 7, Housing as a verb. In: Turner JFC, Fichter R (eds) Freedom to build: dweller control of the housing process. Macmillan, pp 148–175

Turner JFC (1972b) Chapter 6, The reeducation of a professional. In: Turner JFC, Fichter R (eds) Freedom to build: dweller control of the housing process. Macmillan, pp 122–147

Twigg J (2004) Good practice review, disaster risk reduction, mitigation and preparedness in development and emergency programming. ODI. Number 9, Mar 2004. London. www.odihpn.org/download/gpr9/part1pdf. Accessed 12 Dec 2014

Twigg J, Benfield Greig Hazard Research Centre (2006) Technology, post-disaster housing reconstruction and livelihood security—practical action consulting (Website) practical action. http://practicalactionconsulting.org/t4sl_disasterapproaches. Accessed 02 Sept 2011

UN (2010) Evaluation of UNDP contribution to disaster prevention and recovery; executive summary. www.undp.org/execbrd/word/dp2011-4.doc. Accessed 22 Jan 2011

UN (2011) Roles and responsibilities-UNDP. www.undp.org/execbrd/word/dp2011-5.doc. Accessed 22 Jan 2011

UN Intergovernmental Preparatory Committee (2014) Suggested elements for the post-2015 framework for disaster risk reduction. http://www.preventionweb.net/files/37865_06elements withdisclaimer.pdf. Accessed 1 Sept 2014

UNDP Bureau for Crisis Prevention and Recovery (2004) Reducing disaster risk a challenge for development (A global report). UNDP, USA

UNDP Evaluation Office (2010) Evaluation of UNDP contribution to disaster prevention and recovery; executive summary. www.undp.org/execbrd/word/dp2011-4.doc. Accessed 22 Jan 2011

UNDP (2004) Reducing disaster risk: a challenge for development. UNDP: Bureau for crisis prevention and recovery, New York. http://www.undp.org/cpr/disred/rdr.htm. Accessed 24 Oct 2012

UNDP (2008) Donor proposal for disaster reduction and recovery 2009–2010. http://www.preventionweb.net/files/8330_DRRdonorproposal081.pdf. Accessed 12 March 2011

UNESCO (2015) (Meeting) 'Concept note', In: Post-Conflict Reconstruction in the Middle East Context, and in the Old City of Aleppo in particular, Paris, France: UNESCO. http://whc.unesco.org/document/139955. Accessed 2 Nov 2017

UN-Habitat (2017) Trends in urban resilience. UN-Habitat, Nairobi GPO Kenya. https://unhabitat.org/books/trends-in-urban-resilience-2017/. Accessed 06 Aug 2017

UNISDR (2005) World conference on disaster reduction, Kobe, Hyogo, Japan Hyogo Framework for Action 2005–2015: building the resilience of nations and communities to disasters (No. A/CONF.206/6), world conference on disaster reduction. Kobe, Hyogo, Japan. http://www.unisdr.org/files/1037_hyogoframeworkforactionenglish.pdf. Accessed 22 Nov 2010

UNISDR (2009a) UNISDR terminology on disaster risk reduction. UNISDR, Geneva, Switzerland. http://www.preventionweb.net/english/professional/terminology/v.php?Id=7823. Accessed 24 Sept 2011

UNISDR (2009b) Chapter 3, Deconstructing disaster: risk patterns and poverty trends at the local level, global assessment report on disaster risk reduction, pp 59–86. United Nations, Geneva, Switzerland. http://www.preventionweb.net/english/hyogo/gar/report/documents/GAR_Chapter_3_2009_eng.pdf. Accessed 21 Dec 2010

UNISDR (2015) Sendai framework for disaster risk reduction 2015–2030 (No. UNISDR/GE/2015-ICLUX EN5000 1st edn). Geneva, Switzerland. http://www.preventionweb.net/files/43291_sendaiframeworkfordrren.pdf. Accessed 20 Nov 2016

UNISDR (2017) Terminology (Website). https://www.unisdr.org/we/inform/terminology. Accessed 06 Aug 2017

UNISDR (n.d.a) The world conference—UNISDR (Website). http://www.unisdr.org/we/coordinate/wcdrr. Accessed 06 Aug 2014

UNISDR (n.d.b) Making cities resilient: my city is getting ready (Website). http://www.unisdr.org/campaign/resilientcities/about. Accessed 06 Aug 2014 and 03 Aug 2017

UNISDR (n.d.c) Chapter 4, The heart of the matter: the underlying risk drives, global assessment report on disaster risk reduction. United Nations, Geneva, Switzerland

UNISDR (n.d.d.) Sendai framework for disaster risk reduction (Website). http://www.unisdr.org/we/coordinate/sendai-framework. Accessed 03 Aug 2017

United Nations Intergovernmental Preparatory Committee (2014) Suggested elements for the post-2015 framework for disaster risk reduction. http://www.preventionweb.net/files/37865_06elementswithdisclaimer.pdf. Accessed 24 Sept 2014

Van Den Eede M, Dirk Kenis G, Van de Walle B (2004) Combining flexibility and reliability for mainstream organisational learning. In: Proceedings of the 5th European conference on knowledge management. Presented at the 5th European conference on knowledge management, pp 851–860

Vincent S (2004) Participation, resistance and problems with the 'Local' in peru: towards a new political contract? In: Participation: from tyranny to transformation?: exploring new approaches to participation in development. Zed Books, London, New York

WCDRR (n.d.) Conference—third UN world conference on disaster risk reduction, WCDRR. (Website). UN World Conference. http://www.wcdrr.org/conference. Accessed 06 Aug 2014

Wenger D (1977) A review essay on dennis S Mileti, Thomas E Drabek and J Eugene Haas, Human systems in extreme environments: a sociological perspective. Mass Emergencies 2(1), pp 51–59

Whitley R (1984) The scientific status of management research as a practically-oriented social science. J Manage Stud 21:369–390. 10.1111/J.1467-6486.1984.Tb00234.X

Wilford J (2008) Out of rubble: natural disaster and the materiality of the house. Environ Plann D 26(4):647–662

Williams G (2004) Towards a repoliticization of participatory development: political capabilities and spaces of empowerment. In: Hickey S, Mohan G (eds) Participation: from tyranny to transformation?: exploring new approaches to participation in development. Zed Books, London, New York, pp 92–107

Wisner B (2004) Chapter 9, Towards a safer environment. At risk: natural hazards people's vulnerability and disasters. Routledge, London, pp 321–376

World Economic Forum (2015) Building resilience in Nepal through public-private partnerships. Geneva, Switzerland. http://www3.weforum.org/docs/GAC15_Building_Resilience_in_Nepal_report_1510.pdf. Accessed 07 Aug 2017

Wren DA (2005) The history of management thought, 5th edn. Wiley, Hoboken, NJ

Zetter R, Boano C (2010) Space and place after natural disasters and forced displacement. In: Lizarralde G, Johnson C, Davidson C (eds) Rebuilding after disasters: from emergency to sustainability. SponPress, London, New York, pp 206–230

Zargar A (2004) Afterthoughts from an Iranian perspective In: Mumtaz B, Noschis K (eds) Development of Kabul: reconstruction and planning issues. Comportments, Lausanne, pp 125–129. http://www.kunstgeschichte.uni-muenchen.de/forschung/publikationen/pdf_publ/schadl_dev_of_kabul.pdf. Accessed 20 Feb 2014

Part II
Case Study

Chapter 3
The Relationship Between Urban Governance, Disaster Management and Reconstruction Trends in Iran

Abstract Similar to many other developing countries Iran experiences rapid urbanisation but within a highly earthquake-prone geographical context. As a developing country, Iran follows the pathway towards development in various fields. However, the 8-year war with Iraq not only postponed the development process, but also legitimised and prolonged a centralised government. This chapter provides an overview of Iran's institutional context and experiences in reconstruction and the interplay of underlying developmental aspects influenced strategising the recovery and organising reconstruction of Bam. Access to knowledge and increase of urban professionals contribute to old-fashioned urban development programmes in the existing urban governance system being criticised for their quantitative ignorant roles and their damaging architectural values of the country's ancient history on organised settlements. Post-war reconstruction experienced different approaches to reconstruction. It started from idealism, radical changes and not counting on people for reconstruction, but it shifted towards realism and delegating the task of reconstruction with locals. Manjil reconstruction was a turning point for disaster management and a milestone for institutional professional context. It cumulated the existing know-how and lessons learned as the broad national reconstruction policies. However, such valuable lessons for the country, in general, were mainly based on rural reconstruction experiences. Dealing with the complexities of urban reconstruction had preciously proved to be a challenge. The Bam earthquake on 26 December 2003, was the first large-scale disaster after such advances in disaster management, happening in a historic landscape with an internationally famous ancient citadel. It created a field which was a transversal intersection of various developmental patterns.

3.1 An Overview of Iran

Located in southwest Asia, Iran is an ancient country of 1,648,000 km^2, the second biggest country, after Saudi Arabia, in the Middle East–West Asia region. With a population of 71,830,091 in 2007, and nearly 78 million in 2014, Iran is the second

© Springer International Publishing AG, part of Springer Nature 2018

F. F. Arefian, *Organising Post-Disaster Reconstruction Processes*,
The Urban Book Series, https://doi.org/10.1007/978-3-319-70911-6_3

most populated country, after Turkey, in this region. This population mostly live in urban areas. In 2006, the percentage of population living in 1016 towns and cities was 68.45%. With seven cities, with over one million inhabitants, Iran has the largest number of megacities in this region (The World Bank 2008 National Statistic Centre of Iran n.d.).

Iran experienced two major revolutions during its contemporary history: the first was a constitutional revolution in 1907 that limited the role of the King; the second one was the 1979 revolution (Meskinazarian 2011). Iran's political structure changed radically after 1979 from Kingdom to the Islamic Republic. The introduction of the Constitution of the Islamic Republic of Iran (adopted in 1979 and amended in 1989) explains this: On 3rd April 1979, after 2 years of demonstrations that led the Shah to leave the country, 98.2% of people voted for a change of the political regime by voting 'Yes' to the Islamic Republic as the new political system (Constitution of Islamic Republic of Iran, 1989 edition). The post-revolution structure stands on two feet, Islamic nature and Republic nature. The Faghih (Supreme Leader), who is a just and knowledgeable cleric elected by a council of high-profile clerics (Guardian Council), has the supreme power to control the system. The Supreme Leader has the duty to prevent any deviation by various organs from their essential Islamic duties (Constitution of Islamic Republic of Iran, 1989 edition; Meskinazarian 2011).

Concurrently, there publican nature of the country emphasises the crucial role of people as the main administrator of the Islamic Republic. The constitution states:

> In the Islamic Republic of Iran, the affairs of the country must be administrated on the bases of public opinion expressed by means of elections, including the election of the President of the Republic, the representatives of the National Consultative Assembly (the Parliament), and the members of Councils, or by means of referenda in matters specified in other articles of this Constitution.'
>
> (Article 6)

Within this structure, the Faghih and people govern the country together by means of three sources of Power: the Legislator (including members of Parliament), the Executive (the government formed of an elected President and the 22 Ministers), and the Judiciary (its head is appointed by the Faghih). Such a structure aims to safeguard both Islamic and Republic aspects of the revolution. As Imam Khomeini, the leader of the revolution stated the republicanism in it Islamic concept means the acceptance of public opinion (The Portal of Imam Khomeini-a, n.d.). As quoted in the official online resource about Imam Khomeini (the Portal of Imam Khomeini), he warned the government:

> As for the future of this Revolution the person who is with the people, who moves with the tide of the people will become successful and forge ahead. If you step aside then you will definitely be ruined. You should go and be with the people and pursue their policy line.
>
> (Imam Khomeini quoted in the Portal of Imam Khomeini-b n.d.)

This unique way of blending two streams creates a complex internal dynamic inside the country at different stages. Despite such emphasis on the role of people in governing the country, post-revolution Iran is mainly characterised through its centralised governance approach. Among the reasons identified for this are: revolution caused big investors and industrial owners to flee from the country, thus, the government had to take their businesses over and to nationalise them; and, the 8-year war with Iraq required the government to centralise military and administrative functions and unite resources (Tajbakhsh cited in Meskinazarian 2011).

Following the discovery of Oil in 1908, the economy of Iran shifted towards an oil-based economy. According to the Human Development Report 2013 conducted by UNDP, Iran is one of the major oil export countries and is now a middle-income country; the income per capita is 10,695 dollars in 2013 (UNDP 2013).

According to the latest Human Development Report on Iran, the country's pathway towards development during the last decades shows significant progress in some areas, such as access to education, higher education and health. Iran's Human Development Index (HDI) has considerably progressed during the last 32 years. For example, Iran's HDI for the year 2012 is 0.742, meaning that the country is at a 'high human development'. The country also achieved the second highest reduction in HDI shortfall—among developing countries—during the 22 year period between 1990 and 2012. Only one country (the Republic of Korea) was able to do better. Between the years 1980 and 2012, Iran's HDI value increased by 67%—or at an average annual increase of about 1.6% (UNDP 2013).

HDI evaluation is based on long and healthy life, access to knowledge, and a decent standard of living; UNDP uses four HDR categories: very high, high, medium and low. The first HDI component is 'a long and healthy life'. UNDP measures this in terms of life expectancy at birth. During the period from 1980, Iran increased this measure from 51 years to 73 years. The second HDI component is 'access to knowledge'. The yardstick is expected years of schooling starting in 1980 (23 years before the Bam earthquake). Iran recorded an increase from 8.7 to 14.4 years this is praised by UNDP as an impressive achievement. This raise indicates the importance of higher education in Iran because obtaining the National Diploma requires 12 years of schooling, therefore any education more than this belongs to higher education. The third HDI component is 'a decent standard of living'. Here again, the increase has been considerable in recent decades and based on the GDR 13,451 gross US Dollars per capita in 2013 Iran has attained the status of a middle-income country.

However, as the Human Development Report 2013 outlines, despite such positive progress, Iran strives to meet a range of development challenges. This includes disaster preparedness and poverty. Unemployment combined with inflation are persistent challenges, as well as gender equity. The country's gender inequality index is 0.510, ranking it 109 out of 148 countries in the 2013 index. This is reflected in access to employment and economic activities, for example, female participation in the labour market is 16.4% compared to 72.5% for men. Women's presence in top jobs and decision-making levels is low, for example, women hold only 3.1% of the parliament entry seats and 1.53% of the municipal council seats (UNDP 2013; UNDP 2014; United Cities and Local Democracy & World Bank

2008). In the 2013 report, Iran is ranked 75 out of 148 countries for the Multi-Dimensional Poverty Index, MPI, however, the actual index was not calculated due to lack of data (UNDP 2013).

The website of United Nations in Iran classifies it within the top ten most disaster-prone countries in the world, with more than half of all natural disasters related to earthquakes between 1986 and 2007. Climatic events, e.g. floods, drought and sandstorms, represented another 47% of disasters during this period. The previous disaster management activities have been developed further through recent national-level initiatives, such as the adoption of new legislation on disaster management and the establishment of the National Disaster Management Organization (NDMO) to oversee coordination. The Government allocates 2% of its national budget to disaster risk reduction (UN, n.d.). Areas for development such as disaster monitoring and early warning, relevant protocols and standardised operating procedures are to be developed further.

As the Global Assessment Report (GAR) 2013 concludes the world is an urban world now; by 2050 more than 70% of the world population lives in urban areas. Rapid urbanisation applies to Iran too. The Global reports on Decentralisation and Local Democracy place urbanisation in Iran at 68.5%. This is moderate in the region and comparable with Turkey, 68.7% and Iraq, 66.6% (World Bank & United Cities and Local Democracy 2008 & 2010). According to the National Statistics Centre in 1956 there were 199 cities in Iran, increasing to 373 cities in 1976, 612 cities in 1996 and 1016 cities in 2006. Amini (cited in Meskinazarian 2011) identifies four factors for this rapid urbanisation and the increase of urban population: increase of the number of births after the post-revolution baby boom, better health facilities and decrease in the number of deaths; migration from villages and expansion of cities and merging the periphery settlements to the cities, and finally classifying large villages as towns and classifying large towns as cities due to the number of their inhabitants.

3.2 Urban Governance and Historic Urban Characteristics

The country is governed at three levels, national, provincial and local. Iran consists of 31 provinces, each governed by a provincial governor, appointed by the Ministry of Interior Affairs. At local level, the country saw a shift in urban governance from 1999. Previously, mayors were directly appointed by the governor. In 1999, finally, the long-delayed town council elections took place, therefore, the approach towards urban governance became relatively bottom-up and democratic. For the town council elections, local people elect members of the town council, which in turn chooses the mayor and informs the Ministry of Interior for its endorsement. The Bam earthquake happened during the first term of the Bam and Barabvat town councils and their chosen mayor (Meskinazarian 2011).

Known as Persia until 1936,[1] Iran is an ancient country with longstanding tradition of urbanism within climatic and geographic diversity. The UNDP website introduces the country as the home of some school of thoughts about city formation that are reflected in what is commonly known as Islamic urbanism. The progressive combination of philosophical approach (i.e. Islam and Sufism), climatic and geographic responsiveness led historically to form urban fabrics that are responsive to the climate, relate to the landscape and offer distinctive townscapes. Some of these cities, e.g. Isfahan, are internationally recognised as urban masterpieces, and some, e.g. Yazd, Kashan, Kerman and Bam have been tourist destinations with international fame. The reminders of such historic tradition are reflected in older parts of the cities and city centres as urban morphology, architectural styles and the way natural and the built environment are related to each other.

Rapid urbanisation, however, does not respect some crucial aspects, for example, the capacity of the infrastructure, shortage of green space, quality of the built environment in general. Urban development programmes are guided and directed through comprehensive plans that in their nature are the inheritance of the first comprehensive plan for Tehran in 1968. As Madanipour (2006) in his study on urban development in Tehran points out the comprehensive plan of Tehran (produced by a joint activity of the first-ever Iranian consultancy[2] and an American consultancy) reflected the spirit of the fashionable land use-based planning approach of the 1960s. It was mostly based on American cities' patterns, including a network of freeways and wide roads to connect disjointed areas, zoning as the bases for managing the social and physical dimensions, and introducing Floor Area Ratios for controlling development. Replicating such an approach for other cities comprehensive plans cover a 20 years period and the municipalities are responsible to implement these inflexible plans. The comprehensive plans treat the traditional urban fabric the same way as the newer parts. Plans determine the bulk and outer volume of the buildings, including houses. However, they do not take account of the existing architectural style and spatial characteristics, e.g. urban morphology. The architectural design regulations are limited to the Floor Area Ratio; and the building, regardless of its plot's orientation, surroundings and position in urban morphology has to be located in the north of the plot.

Barakpou and Keivani (2016) link urban governance and unsustainable urban development programmes in Iran. During the application of such planning systems over 40 years its quantitative approach has damaged Iranian cities, especially the city centres and traditional fabrics in older cities. Simultaneously, it could not offer a new contemporary urban identity that reflects the traditional architectural styles. Since the mid-1990s there has been a growing new wave of highlighting the need for qualitative approach and respecting the historical context of urban plans. This new trend can be seen as related to the higher education system that produced a new generation of urban design professionals.

---

[1]In 1936, the King, Pahlavy the first changed the name of the country to Iran.

[2]Abdolaziz Farmanfarmaeian Consultancy and Victor Gruen Accociates.

Before 1990, after having graduates who have studied urban planning in the country, their perceptions of the city were based on quantitative measures, whereas urban development projects were in need of qualitative approaches in city planning and design. Within this existing planning system, regardless of the existing traditional/contextual characteristics, design codes and planning regulation did not offer enforceable regulations and guidelines respecting the context or the appearance of the buildings. The quantitative calculations of the needed homes, schools, hospitals, and so on were the only main considerations for design codes. Also, wide roads, geometric grids and prioritising cars to pedestrians have been the mainstream of the urban transport system (Etessam and Pakzad 2012).

Respecting and safeguarding the historic urban fabric was limited to a few flagship conservation projects for famous cities, e.g. Yazd and Isfahan, introduced by the Urban Conservation Department (which had been formed at some stage along the way), and the conservation of buildings by the Cultural Heritage Organisation.

A university professor at the Centre of Excellence in Urban Design at Shahid Beheshti University[3] (SBU), a prestigious university in Tehran, elaborates on the changing trend. In 1990, the country had 900 big and small cities needing 18,000 urban professionals in different fields of urban studies. However, the total number of active professionals in urban studies was only 1800, which was 10% of the actual need. After the Iraq–Iran war (also known as the first Persian Gulf War, 1980–1988), an upsurge began from 1989 in launching urban development courses in universities. Concurrently, trained professionals were needed to advocate the forgotten qualitative aspects of city planning and design to tackle the problem of the mainstream quantitative approach to urban development. Therefore, based on the suggestion of a group of university professors (including him) two-course syllabuses for urban design and planning were prepared. In 1989, Urban Planning and Urban Design were presented as two separate but complementary majors (interview 2013; Etessam and Pakzad 2012).

Initiating urban design as an independent discipline in universities was a milestone to start recognition of the need for a qualitative approach towards the cities. Hence, safeguarding the remainder of the historic urban fabric that still existed was not just a matter of architectural nostalgia. The new generation of urban designers (including me—the researcher) gradually spread in the public and private sector, triggering a new trend of highlighting the need for a qualitative approach to the cities and the role of urban design as a strategic urban development tool. The result was the introduction of urban design projects and programme at various scales that were commissioned by the Ministry of Housing and Urban Development (MHUD). Many of them, however, remained on bookshelves. The mainstream comprehensive planning as reference point for housing development practice has remained intact.

Concurrently, the relatively open political space during the presidency of Khatami, 1997–2005, (Meskinazarian 2011) meant more political tolerance in

[3]The SBU was called the National University before the 1979 Revolution.

general throughout the country. The earthquake happened in the second term of his presidency of and the housing reconstruction programme was formulated and its delivery organisation was formed in this period that broadly and relatively nourished the presence of public/professional voices. His moderate approach emphasised on culture and the importance of dialogue that in turn resulted in the emergence of social initiatives which gave voice to people (including, architectural, urban design and planning professionals) in various directions. For example, new advocacy professional societies and institutes (such as the 'Society of Iranian Urban Planners' that pursued urban design and planning matters) were founded and the existing ones became more active. Those professional societies and institutes facilitated sharing ideas, and they, in fact, were platforms that connected the professional body concerned with the quality of the built environment regardless of their official positions (interviews 2013). For example, at the time of the Bam earthquake, the deputy minister and a number of high-profile managers at the MHUD were among the first generation of the above urban designers. Such a professional context meant readiness to explore the potentials of a qualitative approach and ideas for the reconstruction of Bam that had famous historical and traditional urban characteristics before the earthquake. This built the foundation that contributed to the introduction of an objective on safeguarding the historic and traditional urban architectural fabric for the reconstruction of the Bam area and practical considerations for approaching this objective, as will be discussed in the next chapter.

3.3 Earthquake Vulnerability

The links between urbanisation and increased vulnerability to disaster is acknowledged in disaster development studies (Pelling 2003). Being part of the Alp-Himalaya belt with high level of seismic activities, Iran is one of the most earthquake-prone countries in the world. As Iran's national report for the Hyogo Conference in 2005 highlights during the last 90 years deadly earthquakes claimed 180,000 lives in the country; and UNDP (2004) identified Iran as the fourth most disaster-prone country in the world.

Iran started introduction of seismic building codes within the first two chapters of its National Building Codes in 1989. However, it was the deadly earthquake in Majnil in 1990 that exposed the need for enforcing seismic building regulations. The Manjil earthquake was a turning point, and the seismic regulation for all buildings, including houses, became compulsory from then on. The law of Engineering Organisation and Building Control (that was put to test in 1992) has been in charge of ensuring the National Building Codes since then (Ghafory-Ashtiany and Hosseini 2007; IFRC 2004).

It can be observed that the country suffers from high demand but an unhealthy housing industry (IFRC 2004), where many developers and contractors aim to increase their profit margin, by shortening the overall time of return on their capital and compromising on construction materials wherever possible. As IFRC (2004) reports, growing demand for housing as well as the lucrative building boom, lack of appropriate inspection of new buildings and whether they are conforming to earthquake norms and corruption in this field are the diseases in the country's housing industry. Detailed seismic regulation, (article 2800) is compulsory.

Municipalities in Iran are the executive body for issuing planning permissions, supervision and enforcement of application of the national building codes. The housing development activities in Iran bring together the following: developer and owner, contractor (it can be the developer himself), technical professionals (architects and engineers), engineer supervisor (commissioned/paid by developer or an engineering office, and responsible before the Engineering Organisation) to provide supervisory reports on the construction operation. The inspector from the municipality has to check the accuracy of the supervisor's reports and compliance of the construction operation with the approved plans. There can also be an optional resident supervisor (usually for high-end developments, who are commissioned by the developer and provide in-house supervision of the contractors' work, usually not the case for ordinary housing projects) (member of Tehran municipality, interview 2013; local consultancy, interview 2013).

Engineering Organisation Law requires control mechanisms for housing construction by which the engineering supervisor must control the quality of construction operation at the minimum at five stages and report back to the municipality. If the report indicates non-compliance with the National Building Codes the municipality should enforce the law. These five stages are as follows: foundation, structure, erecting walls and roofs, plastering and end of construction operation (Deputy of Housing and Construction, MHUD. 2011). However, as IFRC (2004) highlights its legislation often is not applied. Regulations hold engineers responsible, but prosecutions of individuals were almost non-existent before the Bam earthquake. There are no restricting laws against negligent municipalities which fail to retrofit infrastructure. Municipalities are blamed for their inability to apply controlling mechanism on the construction operation in the process of housing development by allowing loopholes within the practice of housing development that provides room for 'friendly inspections'.

Moreover, houses in Iran are expensive equities for the lower or even middle-class income of families in Iran. Inflation is high, e.g. the average price for a home per square metre in Kerman province (the provinces which Bam is located in) was 1,901,000 Rials (around $250) in 2003, rising to 7,877,000 Rials (around $710 in 2008), according to the data National Statistics Centre of Iran in 2003 and 2008 (National Statistics Centre of Iran, n.d. a and b), causing uncertainty and constant pressure to enter the property ladder as soon as possible. This has transformed houses from safe living needs to capital investments, as IFRC (2004) states there is a housing boom. This, in turn, encourages unprofessional developers and contractors who target low-income or middle-class buyers. Dodgy engineering practices are always ready to 'professionally' support these developers and contractors. This also tempts developers to target higher income buyers who are looking for the best 'Investment'.

3.4 Disaster Management System

The disaster management system in Iran has an evolutionary nature. Although historically it was more focused on post-disaster activities for which disasters were referred to as unexpected incidents. The importance of DRR gradually and recently has entered the disaster management system in Iran.

After the Manjil earthquake in 1990, a new law on disaster management was passed, in 1991, which appointed the Ministry of Interior Affair as the overall responsible body for disaster management through two newly established agencies: The Bureau for Research and Coordination of Reconstruction Affairs (BRCSR) and the Natural Disaster Headquarters (NHD). BRCSR researches safety measures, formulates preparedness and mitigation plans; collects, analysis and disseminates disaster information; provides coordination services for relief, rehabilitation and reconstruction; monitors activities for disaster management including budget forecasting; provides logistical and procurement support services for provinces (Meskinazarian 2011; Ghafory-Ashtiani 1999).

NHD co-ordinates and supervises between organisations involved in disaster management or post-disaster activities. This includes rescue and relief operations, temporary settlement and reconstruction with the cooperation of all respected governments agencies. The direct specific responsibilities under this supervision and coordination are for and between the three stages of disaster management. The first is the rescue and relief stage that engages the Red Crescent of Iran and the Armed forces for 72 h or more if needed. The second is the temporary settlement stage that engages the Red Crescent of Iran and the Ministry of Interior. Third is the reconstruction stage which engages HFIR and MHUD with emphasise on seismic structure after reconstruction (Ghafory-Ashtiany 1999). As HFIR describes, immediately after a catastrophic incident a dedicated committee, including representatives of other ministries and related organisations is formed to discuss and make decisions (HFIR 2012).

According to Iran's national report for the Hyogo conference on disaster risk reduction, Iran, in line with the International *Decade For Natural Disaster Reduction* (IDNDR), approved the formation of the National Committee for Natural Disaster Reduction in 1991 headed by the Ministers of Energy, Agriculture, Health, Commerce, Jihad of Construction, and MHUD. The Committee was designed as a policy making body to provide exchange of information and to allow the government to have the authority to support and follow up the related activities. Legislation introduced and enacted in relation with NDR during INDR included the Act on the formation of National Committee for NDR, the approval of the executive bylaw for the above mentioned act by the cabinet of Ministers, as well as instructions for particular cases: Technical standards for sand and gravel exploration in river channels; Criteria setting for construction of coastal and intersecting structures, parallel to or over rivers; Consideration of construction limit along river channels and the like; Act on the compensation of damages resulting from floods; Regulation for general directors for NDR; Enactment of building code under standard no. 2800 as an obligatory standard for construction companies and institutions at the national level; Regulation concerning the resistance of buildings against earthquakes; Setting instruction documents to be supervised by municipalities and other executive organisations for incorporation of technical and safety measures; Enactment of the Comprehensive Crisis management plan (Government of Iran 2005).

The Council for Earthquake Disaster Risk Reduction under the responsibility of the Planning and Budget Organisation (the supreme approving body for all public sector development plans and programmes) was formed in 2000. It aimed to provide plans for the reinforcement of important public buildings, infrastructures and the main lifelines of the country.

The Bureau for Natural Disasters Preparedness and Management, under the responsibility of the vice president, was formed in 2004 (after the Bam earthquake). A new National Disaster Management Law was passed by the Parliament, in 2008, to be provisionally implemented for a 5 years period. The National Disaster Management Organization (NDMO) was formed in order to utilise the national, regional and local capacities to cope with the disasters triggered by natural hazards, as well as creating an integrated management system for planning, and coordinating executive activities in a cohesive manner. As NDMO declares, the Hyogo Framework is the reference point for its activities. The National Disaster Management Law extends itself to DRR as it focuses on four stages, including, disaster risk reduction, preparedness, emergency response, reconstruction and recovery (National Disaster Management Organisation, n.d.).

According to the Law, a Disaster Management Coordination Council at national and provincial level is formed and all governmental organisations should have dedicated organisational posts for disaster management and run related activities for disaster risk reduction. However, such steps are at the experimental stage at present. As responsible people for this newly established department in a Ministry and a manager at the municipality of Tehran conclude it might take a generation to move towards a culture of safety (member of a Ministry interview 2013; member of the Tehran municipality interview 2013; National Disaster Management Organisation, n.d.). More recently, the issue of disaster risk reduction and management have been included in the Fifth National Development Plan (2011–2015), including enhancement of disaster preparedness and response, upgrading building and construction codes and standards and earthquake prevention measures, improving safety in rural settlements and allocation of 2% of the annual national budget to this issue (UNDP-d 2013).

Meskinazarian (2011) concludes that the following governmental and/or research organisations are directly concerned with disaster research and management: Army of Islamic Republic of Iran, Planning and Budget Organisation, National Welfare Organisation, Meteorological Organisation, IRIB, Geography Institute of University of Tehran, Disciplinary Forces, Building and Housing Research Centre, Earthquake Committee of National Research Council, International Institute for Seismology

Earthquake, Natural Disaster Prevention Centre. It must also be noted that other organisations, e.g. HFIR that are directly involved in disaster reconstruction are presented as non-governmental but they are not quite independent from the state.

3.5 Previous Reconstruction Experiences: Buyin-Zahra Earthquake 1962

The catastrophic earthquake in Buyin-Zahra, in a rural area in northern Iran, occurred on 1 September 1962 (10/06/1341 in Persian calendar), and killed 12,225 people, injured 2776 people and severely damaged 21,310 houses beyond repair. In total 91 villages were totally destroyed and affected 199,500 people's lives (Ambraseys cited in Meskinazarian 2011). Buyin-Zahra earthquake happened well before the revolution, during the Shah's time, when quick modernisation and development of the country was on the agenda. According to country's reconstruction professionals such as Alizamani (2012), the positive aspect of this reconstruction was that it was linked with developmental goals and strategy of the country at that time, and was seen as an opportunity for improvement. A national newspaper, dated 09/09/1962 (1/07/1341 in Persian calendar), quoted the Shah of Iran that the villages, which will be reconstructed, will be exemplary and provide role models for other villages. A mere reconstruction of houses is not considered; provision of water in any form from Qanats and deep wills is also considered (Etelaat Newspaper reprinted in Alizamani 2012).

However, in practice such opportunistic developmental ambition led to the government decision to relocate 150 villages, promising that it will undertake all developmental activities in these villages, for example building health facilities (Etelaat Newspaper reprinted in Alizamani 2012). The strategy of this reconstruction involved radical changes in the built environment and the society. However, it was not what villagers favoured. In many cases, these new villages were merely housing settlements. In many cases, people did not welcome those new settlements and did not move into the new houses. These constructed settlements after passing decades are still empty (Havaie, n.d.). Figure 3.1 presents one of these new settlements that are still empty after over 40 years.

The extent of the Buyin-Zahra disaster mobilised both international and national help. For example, the Japanese government deployed earthquake experts to help with reconstruction. It also contributed studies on establishing a seismic studies department at the University of Tehran, with a research centre for construction materials, regulations and standards on earthquake resistance buildings (Etelaat Newspaper reprinted in Alizamani 2012). National and religious leaders, sport role models and various social groups helped earthquake-affected people. Buyin-Zahra became the first of such natural disasters in the modern history of Iran.

Fig. 3.1 Buyin-Zahra reconstruction experience in 1962. *Source* HFIR, (Reprinted with permission)

3.6 Previous Reconstruction Experiences: Tabas Earthquake 1978

The catastrophic earthquake in Tabas, occurred on 16 September 1978, killed around 20,000 people and severely damaged 90 villages and the small town of Tabas. It is reported that the earthquake killed 85% of the population of the town of Tabas (Baker et al. cited in Meskinazarian 2011). As Ghafory-Ashtiany (1999) reports in total 5334 housing units were reconstructed including 2436 houses in the town of Tabas and 2774 houses in villages. The Tabas reconstruction contains both positive and negative experiences. The Tabas earthquake happened during the revolution that sought to establish new systems in the country. Many volunteer groups entered the region to build houses for earthquake-affected people, especially the poor. New houses were located next to the old damaged ones and the general approach was to use earthquake resistant construction systems (Ghafori-Ashtiany 1999).

The government started reconstruction and used modern designs, materials and construction techniques. For example, most of the houses were built by reinforced masonry systems, concrete frames or in some cases pre-fabricated units. However, not all these extravagant considerations suited people's lifestyle and taste, and the hot climate of the region. For instance, in one of the villages, where pre-fabricated units were erected, these new buildings were not compatible with people's rural lifestyles. Many villagers left these so-called 'modern homes', repaired their old damaged homes and then used these new buildings as storage or as animal sheds, even though these new buildings included seismic measures and were earthquake resistant buildings (ibid.).

The earthquake, however, created a new understanding of the need to build houses safer than before that influenced the construction system afterwards. Reconstruction took four and half years. It is reported that most of the buildings after the earthquake were safer than before; and people kept the ruins as a memory of prevention (ibid.).

3.7 Previous Reconstruction Experiences: Post-War 1980–1988

Soon after the new revolutionary government took over in Iran, in September 1980 (1359) the Iraqi army evaded Iran, and the 8-year war began that was addressed by the Revolution's Leader, Imam Khomeini, as the 'Wholly Defence'. Along a front of 1352 kilometres penetrating at certain points as deep as 80 kilometres, five provinces were directly affected and under attacks. During the course of 8 years, 3891 villages were damaged. Around 5 million people lost their homes. For instance, in one single province, 236,000 families had to leave their homes to find refuge in other parts of the country (Amirahmadi 1987). Post-war reconstruction started during the war. Soon after overcoming the initial shock of Iraqi attacks the Iranian government found strength to reconstruct the war-affected areas started. From early 1982 (1361 in Persian Calendar) after liberation of a good part of occupied Iran, the Parliament approved the start of reconstruction activities. Reconstruction during the war was a symbolic sign of defiance and self-reliability in the country. This was despite the fact that some of the reconstructed areas were bombarded again and damaged again and parallel reconstruction in war-affected areas was sometimes extremely dangerous (two post-war reconstruction veterans, interviews 2013).

> Take the reconstruction in the same way as the war and strive to solve its problems with the assistance of the people.
>
> (Imam Khomeini 1982, quoted in Meskinazarian 2011, p. 137)

Construction Jihad and HFIR were appointed jointly to this (Miri and Shakerizad Abyaneh 2011). These organisations were revolutionary, the former in charge of Rural Development and the latter in charge of housing for the poor.

3.7.1 Mobilising National Capacity

The extent of the war-affected areas and the scale required for financial resources of reconstruction activities triggered an innovative way of sharing the burden of reconstruction costs. It was through commissioning cities in non-affected areas to support reconstruction of a certain city or area. For example, Tehran in general and especially Tehran Grand Bazaar supported the reconstruction of Khorramshahr, an important war-torn port in the south of Iran. A somehow similar approach for organisational arrangements was taken towards dividing the workload. Given the great number of villages and settlements in a big area, the two organisations in charge, Construction Jihad and Housing Foundation of Islamic Revolution (HFIR), grouped a number of villages together. Each group was attributed a special task organisation for reconstruction. These task forces called Setad-e-Moein (Assistant Organisation) were associated with a specific province or city in the country (Miri and Shakerizad Abyaneh 2011).

Furthermore, the extent of required technical human resources for reconstruction exceeded the capacity of the responsible organisations. Arrangements were made by these organisations and faculties of Architecture and Urban Development in three main universities, i.e. SBU, University of Tehran, and University of Science and Technology, to technically support the reconstruction. Professional technical offices were established in these universities (post-war reconstruction veteran and academic, interview 2013). In addition to voluntarily deployments, architectural students (only boys) were deployed in groups, led by their university lecturers to technically assist HFIR and Construction Jihad in the war-affected areas, which were sometimes still unsafe. Interestingly, such post-war practical strategy influenced the future of the organisations in charge of reconstruction and contributed to the continuity of learning about reconstruction. One obvious example is the HFIR that after the post-war experience became in charge of post-disaster reconstruction and hosts the Office of Reconstruction Affairs (ORA). Many of those architectural students were recruited by these organisations; and as this research finds, moving up the career ladder, for example in 2013, at least four middle managers or high-level managers at HFIR and ORA are from among those students.

3.7.2 Experiencing a Variety of Approaches

Post-war reconstruction saw a variety of approaches to reconstruction that form a spectrum. The early stages of attempts to coordinate and organise reconstruction in 1983 (1362 in Persian calendar) started with idealism and were influenced by revolutionary and Islamic social values. They desired to create a new order for the previous relations between human and nature, human and human, human and livestock and finally human and automobile in the reconstructed settlements. The essence of the reconstruction and design policies as the Construction Jihad portraits it in its guideline for reconstruction is as follows:

> Returning to the pre-war situation is not considered; but access to the methods that can induce change through creating new values in interactions and rural areas is considered.
>
> (Construction Jihad 1983, reproduced in Miri and Shakery 2011, p. 311)

The practical interpretation of such an approach was radical interventions in settlements to incorporate such new orders. An extreme example is regulations for segregating roads for humans and livestock. However, the slow trend of reconstruction compared to the overall volume of the areas which required reconstruction, a lengthy time consuming and costly process, and most importantly the fact that people did not welcome such radical changes in their settlements became apparent. People were changing the layout of the buildings, which professionals had built for them, a waste of already scarce financial resources. Insights towards reconstruction shifted towards acknowledging people's preferences, needs and delegating the task of reconstruction to them. Limiting any physical intervention that might cause unfamiliarity with the new settlement and share the task of reconstruction with people also speeds up the reconstruction process. People are more likely to welcome new places and appreciate reconstruction efforts. Professionals also noticed the educational role of underlying social interactions for changing people's attitude and preferences. A post-war reconstruction veteran and academic (interview 2013) gives an example from early reconstruction stages which tried and failed to change villagers' lifestyle by changing the location of toilets for health reasons. However, they later observed that people had learned such things from their informal interactions with a number of builders who worked both in urban and rural areas.

3.7.3 Experiences of Urban Reconstruction Versus Rural Reconstruction

In Iran, post-war reconstruction was a point of recognition for Iranian practitioners and scholars that urban reconstruction is more complex than rural reconstruction. As Zargar (2004), a prominent practitioner and scholar in Iran's post-war reconstruction, in his recommendations for reconstruction of Kabul (the capital of Afghanistan) explains, classic architecture and urban studies can only be useful in the context of normal societies, not in a city that is badly affected by war. Masterplans or other physical plans for a city are necessary but not sufficient to accomplish reconstruction activities. Reconstruction plans should incorporate a number of other policies that are not usually part of the master plan in order to create a rehabilitation plan. This includes a vision of modernisation of the city and the community due to reconstruction activities, instead of traditionalism, damage assessment and finding financial resources. What adds to the complexities is the fact that urbanites are different from villagers, as rural people require fewer resources, have lower expectations, are more acquainted with and tolerant of a harsh environment, are more ready to participate in and contribute to the reconstruction process, and are often familiar with vernacular construction skills. Importantly the rural economy relies of land and irrigation, both of which can often be restored in a short time after a war (member of HFIR, interview 2013; Zargar 2004).

An illustrative example was Khorramshahr in the south of Iran was an important modern port and has a special place in war-literature/history. It was occupied by Iraqis during the early days of the war, and its liberation is still remembered in a national celebration day. Reconstruction of Khorramshahr as a flagship project was of national importance. For urban reconstruction, the approach was to rely on new urban development approaches within the urban development programmes. Although Khorramshahr was not relocated, its reconstruction saw large-scale physical interventions, without attention to land ownership, unable to acquire required land to implement the new urban development plans. Khorramshahr could only absorb poorer groups of its previous population (post-war reconstruction veteran, interview 2013; Alizamani 2012; Meskinazarian 2011).

3.8 Previous Reconstruction Experiences: Manjil Earthquake 1990

The Manjil reconstruction linked the valuable experiences of post-war reconstruction to the sphere of post-disaster reconstruction. It is also considered as a catalyst for moving towards earthquake risk reduction and a turning point for the disaster management system in Iran. On 21 June 1990, soon after the end of the war, an earthquake of 7.5 magnitudes damaged a vast area of 1500 square kilometres in northern Iran with round 14,000 fatalities, 60,000 injuries and 500,000 homeless. Around 217,000 houses in 700 villages and 3 urban nodes required reconstruction (Akhoondi and Bahraini 2000; Ghafory-Ahtiany 1999).

The government appointed HFIR, based on their previous experiences of both housing and reconstruction, as the executive body for reconstruction in the area. As post-war reconstruction veterans and academics recall, it was natural to offer help for post-disaster reconstruction. This was the time when the fresh cumulated know-how on post-war experience was transferred to post-disaster reconstruction. Technical offices of the universities that previously worked on post-war reconstruction started to work on post-disaster reconstruction (two post-war reconstruction veterans, interviews 2013).

The Manjil reconstruction was a catalyst for moving towards a policy of earthquake risk reduction at the national level that triggered institutional improvements and housing development practices at the national level. At the time of the incident, there were seismic regulations in place but they were not compulsory. The Manjil earthquake revealed that having optional regulations are not enough. It, therefore, triggered the introduction of the Law of Engineering Organisations, which aimed to regularly update the national building codes and assure their application through professional activities in the construction industry. A national professional body, called the Country's Engineering Organisation (CEO), was formed to implement the

Law and to regulate engineering activities (including structure, civil, electrics and mechanics, architecture and later urbanism) and to protect those professionals' rights (Ghafory-Ashtiany 1999; Deputy of Housing and Construction, MHUD 2011).

The Manjil experience was also the point at which the lessons learned from previous reconstruction programmes, both post-disaster and post-war, led to the establishment of a number of broad policies on various aspects of organising reconstruction to avoid previous mistakes and to harmonise activities. Soon after the earthquake, HFIR introduced various macro and micro policies that drew on the previous experiences and established strategic and practical policies on a number of critical topics. For example, the question was not anymore whether people should be engaged in reconstruction or not; or whether the government should deliver ready to move in houses for people or not. All had learned that the active participation of beneficiaries creates a win-win scenario for both people and government that saves time and money, delivers houses that are acceptable by people and helps people to return to normality quicker. The broad national reconstruction policies included the role of government and people for reconstruction, construction techniques, financial support for reconstruction and so on. Following the satisfactory performance of the large-scale reconstruction of Manjil, they were used at various scales of reconstructions. These policies are flexible in their details to reflect every single situation. They were also applied in the Bam case, as will be discussed later (Akhoondi and Bahraini 2000; member of HFIR interview 2013; Ghafory-Ashtiany 1999).

Knowledgeable people in the reconstruction field and HFIR give credit to the president of HFIR at the time, Dr. Akhoondi, as the architect who had established such reconstruction policies towards integrating, coordinating and organising activities. He was the president of Construction Jihad (responsible for rural development programmes during post-war reconstruction), and is regarded as the architect of the core reconstruction policies and an avant-garde manager (post-war reconstruction veteran 2013; member of HFIR 2013).

The overall goal in the Manjil reconstruction was to revitalise life in human settlements. The main fields of attention are to address occupation, build permanent houses and re-establish the needed services with regional and national capacity. The reconstruction programme in Manjil acknowledged that addressing all developmental goals through reconstruction activities is not possible. However, the chance of building grounds for future developmental activities should not be lost. Understanding the region, recognising local and regional potentials and engaging people and attracting people's participation, while examining the potentials for support at the national level, can be helpful for building the ground for future developments. Therefore, long-term idealist comprehensive plans should be avoided. Practical plans should be devised instead, based the existing realities while avoiding quick and unexamined decisions (Akhoondi and Bahraini 2000).

The Manjil earthquake is praised for a successful large-scale rural reconstruction. As Akhoondi and Bahraini (2000) reflect on this experience in their reflective book published 3 years before the Bam earthquake, people's participation is seen as the most distinct characteristics of the Manjil reconstruction. People were in charge of the reconstruction of their houses; and government supported them and delivered what people cannot do by themselves. This took place by building the infrastructure, providing construction material and free technical support and overall supervision of the process to complete the reconstruction activities within the shortest possible time, at best possible quality and with minimum social tensions and problems. People were supported by Setads, technical reconstruction groups for their respective area for the design, planning and construction of their houses. During 3 years the operation engaged 17 Setads and 50 technical construction teams. Additionally, local social groups, Rural Development and Reconstruction Committees, in rural areas were formed. They included local trustees and youth. They had a representative and were participating in decision-making and coordinating the overall reconstruction, for example, presenting technical information or the layout of the public roads, etc. Meetings between these committees and local officials were held in mosques and schools. The earthquake-affected area covered two provinces with two different climatic situations, construction technologies and distinct ethnic background, but reconstruction policies were adjusted to specific situations in each province (Sartipi 2006).

The Manjil reconstruction also confirmed intrinsic challenges in urban reconstruction activities. Despite being characterised as a rural reconstruction the earthquake-affected area also included three small towns, Manjil, Roodbar and Loushan. As Akhoondi and Bahraini (2000) reflect, the urban housing reconstruction compares to the rural one, encountered difficulties. The initial reconstruction programme indicated that all rules and procedures for urban housing procedures had to be followed. Therefore, engineers and architects were needed to prepare technical maps and supervise the construction operation to ensure the application of existing seismic codes—as mentioned in the programme. However, those small cities did not have their own engineers and architects to work with people, follow the design and supervise the procedures in the early 1990s. The responsibility was delegated to the Ministry of Housing and Urban Development, which called on individual engineers and architects and commissioned them from nearby bigger cities and Tehran. In many cases, the engineers did not reside in these cities and some of them provided maps without visiting the sites and there was a minimum level of construction supervision. Akhoondi and Bahraini suggest that perhaps it is worth to examine the idea of having engineering companies in these cities, similar to the approach HFIR used for retrofitting cases. To retrofit urban houses HFIR commissioned one engineering company to establish a local branch and assist people in assessing the damages of their properties and propose retrofit plans. This was the first experience of this kind and proved successful.

Furthermore, the Manjil reconstruction policies emphasised on in situ reconstruction unless strong geological reasons justify relocation. Based on a quick examination 600 villages out of the 700 in need of reconstruction, were decided to be in situ reconstruction, while the other 100 villages, based on geological information were decided to be relocated, integrated and regrouped. A study by Badri et al. (2006), conducted more than 10 years after the reconstruction, showed lower level of quality of life and economic situation of the relocated community compared to the host community.

One of the contextual aspects that positively supported a successful reconstruction in Manjil was that it benefited from stable political decision-making posts during reconstruction. A key policy of the reconstruction was to engage and work with the existing local organisations and institutions for local governance in the reconstruction process. The Manjil earthquake happened during the presidency of Hashemi-Rafsanjani and his government with the slogan of 'Construction Government', which was an indication of the need to compensate the lack of developmental programmes during wartime. The reconstruction activities started and finished during his government (Madanipour 2006; Akhoondi and Bahraini 2000).

3.9 Conclusion

This section provided an overview of the national institutional context and experiences in reconstruction. As a developing country, Iran follows the pathway towards development. Various fields discussed in this section showed a progressive pattern towards development. However, the 8-year war with Iraq not only postponed the development process, but also legitimised and prolonged a centralised government. Town councils, as the manifestation of people's participation are fresh experiences.

Access to knowledge and increase of urban professionals contribute to old-fashioned urban development programmes being criticised for their quantitative ignorant roles and their damaging architectural values of the country's ancient history on organised settlements. Safeguarding historical sites in older cities became a concern. The relatively open social space favoured social and professional activities.

Post-war reconstruction acted as a wide-ranging laboratory, where experiments were undertaken and reflections were observed, to examine different approaches to reconstruction. It started from idealism, radical changes and not counting on people for reconstruction, but it shifted towards realism, limiting unnecessary changes and delegating the task of reconstruction more realistically. Success stories come more from rural reconstruction experiences. It proved that engaging people in reconstruction is advantageous and revealed the importance of the underlying dynamic of social organisations within the community in need of reconstruction.

Like many other developing countries Iran experiences rapid urbanisation but within a highly earthquake-prone geographical context. Manjil reconstruction was a turning point for disaster management and a milestone for institutional professional context. It cumulated know-how and knowledge and lessons learned as the way forward for future reconstruction. However, despite such valuable lessons for the country, in general, they are based more on rural experiences. Dealing with the complexities of urban reconstruction proved to be a challenge.

The Bam earthquake on 26 December 2003 is the first large-scale disaster after such advances in disaster management. It creates a field which is a transversal intersection of various developmental patterns.

References

Akhoondi B, Bahraini H (2000) Reconstruction management of the disaster-affected areas, the experience of reconstruction of housing in the earthquake-affected areas of the Guilan and Zandjan provinces, Iran 1990–1992. The Institute of Publication and Print of the University of Tehran, Tehran (in Persian)

Alizamani MH (2012) Lessons from past experiences in temporary housing in Iran. (Unpublished PowerPoint file presented at various conferences) (in Persian)

Amlrahmadl H (1987) Destuction and reconstruction: a strategy for the war-damaged areas of Iran. Disasters 11:134–147. https://doi.org/10.1111/j.1467-7717.1987.tb00628.x

Badri A, Asgary A, Eftekhari AR, Levy J (2006) Post-disaster resettlement, development and change: a case study of the 1990 Manjil earthquake in Iran. Disasters 30(4):451–468

Barakpou N, Keivani RM (2016) The relationship between urban governance and sustainable urban development in Iran. In: Arefian FF, Moeini SHI (eds) Urban change in Iran: stories of rooted histories and ever-accelerating developments. The urban book series. Springer International Publishing, Cham, Heidelberg, New York, Dordrecht, London, pp 153–169

Constitution of the Islamic Republic of Iran (1979) (1989 amendment) (in Persian)

Deputy of Housing and Construction, MHUD (2011) The Law of Engineering

Etessam I, Pakzad J (2012) Urban design education in Iran, urban design, 124 (Autumn 2012), urban design group, pp 32–33. http://www.udg.org.uk/publications/urban-design-journal-issue/urban-design-124-%E2%80%93-autumn-2012. Accessed 15 Feb 2013

Ghafory-Ashtiany M (1999) Rescue operation and reconstruction of recent earthquakes in Iran. Disaster Prev and Manage 8(1):5–20

Ghafory-Ashtiany M, Hosseini M (2007) Post-Bam earthquake: recovery and reconstruction. Nat Hazards 44:229–241. https://doi.org/10.1007/s11069-007-9108-3

Government of Iran (2005) National report of the Islamic Republic of Iran on Disaster Reduction. Tehran. http://www.unisdr.org/2005/mdgs-drr/national-reports/Iran-report.pdf. Accessed 22 Oct 2014

Havaie MH (n.d.) A brief review on Bam reconstruction program. (Unpublished Power Point file provided by HFIR)

HFIR (2012) Documenting the reconstruction of Bam: design process in reconstruction, (unpublished report) (in Persian)

International Federation of the Red Cross and Red Crescent Societies (IFRC) (2004) Chapter 4, Bam sends warning to reduce future earthquake risks. In: World Disaster Report 2004 Focuses on Community Action, Report no. 7757. http://www.unisdr.org/eng/library/Literature/7757.pdf. Accessed 23 Nov 2010

Madanipour A (2006) Urban planning and development in Tehran. Cities 23(6):433–438

Meskinazarian A (2011) Social resilience of post-earthquake Bam. PhD thesis, Department of Geography, King's College London, London, UK

Miri SH, Shakerizad Abyaneh A (2011) From Bardieh to Sarieh. Publication Centre of ShahidBeheshti University, Tehran (in Persian)

National Disaster Management Organisation (n.d.) National disaster management organisation (Website). National Disaster Management Organisation. http://www.ndmo.org/aboutus.aspx?Id=1. Accessed 31 Jan 2014 (in Persian)

National Statistical Center of Iran (n.d.) Publications information base. (Website). http://amar.sci.org.ir/index_f.aspx. Accessed 01 Aug 2011 (in Persian)

National Statistics Centre of Iran (n.d.a.) Detail (Website). http://amar.sci.org.ir/Detail.aspx?Ln=F&no=224096&S=TP. Accessed 01 May 2011 (in Persian)

National Statistics Centre of Iran (n.d.b.) Detail (Website). http://amar.sci.org.ir/Detail.aspx?Ln=F&no=242673&S=TP. Accessed 01 June 2011 (in Persian)

Organisation and Construction Control: Endorsed in March 2005, with Latest Adjustments and Appendixes, Ministry of Road and Urbanism, Tehran (in Persian)

Pelling M (2003) Conclusion. In: Pelling M (ed) Natural disasters and development in a globalizing world. Routledge, London, New York

Sartipi M (2006) Natural disasters and diagnoses of man-made factors, the cases of Manjil and Bam. Environ Sci 12(Summer):59–66. http://www.sid.ir/en/VEWSSID/J_pdf/112920061207.pdf. Accessed 10 April 2013

The Portal of Imam Khomeini-a (n.d.) Imam Khomeini. Can we put the words “Republic” and “Islam” together in the Islamic Republic Government? (website). http://en.imam-khomeini.ir/en/c104_12453/Ask_Question/Cultural_Social_Sciences/Can_we_put_the_words_%E2%80%9Crepublic%E2%80%9D_and_%E2%80%9CIslam%E2%80%9D_together_in_the_Islamic_republic_government. Accessed 10 Aug 2015

The Portal of Imam Khomeini-b (n.d.) Imam Khomeini. You should go and follow the policy line of the people. (website). http://en.imam-khomeini.ir/en/n12701/Memoirs/Islamic_government_and_governance/You_should_go_and_follow_the_policy_line_of_the_people. Accessed 10 Aug 2015

UN (n.d.) Disasters. United Nations in Iran. http://un.org.ir/index.php/issues-development/issues-disasters. Accessed 05 Oct 2014

UNDP Iran (2004) UNDP helps Bam people to play a part in the reconstruction planning of their quake hit city. UNDP Iran. http://www.undp.org/cpr/disred/english/news/archives/2005/april.htm#iran040405. Accessed 07 Jan 2011

UNDP (2013) Iran extract from HDR 2013. http://www.undp.org/content/dam/iran/docs/Publications/Inclusive%20Growth%20&%20Development/IR%20of%20Iran%20Extract%20in%20HDR%202013%20ENG.pdf. Accessed 15 Feb 2014

UNDP (2014) | Human Development Reports [WWW Document]. UNDP. http://hdr.undp.org/en/countries/profiles/IRN. Accessed 21 Oct 2014

UNDP (2014) | Human Development Reports [WWW Document]. UNDP. http://hdr.undp.org/en/countries/profiles/IRN. Accessed 21 Oct 2014

World Bank (2008) Decentralization and local democracy in the world: 2008 first global report by united cities and local governments. Washington, DC: World Bank

Zargar A (2004) Afterthoughts from an Iranian perspective. In: Mumtaz B, Noschis K (eds) Development of Kabul: reconstruction and planning issues. Comportments, Lausanne, pp 125–129. http://www.kunstgeschichte.uni-muenchen.de/forschung/publikationen/pdf_publ/schadl_dev_of_kabul.pdf. Accessed 20 Feb 2014

Chapter 4
Relationships Between Local Urban Development and Disaster Management System in Bam

Abstract This chapter discusses the local-level underlying drivers linked with existing urban development system, broad disaster management policies and severity of destruction, and professional context interplayed towards shaping the Bam recovery strategy and reconstruction objectives. It examines how the historic and traditional urban characteristics of the Bam urban area, the fame of its ancient citadel (Arg-e-Bam) along the severity of the earthquake, death toll and devastating disaster made a special case for the Bam recovery and reconstruction that is relevant to other international cases. The reconstruction of the Bam urban areas was seen as an emerging opportunity to influence the broader urban development system in the country. The Bam recovery positioned itself as optimistic but pragmatic within the contemporary debates on linking reconstruction and urban development. Yet there were challenges of linking short-term activities to longer-term reference documents of urban development plans, as well as existing sanctions on Iran at the time that also challenged the deliverability of key developmental initiatives that were agreed with international agencies. Despite challenges, the Bam case exemplifies reconstruction per se is part of the developmental path of the area. It highlights a fundamental difference between participatory housing programmes and non-participatory reconstruction projects.

4.1 Overview

The Bam area is located in Kerman province in southeast Iran. It has historical importance for the region and the country, as it is on the connecting trade routes of the Silk Road. The city was an important centre for the production of fine silk and dates. Being the only habitable area in the eastern deserts and being close to Sistan made Bam a strategic location and a gateway to India throughout history. The internationally famous citadel Arg-e-Bam is a reminder of this history as UNESCO World Heritage Centre describes on its website:

> The origins Arg-e Bam, can be traced back to the Achaemenid period (6th to 4th centuries BC) and even beyond. The heyday of the citadel was from the 7th to 11th centuries, being

© Springer International Publishing AG, part of Springer Nature 2018

F. F. Arefian, *Organising Post-Disaster Reconstruction Processes*,
The Urban Book Series, https://doi.org/10.1007/978-3-319-70911-6_4

> at the crossroads of important trade routes and known for the production of silk and cotton garments. For centuries, Bam had a strategic location on the Silk Roads connecting it to Central Asia in the east, the Persian Gulf in the south, as well as Egypt in the west and it is an example of the interaction of the various influences.
>
> The existence of life in the oasis was based on the underground irrigation canals, the qanāts, of which Bam has preserved some of the earliest evidence in Iran. Arg-e Bam is the most representative example of a fortified medieval town built in vernacular technique using mud layers (Chineh).
>
> (UNESCO World Heritage Centre 2004a)

According to UNESCO World Heritage Centre (ibid) the cultural landscape of the Bam area (including the urban area) is an important representation of the interaction between man and nature and retains a rich resource of ancient canalisations, settlements and forts as landmarks and as a tangible evidence of the evolution of the area. With the shift of international trade from the old Silk Road towards ports at the Persian Gulf and Indian Sea, the golden days of trade importance passed. However, Bam retained its importance as a regional centre for the eastern part of the country, and still exports its fine dates to the world, and is on the international transit freeway from east to west (Ekhlaspour 2009; UNESCO World Heritage Centre 2004b; Bastani Parizi cited in Meskinazarian 2011). Unlike its dry and desert-like surroundings, the Bam area is mainly characterised by its palm groves, which are irrigated by traditional underground water systems. Figures 4.1 and 4.2 show the geographical location of the area in relation to the historic trade routes, and the Arg viewed from inside the medieval city of Bam. Figure 4.3 shows the aerial map of the Bam urban area. The area resembles an oasis in a desert.

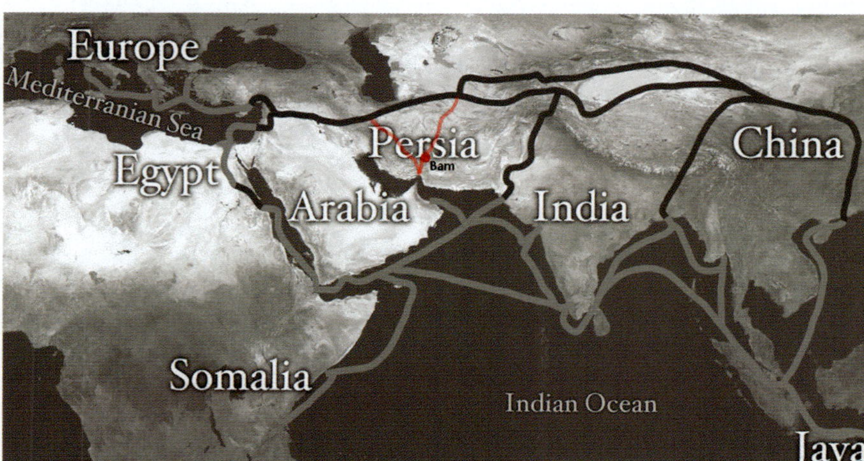

Fig. 4.1 The schematic geographical location of the area on historic trade routes. *Source* Wikimedia Commons, https://en.wikipedia.org/wiki/File:Silk_route.jpg [Accessed 29/10/2014]. (Public domain)

Fig. 4.2 The citadel from a view from inside the medieval city of Bam. *Source* Arad, via Wikimedia Commons, https://commons.wikimedia.org/wiki/File:Arge_Bam_Arad.jpg [Accessed 08/03/2017]. (Public domain)

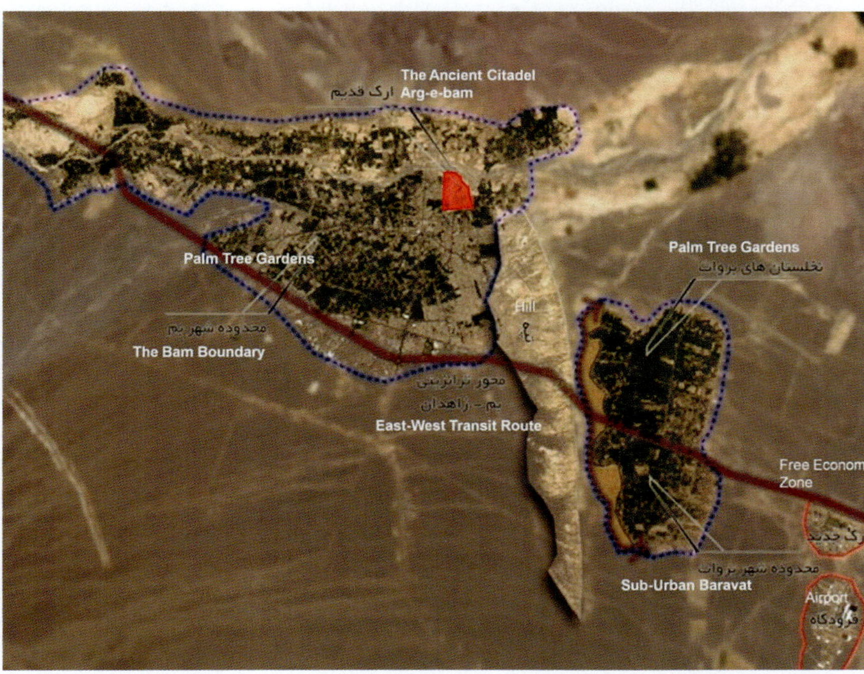

Fig. 4.3 The aerial map of the Bam area in 2003. *Source* HFIR (Reprinted by permission)

4.2 Socio-Economic Characteristics of Bam

The Statistical Centre of Iran (2004), in its special issue on Bam published 2 months after the earthquake, explains that the earthquake damaged region of Bam had a population of 142,376, and that the urban population of Bam and its suburb Baravat were 89,145 and 15,324 respectively. The rest of the population, 37,907 lived in 65 rural centres. According to 1996 census, the closest to the earthquake, 11 to 24 years olds formed 34.66% of the population in the Bam area. This means that at the time of the earthquake, 7 years after the 1996 census, more than one-third of the population were 18 to 31 years old, a young generation who were about to start a family and needed housing (Meskinazarian 2011). The same census indicates that the literacy rate in the Bam urban areas averaged 86.02% (88.10% for males and 83.92% for females) which is significantly higher than the country's average of 74.2 and the province average of 74.8% (Armanshahr 2004). However, in rural areas the literacy rate, 64.2%, falls below the country's average (Statistical Centre of Iran 2004).

The main economic activities in the area include agriculture, industry and the service sector, and the area is a tourist attraction. The historic and modern life of the locals (both rural and urban) emotionally and economically is interrelated to the palm grove gardens. Fine dates are the main agricultural product of the region; and they are exported worldwide. Most families are related to palm grove gardens, even if they formally have another job, through either owning a number of palm groves or working on them seasonally. The Bam area continued its date production during a lengthy drought, in 2002 and 2003, in southeast Iran that affected the rural poor the most (Ekhlaspour 2009). As a historic urban node remote from other cities, Bam is the commerce and service hub in its region. The service sector's share of employment is 46%, of which 26% belong to public services provided by the government, including the military (Ekhlaspour 2009; Khatam 2006). Arg-e-Bam is a national tourist attraction and one of the international tourist destinations in Iran. Despite the country's post-revolution turbulent international relations, it still attracts overseas tourists, with reports of 104,446 visits to Arg-e-Bam in 2002 (Statistical Centre of Iran 2004).While the area has decent hotels and an airport to accommodate such interests, tourism is not the core of the economic life of the city (Ekhlaspour 2009). Historically, the area has agriculture-related industries, e.g. storing, packaging and transport. The Bam area also saw the establishment of a free economic zone (including a car factory, Daevoo Motors) and its related services within commuting distance of Bam that created jobs for locals.

According to the 1996 census the rate of unemployment was 13% for the Bam area (Statistical Centre of Iran 2004). Given the size of the 18–31 age group, we can conclude that this group also had the biggest share of unemployment at the time of the earthquake. Official census reports indicate a mere 10% as the women's share in the total economic activities in the region, however, Ekhlaspour, a sociologist from the region, argues that this does not reflect the real situation and the real rate is higher. According to her, the collected primary data for the census did not consider women's work in their family-related economic activities in agriculture, livestock and in-house light industries, or as seasonal workers (ibid). She also reports of a marginal society at the periphery areas of the city, comprising of Afghan refugees and seasonal workers who come to the city to work on agricultural lands or other industries.

4.3 Bam Urban Characteristics

The present city of Bam is the expansion of the medieval-walled city around the citadel, towards south and palm tree gardens, with establishing urban nodes such as bazaars, mosques, caravanserais and baths (Armanshahr 2004). Emotional and economic interrelations with palm grove gardens are also manifested in the urban fabric, both in present Bam as the regional urban node and in its suburb Baravat. The city is organically formed of dispersed blocks of palm groves and residential houses that are situated among the palm groves. The city is famous as an organic traditional '*garden city*'. The urban fabric of Bam, in general, is dispersed; the only relatively dense urban fabric is the city centre, formed around the main Bazaar. Bam is characterised by its organic urban morphology and garden houses. Narrow roads follow the traditional irrigation system, qanats, which organically pass between garden houses.

The study of architectural characteristic of the Bam area, focusing on housing, based on desk study and fieldwork immediately after the earthquake, later published by the MHUD in 2007, divides garden houses into two main types, traditional and semi-modern. The first type, referred to as traditional garden houses, belongs mainly to older families in Bam and can as much as 200 years old. Various spatial configurations of houses with central courtyards or freestanding can be observed. They present traditional styles of architecture, including traditional materials, dome roofs and wind-catchers. The second type consists of newer garden houses that again have various configurations with a central hall or 'villa' (freestanding) style. This type consists of masonry or steel structure buildings, flat roofs and modern architectural styles. This was the prominent type in recent years (Golpayegani and Einifar 2004). Figures 4.4 and 4.5 show examples of those housing types that appeared in the aforementioned book. Additionally, the recent city expansion followed a grid network urban morphology. Residential blocks were situated in linear forms adjacent to the roads and houses were built of modern building materials, with flat roofs and modern architectural styles, as of other cities in Iran, with a number of two or three-storey buildings (Armanshahr 2004). The suburb of Baravat is a peri-urban district and has a dispersed urban fabric. It is dependent on Bam in almost every aspect of urban life, despite having a separate municipality. It is like a detached district of Bam as a result of a fault-line in between the two. Since it was recently recognised as a town it does not require a comprehensive plan; instead it has a guiding plan which is typically prepared for villages.

Fig. 4.4 Examples of traditional garden houses in Bam. *Source* Research and Documentation Centre, Faculty of Architecture and Urban Development, Shahid Beheshti University (Reprinted with permission)

Fig. 4.5 Example of newer garden houses in Bam. *Source* Golpayegani (Reprinted with permission)

The most recent comprehensive urban development plan followed the same fashion in urban development as the rest of Iran. It was prepared by a consultancy company (Armanshahr), in 2003, and endorsed just before the earthquake. It envisioned Bam as the regional service centre for the surrounding villages and smaller towns. It was a long-term plan which looks at the city for the next 20 years, from 2003–2023 (1382–1402). After the earthquake, the same company was later appointed to prepare the post-earthquake urban development plan, in the format of '*The Special Structural-Strategic Plan for Bam*', as such fresh knowledge on the destroyed city was extremely valuable at the time. The Bam population was estimated to be 150,000 by 2023. According to the Structural-Strategic Plan for Bam, the majority of the buildings (especially houses), before the earthquake were one storey, and the average household size in the 1996 census was 5.23 people; it was reduced to 3.10 after the earthquake and the Plan envisioned reaching 3.4 in 2023. The culture of house ownership was strong, as 73.60% of households were home-owner, while, 8.11% lived in their accommodation for free (e.g. extended family and organisational housing), and 17.03% of households were renters (Armanshahr 2004; Statistical Centre of Iran n.d.).

4.4 When the Earthquake Hit Bam

In the early morning (5:28 am) of Friday, (the weekend in Iran), of 26 December 2003 a major earthquake measuring 6.5 on the Richter scale struck Bam. Within 12 s, according to the UN Flash Appeal in 2004, the earthquake caused around 30,000 fatalities, injured 20,000 people and an estimated 75,000 became homeless. The extent of devastation touched all aspects of survivors' lives, for example 131 schools, all 3 existing hospitals with 225 beds and 24 health centres were destroyed or became unusable. Approximately 85% of the houses, commercial units, educational facilities and health facilities, as well as public and private offices in Bam and its surrounding villages were completely destroyed or became unusable (UN 2004). President Khatami announced the disaster and the government's policy of open skies for the first time since the revolution, eliminating the necessity of requesting a visa to enter the country for the purpose of assisting the affected area. As the news broke, the organisations responsible for the emergency management in the country, their international counterparts, and volunteers from inside and outside the country rushed into the area. As the Oxfam's Humanitarian Programme manager (cited in Meskinazarian 2011) recalled in the first 2 weeks in addition to 8,600 Iranian personnel and 4,500 tonnes of

Fig. 4.6 Destruction of the Bam central area, *Photo* Arefian 2004.

supplies deployed by the government 1,900 international personnel and 120 tonnes of supplies arrived in the area (Havaei and Hosseini n.d.; Ghafory-Ashtiany n.d.). Figures 4.6, 4.7 and 4.8 represent Bam after the earthquake.

Fig. 4.7 The extent of destruction of the Bam urban fabric around the ancient citadel after the earthquake. *Source* HFIR (Reprinted with permission)

Fig. 4.8 Zooming out on the residential urban fabric after the earthquake. *Source* HFIR (Reprinted with permission)

National and international practitioners describe the overall emergency and relief performance as impressive. According to Iranian Red Crescent (cited in Meskinazarian 2011) within the first 12 hours of the earthquake food and water were distributed, and over 20,000 tents were distributed during the first 3 days. There was no significant disease outbreak. Due to the severity of the disaster, the early response teams had to deal with repairing infrastructure such as the water system and health and social services alongside the emergency provision. For example, while one-fifth of the teachers died, a few days after the earthquake schools resumed their work in tents with the help of 1,000 educational volunteers to provide a supportive environment for the surviving students. Mental health councillor volunteers organised supportive activities and group counselling (Tierney et al. 2005; Ghafory-Ashtiany and Hosseini 2007).

On 20th January 2004 (28/10/1382 in Persian calendar) three weeks after the earthquake, the Government formed the Steering Committee for the Reconstruction of Bam (in short, the Steering Committee). According to the government appointment letter on 29/01/2004 (07/11/1382 in Persian calendar), the Committee had the full authority of the government on affairs related to planning, organising, directing operational activities and supervision the reconstruction of the earthquake affected Bam area The Committee consisted of the ministers of the Ministry of the Housing and Urban development (the head of the Committee), the Ministry of Interior Affairs, the Ministry of Culture and Islamic Affairs, the Ministry of Economic and Revenue Affairs, the Ministry of Justice; the president of the Planning and Management Organisation; the president of IFRC, the president of HFIR, the governor of the Kerman Province; and three professional representatives of the president. The Committee members were special representatives of the president for reconstruction (The Cabinet 2004).

4.5 Earthquake Destruction

The epicentre of the earthquake was underneath the city of Bam meaning that more destruction occurred in the urban area. The earthquake directly affected 92,000 people in urban areas and 48,000 in rural areas. Early examinations estimated 25,562 urban residential and commercial units needed to be reconstructed and 828 units were in need of repair. A total of 24,332 rural units needed reconstruction (Joodi 2010).

To examine what had happened during the earthquake that caused such severe disaster, technical research centres such as the Building and Housing Research Centre (BHRC) and International Institute of Earthquake Engineering and Seismology (IIEES), quickly deployed expert teams. The Emergency, Preliminary Report of the Bam Earthquake, published 10 days after the earthquake, showed that the majority of destroyed buildings had either sundried adobe structure or unenforced masonry systems with heavy roofs and weak walls. This included a number of public buildings, for example the Municipality of Bam that was made of brick and was destroyed completely (BHRC 2004). The older parts of the city and villages houses are built using sundried adobe structure. These older buildings used traditional materials extensively were blamed for the extent of the destruction during the earthquake. According to the Bam Special Structural-Strategic Plan (Armanshahr 2004) 38.7% of houses were built from sundried adobe structures, while 61.3% were masonry buildings, built using bricks.

However, the BHRC and IIEES separate examinations also found that the area, especially the urban area which had newer buildings with steel or concrete structural systems, were also destroyed. This included residential and non-residential buildings. Those studies revealed that the majority of the newer buildings would not have been destroyed if they had applied basic seismic engineering considerations (BHRC 2004; IIEES 2004). Despite having seismic codes as part of the National Building Codes, the extent of damage and destruction of these newer buildings, as the Armanshahr report (2004) emphasises exposed the fact that engineering regulations and codes during design and construction were not applied. A few remaining '*Mohandesi-Saz*' (Engineer-Built) buildings became evidence of the effectiveness of complying with engineering codes and regulations as they stayed either undamaged or with slight damage causing no fatalities (BHRC 2004).

In fact it had been easy to receive the Certificate of Completion for a building without meeting the compulsory regulations. Municipalities as the bodies in charge of housing development did not have effective mechanisms for enforcing contractors and owners to apply the seismic regulations and even in some cases the basics of structural engineering. People did not hire them; buildings could be built without appropriate supervision; the municipality did not carry out the necessary technical control that they were responsible for. Although according to the Engineering Organisation Law, the compliance with building codes and regulations were compulsory in practice, they could be easily skipped or examined superficially. An experienced Bami structure engineer, and one of the few engineer supervisors before the earthquake, who left Bam after the earthquake, has an interesting story about this: In the midst of blaming local engineers for earthquake destruction, he proved to the Kerman Engineering Organisation[1] (KEO) that Bami engineers supervised building construction operations whenever people wanted them. A few houses with slight damages belonged to the people who had asked supervisors to do the job:

> The first local (*engineer*) participated in reconstruction was me. It had been said that Bami engineers didn't do their job before the earthquake. The first time (*after coming back to Bam after 6 months*) I went to the Engineering Organisation (*office*) to complain (*about such accusations*). They said: "You betrayed Bam". I just gave them 60 names of owners and addresses of their houses. After a while they (*from Kerman Engineering Organisation*) called me in Kerman and asked me: "how could we know that it was you supervised those

[1]The Kerman Engineering Organisation (KEO) is the province branch of the National Engineering Organisation in charge of qualifying and regulating engineers and application of the National Building Codes in construction industry, established in 1992.

> building? Prove it." They had visited those houses and saw minimum damages on them. Municipality confirmed that I did those jobs. In Khordad 83 (*May 2004, six months after the earthquake*) I returned to Bam. It was proved that if an owner or developer had asked for supervision Bami engineers would not cut the job.
>
> (Bami engineer)

4.6 Strategising the Overall Reconstruction

4.6.1 Fundamental Questions on Reconstruction

High-level immediate discussions about reconstruction and recovery among Steering Committee members and their advisors had to answer fundamental questions and topics that would set the overall direction of reconstruction. This was focused on the question of relocation or in situ reconstruction, and managing the scale of reconstruction. The general goal of the reconstruction was defined as:

> Reconstructing residential and commercial units and other physical assets in cities and villages that were damaged as a result of the Bam earthquake, in-line with regional development plans, within the national and local capacities and based on the available capacity and resources in order to return physical life to cities
>
> (Steering Committee 2004)

The Bam reconstruction is an in situ reconstruction despite the volume of its destruction. The most important considerations were the expenses of relocating the city, the existence of Arg-e-Bam, strong emotional and economic dependency of the people of Bam to their palm gardens, previous negative relocation experiences such as Buiyn-Zahra, and finally the geopolitical importance of the current location of the city in southeast of Iran (HFIR 2012; member of Armanshahr, interview 2013).

The Steering Committee commissioned each national organisation to take the responsibility of the reconstruction of their-sector related facilities and buildings. It was to manage the scale of such overall reconstruction which in fact was reconstructing the whole city on the ruins of the previous one, for example: HFIR for debris removal and the operational management of the urban and rural housing and commercial units; the National Organisation of Schools Maintenance and Development for reconstruction of schools; the Ministry of Agriculture Jihad for dealing agricultural affairs; and the Ministry of the Interior (Municipalities in Iran are part of the Ministry of the Interior) for dealing with urban affairs (HFIR 2012).

4.6.2 Dynamics of the Introduction of Core Principles of Reconstruction

The previous literature review in Chap. 2, showed that within disaster development literature it is acknowledged that natural hazards reveal failures in development activities. Similarly, the Bam earthquake was a sharp surgery knife, which exposed shortfalls of safe construction in housing development and urban planning for appreciating the historical–traditional urban identity. As a result, the Steering Committee translated the aforementioned overall reconstruction goal to the three guiding principles of reconstruction that in turn formed the strategic objectives of the housing reconstruction as the following:

1. Designs would safeguard the cultural identity and architectural fabric that Bam was famous for.
2. Buildings would be earthquake resistant.
3. Beneficiaries would be mobilised and would participate (e.g. HFIR 2004, 2012; Joodi 2010)

The dynamics of such translation was linked with the professional, emotional and social environment in the immediate aftermath that directed everyone towards the three main concerns. The relatively open social space fostered and supported the emergent of such discussions.

First, severe damage to the Arg-e-Bam with its international fame and national importance drew attention to the city itself and the unique characteristics that were unknown in many parts prior to the earthquake. This echoed concerns about the importance of appreciating historic and traditional cities and failure of the existing quantitative urban planning system activities in the Bam context. The future of the Bam urban area, given its previous traditional garden city characteristics and architectural style, was one of the main topics for discussion in the meetings. One simple example was what happens to the garden city characteristics (palm tree gardens) during reconstruction if reconstructed houses were to be built according to the mainstream comprehensive urban development planning norms, which located the building in the north of a plot. Such concerns were supported by the fact that the Bam area (including urban area) was a World Heritage area and there were nostalgic feelings after such disaster for the famous city citadel, Arg-e-Bam.

Shared understanding (e.g. Iran Cultural Heritage Organisations, the Ministry of Housing and Urban Development, the Housing Foundation and the Bam Town Council) indicated the need to safeguard the Bam urban architectural identity during reconstruction, including housing reconstruction. The high-profile brainstorming meetings later led to the establishment of the Bam Architecture and Urbanism Council (BAUC); this was formed to ensure that all activities in the Bam reconstruction were design based. Additionally, from this consensus, an opportunity emerged in order to influence the mainstream of the country's urban planning by bringing in the qualitative urban design and architectural features into the mainstream of urban development and housing development (interviews 2013).

Second, similar to the Manjil disaster, the severity of the Bam disaster triggered discussions on disaster prevention in general and insufficiency of the existing measures and processes for safe construction in particular. At the time of the Bam incident, earthquakes were referred as 'unexpected incidents' within the disaster management system. Such labelling was questioned by high-profile managers such as the president of Iran Cultural Heritage Organisation in consultative meetings (Beheshti 2004). Specialist reports, e.g. the ones by BHRC and IIEES confirmed the general consensus that the extensive destruction of Bam area was the result of problems at a national level such as unsafe construction. Having compulsory seismic codes in place was not enough to achieve safer buildings as they were not applied (BHRC 2004; Moghadam and Eskandari 2004; Eshghi and Zare' 2003). Bam was a small example of the big threat posed by large-scale urban disasters. The vice minister of the MHUD, in a consultative and technical workshop for reconstruction and urban planning of Bam held on 17/04/2004 (27/01/1383 in Persian calendar), called such disaster a death lottery throughout the country:

> If the earthquake had hit any other city in the country the result would has been the same or even worse. Bam was destroyed because it happened in Bam. It was a death lottery. Bam was a sample of what could happen in other cities, especially the bigger cities, and the capital. Having compulsory seismic regulation in place (Regulation *2800*) is not enough. Many of the destroyed buildings were the newly built ones which must have complied with those existing regulations. The extensive volume of destruction was a result of weak and poor quality structures of the buildings that could not resist the earthquake.
>
> (Hanachi 2004, Italic from author)

Third, the extensive previous experiences had proved that engaging people in the reconstruction of their own homes and assets was a win–win situation. Following the commissioning of the task of reconstruction management to HFIR, a quick discussion reconfirmed the Bam reconstruction should be a participatory reconstruction. There was no doubt on the necessity of sharing the task with the beneficiaries. The government was the facilitator-supervisor for the housing reconstruction, which was driven and undertaken by beneficiaries themselves. The operational considerations on rural and urban houses reconstruction process were to be decided (HFIR 2012; Joodi 2010).

Each of the above spheres of the Bam earthquake directly led to defining three core principles for the whole reconstruction activities; they directly transferred to housing reconstruction programme as the strategic objectives. The housing reconstruction programme, therefore, had the task of approaching those strategic objectives. To do that the housing reconstruction programme brought organisations and beneficiaries together within the programme delivery system. As will be discussed later, approaching those objectives influenced the system formation, and in turn was influenced by the system formation and implementation. For example, the organisational configuration of the system, as will be discussed later, contributed to unexpected issues during the programme implementation that required adjustments.

4.7 Housing Reconstruction Programme Policies

Besides the previous core principles framework regarding reconstruction of Bam, there were also a number of established broader policies that were in place for organising housing reconstruction activities in general after the Manjil reconstruction. They addressed a variety of considerations and practicalities for housing

Table 4.1 Defining roles between government and people in Bam

The Role of Government	The Role of People
• Enabling people through providing:	
• Grant and loan facilities	• Building their homes through
• Technical guidance and support	• Management of Construction
• Appropriate plans	• Supervision and pursuing
• Resources	• Participation for provision of their house's plans and approving it
• Special supports for the poor and vulnerable people	

reconstruction related to finance, administration, construction technology, construction materials and manufacturing, organisational, designing and planning, as well as participation of beneficiaries (HFIR n.d.). They are adapted for each individual case, as for the Bam case HFIR (2004) introduces them as follows:

Policies on the participation of beneficiaries indicated that they would be the construction manager of their own house reconstruction. Within this cooperation, HFIR was the representative of the government and the executive body for the overall reconstruction (HFIR 2012). Table 4.1 shows how the role of people and the government were defined.

Financial policies indicated the government would financially support the housing reconstruction in three ways: presenting banking facilities with low-interest and commission rates, providing grants, and financially supporting the technical services that are related to housing reconstruction but people cannot undertake, including debris removal, technical supervision and a special budget for housing reconstruction for people with special needs.

Average size of houses was related to the financial support that was provided—up to 60 m^2 in rural areas, and 80 m^2 in urban areas. If applicants required there would be technical considerations in order to extend their houses if possible.

One of the main construction policies was to improve the quality of construction of houses, compared to the situation before the disaster, by using the financial governmental assistance and beneficiaries' awareness. Rebuilding houses as they were before the earthquake would mean losing the opportunity to make changes that would decrease future risks. Construction methods should be either the existing local methods or had the potential to be accepted and used in the country. Therefore, people would be able to undertake and manage the majority of their housing reconstruction by using the available capacity. Educational activities had to be arranged to enhance the quality of construction during the reconstruction period.

Policies related to construction materials and manufacturing of material dictated that attention should be paid to the availability of the raw material within the local area; avoiding complex technologies for transforming raw to finished materials; that materials be economical; it should be possible for the local people to participate in the work; and avoiding materials that are harmful to the environment. Additionally, the policy indicated increasing the manufacture of construction materials and intervening in their distribution management in order to control the market to prevent the creation of a black market or price increase.

Organisational policies indicated that urban development plans and design for cities and villages to be, respectively, the undertaken by MHUD and HFIR. Organisational policies focused on: utilising other provinces' capacities, including human resource and construction instruments; avoiding administrative bureaucracy, and reducing the number of institutions with parallel activities; and speeding up the procedure for people, and reducing the number of organisations that people must contact.

The design and planning policies required reconstruction planning to be flexible. The design procedure had to be in collaboration with beneficiaries, respect what people had already done and try to establish systems and methods that were familiar. The design procedure must respect beneficiaries' expectations, which were according to their social and economic standing. The design system was to accept the previous framework, and to provide a number of technical and design improvements. Proposals for relocation or merging building sites without strong technical reasons were to be refused.

However, the multi-organisational system for delivery of the housing reconstruction programme, that ultimately must approach the strategic objectives, simultaneously is influenced by some practical policies and in turn, the configuration and implementation of the delivery system influenced addressing these policies in practice. Problems occurred and the system was adjusted to respond to the realities of implementing the housing reconstruction programme. Emerging problems were related to the programme formation and the system configuration. Therefore, the housing reconstruction programme was adapted to the realities of implementing the programme. These will be discussed in detail later.

4.8 The International Organisations and Disaster Diplomacy

International organisations at various levels were among key stakeholders in the response, recovery and reconstruction in Bam. For example, WHO, the World Health Organisation was the first international organisation that entered the country; and the UN issued a flash appeal 12 days after the earthquake, 8 January 2004 (HFIR 2012). The UNDP worked closely with the government, provincial and local authorities as well as affected locals. As was documented later by the Housing Foundation, from the very beginning the UNDP played an active role in preparing a UN strategic document: A United Nations Strategy for Support to the Government of the Islamic Republic of Iran Following the Bam Earthquake of 26 December 2003. According to the UNDP report in 2004, collaboration with the UNDP and the Iranian government included both short-term and midterm collaboration on supporting a coordinated UN response in the immediate aftermath of the earthquake, restoration of water supply infrastructure systems for earthquake affected smallholder dates palm plantation in Bam district; provision of technical support through specialist workshops and consultative meetings, provisions of local and international expertise and experience, demonstration projects, advocacy initiatives, capacity buildings and training programmes; and finally a small-scale pilot project for sustainable housing reconstruction in Bam through community mobilisation and participation. The scale of this pilot project was 130 families and a child-friendly park, jointly funded by the Bam municipality and the UNDP (HFIR 2012; UNDP.I. R. 2005). A report on the field trip, 23–25 January 2008, to visit activities carried out in Bam within the aforementioned projects noted:

> The synergy and partnership between national counterparts and international agencies, including UNDP was remarkable.
>
> (UNDP 2008, p. 3)

For the general housing reconstruction the presence of UNDP was more focused on the earlier stage of reconstruction and preparation level, e.g. jointly (with the MHUD and HFIR) organising specialist workshops on 'The Bam We All Want', in April 2004, technical workshops for locals on the earthquake vulnerability of the houses and architectural styles. Dr. Victoria Kianpour, the programme analyst at UNDP Iran in charge of the Bam project at the time, praised the efforts by the Iranian government and especially the Housing Foundation and its president at the time. She clarified that the role of the UNDP in reconstruction was more one of coordinating international efforts and intellectual support at earlier stages rather than financial support or decision-making role (interview 2013). My timeline analysis supports this as that indicates setting the overall reconstruction approach and strategy as well as the reconstruction programme was proposed and approved in January and February of 2004 before the joint consultative workshops with the UNDP.

However, in addition to those promising collaborations the case of Bam also illustrates complexities of the international diplomacy of disasters in an aftermath when interactions with the broader international political representatives might seem hypocritical. Local development programmes and people are the ones who ultimately suffer from such hypocrisy. The government of Iran from the revolution in 1979 has had a turbulent relationship with mostly Western countries, especially the USA. In addition to a general distrust and suspicion between the revolutionary governments and the big players in international diplomacy, Iran faced US sanctions and a number of UN resolutions at the time of the Bam earthquake. This affected absorbing disaster-related loans and implementing longer-term developmental projects. An illustrative evidence of this is the emergency loan agreement between the government of Iran and the World Bank for the Bam reconstruction and recovery.

According to the World Bank project documents, after lengthy preparations and negotiations starting in March 2004, a loan for the Bam emergency reconstruction project amounting to 220 million dollars from the World Bank was approved on 28 October 2004 and endorsed on 16 November 2004, 11 months after the earthquake.

The loan agreement focused on five components: first and foremost was an estimated 150 million dollars for the provision of construction material and equipment for housing and commercial unit reconstruction. The second was an estimated 22.35 million for repair of the transport infrastructure, focusing on the rehabilitation and preventive works for the main international highway that also links Bam to the provincial capital of Kerman. Third was an estimated 11.45 million for the repair of telecommunication infrastructure. Fourth was 8.39 million for retrofitting works on emergency response buildings in Kerman province and provision of the vehicles and equipment. The fifth was an estimated 5.85 million dollars for project management and technical assistance through establishing the Bam reconstruction office and procuring consulting firms for providing advisory services, monitoring advisory services and technical supervision.

However, as a result of reoccurring problems in the fulfillment of the loan agreement the actual loan given for reconstruction and recovery was reduced to 154.43 million dollars of which 92% was allocated to the housing and commercial units reconstruction. This means in total 70% of the loan was absorbed for housing reconstruction and not for other developmental recovery programmes (HFIR 2012; the World Bank 2010). The project appraisal conducted by the Sustainable Development Department in the World Bank (2010), published in 2010 lists the reoccurring problems as the following:

> Several factors affected the implementation of the project, for example: (i) due to the imposition of financial sanctions by the US, Special Accounts denominated in US$ had to be switched to Euros which resulted in delays in disbursements and limited the ability of local firms to submit bid securities; (ii) multiple UN sanctions discouraged potential bidders from outside Iran, thereby limiting the pool of competitive bids leading to increased project costs - because of the heavy reliance on local bidders, the quality and timeliness of many of the works also suffered; (iii) because of UN sanctions, the Bank put in place additional system to screen purchase of equipment (specifications had to be reviewed against a list of prohibited items and entities identified by the relevant UN Resolutions) thus adding another layer of clearance contributing to delays when this process was first introduced; (iv) decision by the Bank not to extend the closing date of the project left some activities incomplete and under-funded, and as a result the government had to approach other donors (i.e. Islamic Development Bank) to fill in this gap; (v) the inability and reluctance of airport authority to allow contractors and workers to enter into the airport premises due to security reasons caused delays that ultimately resulted in dropping from Bank financing activities related to the Bam airport; and (vi) because of an acute shortage of telecommunication specialists in the Bank, at the time, difficulties were encountered in preparing tender documents to import complex modern electronic equipment (New Generation Network) system.
>
> (The World Bank 2010, p. 7)

This appraisal rates the performance of the Bank as moderately unsatisfactory with moderate risk to development outcomes; it also rates the performance of the borrower moderately satisfactory with an ‘excellent’ rate for the performance in housing and commercial unit reconstruction by HFIR. The Bank links it with the previous experience in reconstruction of HFIR.

Furthermore, despite positive responses from international organisations, the Iranian government could attract only 130 million dollars for the reconstruction stage from international organisations, a fraction of the total 1.8 billion dollar costs of the reconstruction. Some financial promises made immediate after the earthquake were not delivered by either foreign governments or international organisations after the emergency stage passed, similar to other disaster cases. Another problem, as it was documented later by HFIR (2012), was that there was not any organisation or institution responsible for an international talk with the countries and international organisations within the humanitarian movement after a disaster. According to the report by OCHA Financial Tracking Service (FTS), and on the basis of information provided by donors and appealing organisations, the total international aid to Iran for the Bam earthquake was 130,225,462 dollars (FTS n.d.). Intentional aid was mostly spent during the emergency and early recovery stage or reconstruction of schools and hospitals, whereas the total cost of reconstruction reached 1.8 billion dollars (FTS n.d.; HFIR 2012). Years later a member of the Office of Reconstruction Affairs (ORA) (cited in HFIR 2012) provides a comparison between Iran and Indonesia: after the Indian Tsunami, Indonesia needed 8 billion dollars for reconstruction and was able to absorb around 7 billion dollars from international donors. The Indonesian government created a council independent from the government consisting of local trustees and through this could attract more financial help for reconstruction (HFIR 2012).

4.9 NGOs in Bam

One of the distinct characteristics of the Bam earthquake was the extensive presence of national and international NGOs after the earthquake. Following the declaration of the disaster openly and the request of Iranian government all UN-related agencies, e.g. UNDP, OCHA, UNICEF and so on, and the well-known international NGOs such as Oxfam and Caritas went to Bam. It was unlike previous disasters that where Red Crescent and a few number of international organisations were able to help at the emergency stage. During the reconstruction and recovery stage a number of 116 national and international NGOs became engaged mostly in public services, for example, the reconstruction of schools and educational buildings, health centres and sports facilities. The representatives of the government introduced those NGOs to the reconstruction executive body, MHUD or other governmental organisations in charge of rebuilding the above amenities for further coordination and allocation of sites. Allocation of sites was coordinated with and reflected in the ‘Strategic and Structural Plan for Bam’, that was being prepared concurrently, to integrate activities. For housing reconstruction, HFIR directed those interested NGOs to rural areas, where the number of houses could be delivered by the NGO matched with the number of houses in the village that required reconstruction. In urban areas, such NGOs were linked with vulnerable groups, e.g. families that lost the main breadwinner, or families with disabled members (HFIR 2012). In these cases based on an agreement with HFIR, the NGO and acted as the representative of the family for following up the house reconstruction for the family and finance the construction (member of HFIR, interview 2013). For example, Caritas helped vulnerable groups such as those with spinal cord injuries (Caritas, n.d). The reason for such an approach was that giving a number of urban houses to a NGO for reconstruction required setting justifiable criteria for selecting those houses (member of HFIR interview 2013). An example of agreement between HFIR and the ADF Foundation in Turkey for reconstruction of 213 houses (dated 13/02/2004) shows that in such cases IFIR had to introduce vulnerable families to the NGO which acted as the representative of the beneficiaries for follow-ups of the reconstruction of their house within the housing reconstruction process and fund the construction operation.

The Bam incident also saw the emergence of local NGOs immediately after the earthquake and each took initiatives for specific groups during early recovery and afterwards up to the present. According to Ekhlaspour (2009) NGOs in Bam were mostly involved in social affairs and improving the livelihood of the most vulnerable, e.g. women and children. Some locals tried to heal their own grief through helping others. An example is the founder of a local NGO, Hamrahan-e-Bam (Friends of Bam). He lost his whole family, namely wife, children and extended family, in the earthquake and recognised that to coping with his own grief required helping others (founder of a local NGO after the earthquake, interview 2013).

4.10 Dynamics of Different Approaches to Temporary Accommodation

Initial decision in the aftermath of a disaster influences the later stages of reconstruction. Therefore, decision makers should think about the potential longer-term effects of their decisions, for example the provision of temporary accommodation for people while reconstruction of the permanent homes is carried out. Temporary housing links emergency shelters to the permanent houses and is a distinctive stage to the reconstruction process (Johnson 2007). The Bam example provides an interesting case for (a) ignoring people's lifestyle during the provision of temporary accommodation, and (b) respecting people's wish and the longer term effects of decisions made about temporary housing.

Camp compounds were quickly erected, using prefabricated units, on a roadside between Bam and Baravat that is a main international transit road from east to west of the country. Figure 4.9 shows a compound Initial decisions on temporary housing in Bam was taken without taking the local people's lifestyle and culture into account, and in fact was an easy answer to providing accommodation for homeless people. Many of these prefabricated units were part of the international aid which came to the region as ready-made units. Each unit was 18–20 m^2 and had water, electricity and air conditioning. Each compound had shared sanitary services (HFIR 2012).

Fig. 4.9 Initial temporary accommodation in form of camp compounds. *Source* HFIR (Reprinted with permission)

However, disaster-affected people did not welcome to the idea of living in camp compounds. They preferred to return to their yards and live by the remains of their own houses rather than move to officially run camps. Locals and their representatives in town council raised various complaints, such as safety issues; inconsiderate to the local social norms within that girls and women would not leave their allocated units to use public toilet services, and potential future damage to the town after removing these camps. These camps also became a heaven for numerous outsiders who descended on the Bam area immediately after the earthquake, claiming to be disaster-affected Bami (Ekhlaspour 2009). Esmaeili, head of the Bam Town Council echoed people's demands in a consultative workshop, 3 months after the earthquake (March 2004):

- Prioritising erection of urgent temporary housing of people in their own yards;
- Discontinue erection of temporary camping compounds;
- Installation of temporary habitable structure within the garden houses such that the building of permanent houses would not be hampered;
- Concurrent erection of boundary walls for each site to provide safety and protection for families in their own yard (Esmaeili 2004).

People's demands were considered, and the provision of temporary accommodations saw a shift towards erecting prefabricated units for households in their own yards. This change of approach on temporary accommodation in Bam played a significant role on the overall outcomes of the Bam reconstruction. As Fayazi and Lizzaralde (2013) conclude it facilitated the recovery process and positively influenced community resilience. It in fact created opportunities for beneficiary households to increase their social capital by settling in their own communities, sharing their understanding of reconstruction realities and their experiences, while having close relationships with local organisations and other participant organisations (ibid). In autumn 2004 when the housing reconstruction process commenced, families had already moved to their own yards and lived in their individual temporary accommodation, which indeed became a permanent investment for many of the applicants' families. Figure 4.10 shows temporary accommodation in form of in situ individual units.

Fig. 4.10 Individual in situ temporary housing what was people wanted. *Source* HFIR (Reprinted with permission)

Design guidelines for reconstructing permanent houses requested to consider these supposedly in situ temporary accommodations. They were seen as the subsidiary supporting space for family lives even after the completion of permanent houses. The reason was to compensate the gap between the large size of the houses before the earthquake (around 200 m^2) and small size of reconstructed houses (80 m^2)—according to Armanshahr the consultancy company prepared the Bam master plan just before the earthquake and the structural-strategic urban plan after the earthquake. These 24–32 m^2 'temporary' accommodations proved to be permanent assets for many families to meet a variety of household needs. At the time of this research, in 2013, a considerable number of 'temporary' accommodations still existed and functioned, wherever the size of the yard was big enough to keep the permanent house, temporary accommodation and palm trees. Figure 4.11 illustrates an example of this.

Fig. 4.11 The old temporary houses in many cases still exist in the yard and are in use for the household needs. *Photo* Arefian 2013

The new approach to the provision of temporary accommodations also played a role in changing the future construction scenarios in Bam. As will be discussed later, living in their own construction site whilst their permanent house was being built in the same site in reality entailed all family members observed the reconstruction process of their own houses, and they were in direct contact with contractor, engineer supervisors, construction workers, and at some point they even could help on this. As this research explores this was influential on the outcomes of housing reconstruction programme and the way it changed previous norms in the Bam area. Those massive camp compounds for temporary accommodation were removed one by one. Some had become places for antisocial behaviours and removing some of them occupied by outsiders proved to be a battle for local management and reconstruction management (members of provincial government, and Bam and Baravat municipalities, interviews 2013).

4.11 Recovery Approach

The literature review showed reconstruction as an agent for physical recovery is an important platform that potentially can be beneficial for psychological and economic recovery and people's livelihood (Davis 2007). Reconstruction also creates an economic boost for the area. Thus, depending on the extent of destruction and intensity of construction and developmental projects and programmes are undertaken, thus it provides an opportunity for blending and addressing people's livelihood and economic recovery (Alexander 2002). The Bam case provides an example of how that opportunity was used or missed to assist the economic recovery.

From early days steps were taken for the stimulating economic recovery in already-recognised sectors and to boost the local businesses. Bam also became a great marketplace for construction workers, contractors and service-based businesses. For the longer-term physical recovery the Steering Committee quickly divided the tasks of overall reconstruction with related organisations. With the financial help of government, international agencies, Iranian expats, local and international NGOs and foreign governments numerous projects on schools, hospitals, mosques, libraries, infrastructure, and recreational facilities and so on were carried out.

As the regional agricultural centre, one of the most important economic resources in the area was palm grove production. Recognising the importance of the palm gardens special attention to those gardens and repairing qanats became a major priority for various governmental organisations, non-governmental ones and international agencies. Reports show that, although the earthquake damaged the irrigation system the palm grove production was not stopped, thus creating hopes among many families who still could rely on their agricultural incomes. Before the earthquake, according to Tierney et al. (2005), 80% of palm garden irrigation system relied on 64 qanats, out of which 30 were damaged. It is reported that 22 qanats out of the 30 were repaired by May 2004. Many of the families (inside or outside the urban areas and villages) owned or worked in palm gardens while they might or might not have another formal job (e.g. Tierney et al. 2005; Khatam 2006; UNDP 2004).

The earthquake also, as Fallahi and Arzhangi (2011) report, destroyed 90% of small businesses and shopping centre structures, including the Bazaar. This affected the livelihood of people who relied on small businesses and services. Recovery of small businesses started quickly after the earthquake by fencing and allocating prefabricated units and containers with the help of the governor, municipality and reconstruction executive body, HFIR. Interestingly such prefabricated business units quickly spread throughout the city and mainly at the edge of main and/or commercial streets. They provided all services and shops, from clothes to beauty salon, from hairdresser to bakery, to electrical equipment and to lawyers' offices. Real-time observations by me and others in 2004 and 2005 suggest '*Life goes on for the residents of Bam. Shops and stalls have already started business*' (Lateef 2004, p. 164). Figure 4.12 shows an example of those small businesses.

Fig. 4.12 Example of working small business and shops in 2005. *Photo* Arefian 2005

Employees at service sectors in Bam, as Khatam (2006) states, were less vulnerable to social instabilities and economic disruption as they still benefited from stability in their employment and were able to return to their jobs. This included around 26% of the workforce including military personnel.

And finally, the reconstruction period as the major step for long-term recovery created a job market both for construction industry related industries and service providers. As a member of HFIR explains (interview 2013), in order to boost the local economy distribution of construction material was only allowed through local shops and businesses. A construction Bazaar was formed. Wholesale traders of construction materials exhibited in the construction Bazaar but they had to set representative arrangements with local business for selling and distributing their products. The government provided low-cost loans for setting up construction material production workshops and factories.

However, neither reconstruction policies nor such initiatives addressed low skilled labourer and unemployed youth at the beginning. According to Khatam (2006) unemployed people, especially youth and low skilled labourers, are the ones who suffer most from the disaster and statistics indicated 20% unemployment within the area. As a number of researchers and observers point out (e.g. Meskinazarian 2011; Tierney et al. 2005) many of local workers felt marginalised at the early stages of reconstruction as they did not meet the criteria to work as local contractors or skilled labourers. Since outsider contractors did not recruit them, they became drivers, serving the numerous operational organisations in the field, e.g. Setads and consultancy companies. As will be discussed later, such approach towards local contractors and construction workers changed during the implementation phase. Training courses were later arranged for local contractors and construction workers in order to help them receive the licence for contractor works (member of HFIR interview 2013; local contractor interview 2013).

Interestingly, although there was the above hindrance, many of operational organisations organically recognised the value of local knowledge for their services, for example, knowledge about neighbourhoods and residents. Many organisations including my own architectural consultancy employed local youth, e.g. for technical assistance or office-based tasks, providing informal training for their local employees. It appears that such approach not only provided employment during reconstruction period it also had longer-term impacts. During this research, I came across two engineers and an architect who during reconstruction worked for those organisations as unskilled trainees. After the reconstruction period, each decided to pursue his university degree and continue in such disciplines, now working at the municipality, Engineering Organisation and the Bam branch of HFIR.

Meskinazarian (2011) and Khatam (2006) indicate that although a number of initiatives for economic recovery were introduced the overall longer term recovery prioritised the physical recovery to the livelihood of people. Thus long-term recovery leaned towards construction projects, and not systematic planning and programmes for increasing people's livelihoods and job opportunities for unemployed youth and low skill labourers.

The lack of comprehensive planning and programmes to systematically address all groups' livelihoods for the longer-term recovery in Bam is obvious. However, it must be noted that around 30,000 people soon after the earthquake moved to the earthquake affected area, claiming to be Bami in search of opportunities as reported by Ekhlaspour (2009). While Ekhlaspour refers to them as the rural poor of the larger region and links them to the lengthy drought in the region in years prior to the earthquake, there is no detailed data on them. But they are most likely among the ones complaining about unavailability of jobs in Bam. By the same token, it is possible that many of non-genuine Bami applicants received small temporary business units. At the end of the reconstruction period removing many 'temporary business units' proved problematic. Municipalities could only remove the majority of the units in 2011 and 2012 after 7 years.

However, the reconstruction of permanent business premises encountered problems and negatively affected the recovery of small businesses. Reconstruction of permanent shops and small businesses had to follow the same procedure as the housing reconstruction process. Each shop owner could receive grants and low-interest loan for reconstruction. The amount was based on the shops' floor area before the earthquake. Fallahi and Arzhangi (2011) in their research on business continuity after the earthquake found a lack of data and legal issues became of two major challenges for reconstruction of Bam. The lack of data meant that there was no sufficient data on small businesses and shopkeepers desired to receive more from reconstruction funding. Another Reconstruction of shops also was linked with the implementation of the city's urban development plan that proved problematic. This will be discussed in the next section.

4.12 Reconstruction Within the Context of the Urban Development of Bam

A destructive disaster is seen as a disruption in development path of the area; recovery and reconstruction help to overcome this disruption so the disaster-affected area will be returned to its development path again. There is a spectrum of a seeing reconstruction as a window of opportunity to deliver radical changes to more pragmatic ones that acknowledge that in any situation there is room for improvement. Moreover, there are questions on how and to what extent a developmental approach could be linked with reconstruction. The case of Bam provides an illustrative example for such debates, as it took a positive but pragmatic approach for linking reconstruction and development.

Various presentations and documentation by the HFIR members and Office of Reconstruction Affairs, ORA, state the Bam reconstruction, as a whole, had a developmental approach (e.g. Alizamani 2012; Joodi 2010; Havaei n.d.). Although they do not explain what they mean by 'developmental approach', given their presentations it appears that they refer to strategising the reconstruction, previously discussed. Their focus also is on housing reconstruction, its objectives and practical considerations, as well as, to some extent, the revised urban development plan that set the roadmap for the future of the disaster-affected area. They highlight the contemporary international disaster discourse and lessons they learned from previous national reconstruction experiences, providing three propositions:

- A reconstruction without developmental approach was not acceptable because it would have been a repetition of what had happened before and what had collapsed during those 12 s of the Bam earthquake.
- An idealistic ambitious reconstruction that required radical changes in people's lifestyle, lengthy operational period and was disrespectful to available resources was not deliverable.
- A reconstruction is an opportunity to re-think about the development of the area and address previous shortfalls and problems, and have a longer-term plan aligned with regional and national development (ibid).

Thus, the third proposition for the reconstruction of Bam was selected: an opportunity to enhance the development of the area and address existing shortfalls, whilst eliminating undeliverable and idealistic objectives (Alizamani 2004). This approach was optimistic yet pragmatic. According to this view, the Bam reconstruction activities as a whole would have an enabling role for the destroyed area to overcome this massive disruption that occurred in the city's previous developmental pathway alongside other parts of the country (Joodi 2010). A member of ORA uses the following metaphor of a patient to explain:

> Imagine there is a patient with chronic heart problems. He just had a car accident, resulting to his broken hand and severe wounds. What do you do as his medical doctor? Will you undertake operation for his broken hand, wounds and heart at once? Or you treat his wounds and hand, and then look for solution for his chronic problem?
>
> (member of ORA)

The Bam case provides valuable examples of operational challenges and opportunities of the enabling role of urban reconstruction, and linking it with routine urban development processes. In Bam the underlying socio-economic dynamics of such reconstruction which in fact was the speedy implementation of the city's urban development plan prolonged business recovery. Long-term business recovery—linked to the reconstruction of shops and independent business premises—was strongly linked to the implementation of the new Bam urban development plan, called '*Strategic and Structural Plan for Bam*'.

The Strategic and Structural Plan for Bam that was approved in October 2004 was, in fact, a revised version of the recently approved detailed master plan of the city that was adjusted to the post-disaster situation. The old master plan followed the fashionable trend of master planning across the country and widening roads for prioritising cars. Similar to the old master plan, the new Plan suggested a number of commercial roads to be widened to assure the capacity of these roads for the longer term view of the city. The Plan also reflected on damages to the urban fabric and included dispersed recovery projects by NGOs and others. It also introduced additional urban design guidance to reflect the core principles of the Bam reconstruction (Armanshahr 2004). The preparation of this new urban development plan was undertaken in parallel to reconstruction activities, with coordinating meetings and on hand advice for the reconstruction when questions arose.

In practice, such proposal for widening commercial roads meant that all shops and businesses premises at the edge of these roads required a setback from their previous building lines. This, in turn, meant the municipality as the implementer of the Plan had to financially compensate the shop owners. The amount of this compensation became a source of dispute between business owners, the town council, municipalities and government. The total allocated fund for the purpose was \$200,000 for Bam and \$80,000 for Baravat that was based on initial estimations. However, the town council pushed the government towards higher amount of total budget and increased financial compensation for each case (at some cases reaching to \$12,000 for an individual case) as the town council saw this fund as a source of income for disaster-affected residents. Consequently, the allocated funds became insufficient to cover all the affected businesses. Uncertainty and inconsistency in financial compensations meant many business owners were reluctant to do the requested setback in the hope of receiving more financial compensation but neither could they reconstruct their permanent business premises. Such issue delayed prolonged overall business recovery. Finally, as minutes and presentations at the BAUC show, the Bam Architectural and Urbanism Council intervened in December 2005 and ruled that no road should be widened (BAUC and HFIR 2005).

Such challenges are linked to the fact that reconstruction, which is urgent, short-term and fast pace in nature, should deliver urban development plans, which are long-term and slow pace towards an ideal vision. The implementation of such visions exceeds the responsibilities and scope of reconstruction. As a member of ORA reflected on the challenges of implementing urban development plans during a limited time of reconstruction:

> Our professionals and consultancy services should understand that urban development plans in post-disaster reconstruction is different than the ones in normal situation.
>
> If in normal situation an urban development plan should be implemented over the course of 20 years for example, in reconstruction we are on a fast pace to do so.
>
> (member of ORA)

Moreover, the devastating extent of physical destruction in Bam meant that almost all other non-residential and non-commercial buildings required reconstruction, for example 130 schools, two hospitals, and all mosques. Like other reconstruction cases, there were also new/promised non-residential public or private sector projects, many of which were sponsored by national or international donors. The Plan was flexible enough to embrace generous proposals by donors. Notably, those projects also are part of the developmental path of the disaster-affected area. They not only influence the physicality of the city but social aspects, and business sense of the people because they bring fresh ideas to the disaster-affected area. The Bam earthquake, like other devastating disasters, attracted national and international attention for undertaking avant-garde projects and programmes, compared to pre-disaster status quo. Such projects enhanced the vision for Bam as a regional centre for the surrounding area within the Plan. Examples of such projects include a music centre by veteran Iranian musicians in the memory of a famous singer—Iraj Bastami—who died in the earthquake, and the child-friendly neighbourhood project funded by UNDP. A local teacher reflects on this:

> Office buildings are now new, modern and cleaner (*than before the earthquake*). Previously we had only one traffic light in the city, Imamzadeh Asiri, and no one cared about it. Now we have more traffic lights. Imam Khomeini road became a boulevard and is wider. Bam now represents a more developed city. We have a well-equipped sport stadium now that we didn't have before. The Education Organisation Office didn't have a conference venue before. Now its new building has a big conference venue. It is very good. We have a number of reception venues for celebrations (*weddings and ceremonies*) whereas previously there were only Azadi hotel and Arg-e-e Jadid hotel. Our houses are not as big and a number of wedding reception venues throughout the city exists.
>
> They (*the Bam municipality and town council*) brought a funfair park here but it is useless. They bought equipments but they are not used as they were second hand. It does not even have a roofed space for the children's playground.
>
> There is a big library now. Universities before offered a few subjects but now there are more subjects.
>
> (local teacher; italic from author)

Such non-residential projects too had to respect the overall core principles of the Bam reconstruction set by the Steering Committee on safeguarding the architectural identity of the old city and building earthquake resistance buildings. In this regard, the Bam city reconstructed area presents interesting examples of construction projects for schools, mosques, universities and governmental offices. However, one must recognise there is a core difference between developmental projects undertaken by a single organisation and participatory programmes from a complexity perspective. For the former the single organisation is one big client or investor with its own professional team; but in the latter participatory programmes are a portfolio of large number of individual projects and their local clients. Those single organisation construction projects are undertaken by informed clients and skilled professional teams on design and construction, whether to be small or large scale. Those projects are not are not similar to large-scale housing reconstruction when the housing reconstruction programme is to embrace people's participation. The challenge for the housing reconstruction was how to embed such optimistic but pragmatic developmental improvements into the housing reconstruction process that was supposed to be based on individual reconstruction cases, driven and administrated by the owners or beneficiaries. The reconstructed Bam, more than 10 years after the earthquake, shows the aforementioned core difference as it represents a two-folded picture of reconstruction. Such widespread finely finished construction projects contrast to modest houses and the existence of uncompleted house structures and some roads that the municipality still has not completed.

4.13 Conclusion

This chapter examined how the strategy to reconstruct the Bam area was directly linked to local historic importance and Bam being a unique traditional garden city, prompting new discussions on the importance of safeguarding historic traditional urban characteristics that Bam was famous for. Simultaneously the Bam case directed general consensus towards shortfalls in safe construction in the housing sector at both local and national levels as compulsory seismic codes were not applied and enforced in practice and such a shortfall was not detected. The practical lesson of the advantages of engaging beneficiaries within reconstruction of their houses was proved.

Repair of qanats, facilitating creating marketplaces for local businesses and requesting construction material producers to partner with local businesses for distributing them were among the steps for economic recovery in recognised sectors. However, such considerations did not extend to the unemployed or low skilled labourers to enter the job market, although the reconstruction period created a big construction market. Since local knowledge was organically appreciated by some organisations they recruited local youth.

Despite involvement and overall arrangements of international agencies, there were contradictory practical problems for the delivery of key developmental projects. Bam exemplifies prevailing international politics and relations prior to disaster influence the deliverability of post-disaster development key initiatives that might be promised immediately in an aftermath. It is the local people and the already disaster-affected area that misses out as a result of such hypocrisy.

The initial approach for temporary accommodation ignored the social norms and created unease for Bami people. Reflecting the people's request to return to their own sites and live there while their house was to be reconstructed, was an important change of approach by HFIR. This proved more socially acceptable temporary accommodation and also, as the future showed, it had a long-term positive impacts as will be discussed later.

The Bam reconstruction was an optimistic yet pragmatic effort within the contemporary debates on linking reconstruction and development policies and processes. It tried to link urban reconstruction and developmental visions together. In practice, challenges included limited financial resources, the short-term reconstruction time span and the difference between immediate and long-term priorities for fully implementing ambitions activities with longer-term visions of the urban development plans. Too many changes in the existing reference documents of urban development plans, in turn, might prolong the overall recovery, for example short-term business recovery as it causes more uncertainties.

New large or small projects and programmes implemented by single organisations for physical recovery are part of the developmental path of the area. They bring fresh ideas and stimulate further development. However, they are less complex than participatory programmes, which have developmental objectives. A participatory reconstruction programme must bring various strategic and practical considerations together and create a reconstruction process for ordinary people. This research focuses on the latter, specifically the participatory housing reconstruction programme in the Bam urban area, examining how it tried to embed such optimistic vision into the sociotechnical delivery system of the programme.

References

Alexander DE (2002) Principles of emergency planning and management, Illustrated edn. Oxford University Press, England

Alizamani MH (2004) Reconstruction approach aligned with development in reconstruction of Bam. In: Proceedings of the specialist workshop for Bam Sustainable Development Manifesto, held on 08/09/2006, Centre for Architecture, Urban Studies and Research, Tehran (in Persian)

Alizamani MH (2012) Contemporary approaches in post disaster housing reconstruction (Unpublished PowerPoint file presented at various conferences) (in Persian)

Armanshahr Consultancy Company (2004) Special structural and strategic plan for Bam (Unpublished urban development project document), MHUD, Tehran (in Persian)

BAUC and HFIR December (2005) Review of the detailed masterplan proposals for widening the roads (Unpublished project document, provided by MHUD), (in Persian)

Beheshti SM (2004/1383) Bam, the world heritage (Historical and cultural specifications of the Bam City in urban planning). In: Proceedings of the technical and professional workshop for reconstruction and urban planning for Bam City. 25–26 February 2004 Tehran, Ministry of Housing and Urban Development, Tehran (in Persian)

BHRC (2004) Primary urgent report of the Bam earthquake (26/12/2003); 06/01/2004, BHRC, Tehran (in Persian)

BHRC Catalogue of Earthquake Strong Motion Records (2003) http://www.bhrc.ac.ir/ismn/catalogue/pdf/2003.pdf. Accessed 17 Oct 2010

Christopolos I, ALNAP, Glemminge Development Research (2006) Provention consortium forum 2006. Strengthening global collaboration in disaster risk reduction. The Elusive ‘Window of Opportunity’ for risk reduction in post-disaster recovery. ipcc-wg2.gov/njlite_download.php?id=5282. Accessed 20 Dec 2010

Davis I (2007) Learning from disaster recovery; guidance for decision makers. International Recovery Platform (IRP). http://www.undp.org.cu/crmi/docs/irp-decmakersleslearned-td-2006-en.pdf. Accessed 17 Nov 2010

Ekhlaspour R (2009) Bam, review of society and economic conditions of Bam during the last two centuries. Iran History, Tehran (in Persian)

Eshghi S, Zare' M (2003) Bam (SE Iran) Earthquake of 26 December 2003, Mw6.5: A preliminary reconnaissance report. International Institute of Earthquake Engineering and Seismology (IIEES), Tehran

Esmaeili A (2004) Advocating community demands in the process of reconstruction and urban planning for Bam. In: Proceedings of the technical and professional workshop for reconstruction and urban planning for Bam City. 25–26 Feb 2004, Ministry of Housing and Urban Development, Tehran (in Persian)

Fallahi A, Arzhangi S (2011) Business continuity after the 2003 Bam earthquake in Iran. In: Kim et al (eds) ASEA/DRBC/EL 2011, CCIS 257. Springer, Berlin, Heidelberg, pp 532–540

FTS (n.d.) (Website), Iran, Islamic Republic of in 2004—related emergencies, financial tracking service. https://fts.unocha.org/reports/daily/ocha_R10c_C99_Y2004_asof___1509181234.pdf. Accessed 18 Sept 2015

Fayazi M, Lizarralde G (2013) Temporary housing: an obstacle or an opportunity for enhancing community resilience in Bam, Iran? In i-Rec conference 2013, sustainable post-disaster reconstruction: from recovery to risk reduction. Ascona, Switzerland. http://www.academia.edu/5064290/Temporary_housing_An_obstacle_or_an_opportunity_for_enhancing_community_resilience_in_Bam_Iran. Accessed 09 Aug 2015

Ghafory-Ashtiany M (n.d.) Bam Earthquake of 05:26:26 of 26 December 2003, Ms6.5. JSEE special issue on Bam earthquake. www.iiees.ac.ir/iiees/English/Publication/jsee/Editorial%20Summary.pdf. Accessed 04 Feb 2014

Ghafory-Ashtiany M, Hosseini M (2007) Post-Bam earthquake: recovery and reconstruction. Nat Hazards 44:229–241. https://doi.org/10.1007/s11069-007-9108-3

Golpayegani A, Einifar A (2004) Typology and design guide for housing in Bam. Ministry of roads and urban development-office of architecture and urban planning, Tehran (in Persian)

Hanachi P (2004) New policies in urban planning and existing references in reconstruction of Bam. In: Proceedings of the technical and professional workshop for reconstruction and urban planning for Bam City. 25–26 Feb 2004, Tehran, Ministry of Housing and Urban Development, Tehran (in Persian)

Havaie MH, Hosseini M (n.d.) Bam earthquake: from emergency response to reconstruction. JSEE special issue on Bam earthquake. http://www.sid.ir/en/VEWSSID/J_pdf/86020040418.pdf. Accessed 07 Jan 2011

Havaie MH (n.d.) A brief review on Bam reconstruction program. (Unpublished Power Point file provided by HFIR)

HFIR (2004) Housing reconstruction programme for damaged regions in the Bam earthquake (Unpublished report) (in Persian)

HFIR (2012) Documenting the reconstruction of Bam: design process in reconstruction (unpublished report) (in Persian)

HFIR (n.d.) Deputy of reconstruction and rural housing (Website). Housing Foundation of Islamic Revolution. http://www.bonyadmaskan.ir/En/sitepages/Deputy%20of%20Recontruction%20and%20Rural%20Housing.aspx. Accessed 09 Sept 2013

Johnson C (2007) Strategic planning for post-disaster temporary housing. Disasters 31(4):435–458. https://doi.org/10.1111/j.1467-7717.2007.01018.x

Joodi, M. 2010. A brief review on Bam reconstruction program. Presented in international recovery forum 2010, organised by International Recovery Platform (IRP), 16/01/2010, Kobe Japan. http://www.recoveryplatform.org/resources/meetings_and_trainings/514/international_recovery_forum_2010. Accessed 23 Dec 2010

Khatam A (2006) Viewpoint: the destruction of Bam and its reconstruction following the earthquake of December 2003. Cities 23(6):462–464

Lateef F (2004) As the city heals, post-earthquake reconnaissance to Bam, Iran. SHG Proc 13(3):162–167

Meskinazarian A (2011) Social resilience of post-earthquake Bam. PhD thesis, Department of Geography, King's College London, London, UK

Moghadam AS, Eskandari A (2004) Post-earthquake quick inspection of damaged buildings in Bam earthquake of 26 December 2003. In: Special issue on Bam earthquake, vol 5, no. 4 (Winter 2004)—vol 6, no. 1 (Spring 2004), International Institute for Earthquake Engineering and Seismology (IIEES), 2004

Statistical Centre of Iran (n.d.) Detail (Website). http://amar.sci.org.ir/Detail.aspx?Ln=F&no=224096&S=TP. Accessed 01 May 2011 (in Persian)

Statistical Centre of Iran (2004) Results of listing places and families in earthquake hit area of Bam. Statistical Centre of Iran, Tehran

Steering Committee for Reconstruction of Bam (2004/1382) Decision by the president's special representatives (A copy of the original appeared in the documentation of the reconstruction of Bam after the earthquake: Design Process in Reconstruction), Tehran (in Persian)

The Cabinet (2004) Decision by the Cabinet (A copy of the original was appeared in the documentation of the reconstruction of Bam after the earthquake: design process in reconstruction), Tehran (in Persian)

The World Bank (2010) Islamic Republic of Iran—Bam earthquake emergency reconstruction project (Implementation Completion and Results Report No. ICR1207). http://www-wds.worldbank.org/external/default/WDSContentServer/WDSP/IB/2010/09/28/000356161_20100928235953/Rendered/PDF/ICR12070P088061C0disclosed091271101.pdf. Accessed 09 Sept 2013

Tierney K, Khazai B, Tobin T, Krimgold F (2005) Social and public policy issues following the 2003 Bam, Iran, earthquake. Earthq Spectra 20(1):S513–S534

UN (2004) United Nations launches appeal for Iran earthquake (Website). http://www.un.org/News/Press/docs/2004/iha852.doc.htm. Accessed 10 Jun 2014

UNDP Iran (2004) F helps Bam people to play a part in the reconstruction planning of their Quake Hit City. UNDP Iran. http://www.undp.org/cpr/disred/english/news/archives/2005/april.htm#iran040405. Accessed 07 Jan 2011

UNDP Iran (2005) UNDP in Iran: 40 years of development partnership. http://www.undp.org.ir/DocCenter/reports/brochure2005.pdf. Accessed 23 Nov 2010

UNDP (2008) Report of mission to Bam; 23–25 Jan 2008. erc.undp.org/evaluationadmin/downloaddocument.html?docid=3120. Accessed 27 Mar 2013

UNESCO (n.d.) Bam and its cultural landscape—UNESCO World Heritage Centre (Website). http://whc.unesco.org/en/list/1208. Accessed 22 Feb 2014

UNESCO World Heritage Centre (2004a) World Heritage Newsletter, 43. http://whc.unesco.org/en/newsletter/43/. Accessed 07 Jan 2011

UNESCO World Heritage Centre (2004b) Bam and its cultural landscape (Website). http://whc.unesco.org/en/list/1208. Accessed 14 Aug 2017

Chapter 5
Initial Organisational Formation of the Housing Reconstruction Programme in Bam

Abstract Disaster exposes failures in previous development activities; there are expectations from reconstruction for improvements. However, the difficult task is how to organise the necessary processes and procedures, e.g. regarding participants and their roles. Practical decisions have to be made towards approaching initial objectives. This chapter examines how the housing reconstruction programme in Bam that covered most of the urban area was initially organised. It enables an understanding of the organisational configuration of the programme delivery system. The interplay of programme size, urban process and existing reconstruction policies led to introducing practical considerations for the formation of a sociotechnical delivery system for the programme that was innovative, purposeful, multi-organisational. It connected financial, technical, social and practical considerations in the reconstruction programme, and connected disaster-affected beneficiaries/applicants and various organisational positions together. Each of these 13 positions including applicants was a system element, from local individuals to local urban governance, provincial professional bodies, private sector consultancies, national organisations and revolutionary organisations. This multi-organisational delivery system created a temporary housing development process across the city. The initial workflow process included two main phases, design and construction phases with controlling points for assuring the works were approaching the primary strategic objectives; and brought the sociotechnical system to life. Various influential considerations underlying the complexities for organisational configuration in housing programmes are explored.

5.1 The Sociotechnical Delivery System

As theoretical chapter showed how one of the most difficult tasks in post-disaster reconstruction is organising the necessary processes and procedures, particularly regarding the participants and their roles in housing reconstruction. Practical decisions have to be made, in a context of competing interests (Davidson 2009) in an emotionally overwhelming post-disaster situation. While the disaster exposes

© Springer International Publishing AG, part of Springer Nature 2018

F. F. Arefian, *Organising Post-Disaster Reconstruction Processes*,
The Urban Book Series, https://doi.org/10.1007/978-3-319-70911-6_5

failures in previous development activities, there are expectations from recovery and its strategic agent reconstruction for improvements. Such expectations add to the complexities of organising reconstruction programmes because adding any qualitative principle or strategy to the existing practicalities and policies adds layers of practical considerations which must be blended together through the organisational design and configuration of the programme. Davidson (2009) defines organisational design as identifying the project's participants and the relationship between them. The success of post-disaster reconstruction largely depends on the complex relationships between the multiple actors, as each of them has different organisational size, capability and organisational norms. The challenge is 'how' to 'design' these relationships between participants in the best interest of the recovery effort (ibid). The Bam's reconstruction case included strategic core principles of people's participation and constructing earthquake resistant buildings. Comparing this to the contemporary debates in theoretical chapter, part I positions Bam as a contemporary case with high level of complexities. Moreover, the Bam reconstruction had to take into account the established overall policies and practicalities in a way to be usable for 25,000 housing and commercial units that was the initial estimation. All these added to the complexities of organising housing reconstruction in the Bam area.

At the initial stage of debates on how to organise post-disaster urban reconstruction in Bam, the BAUC highlighted the need for having a system and a defined mechanism in place for housing reconstruction. The reason was that there was an expectation that a large number of people would want to reconstruct their homes at once. A system was needed to avoid chaos during the implementation of the housing reconstruction programme, an estimated 25,000 cases, on participatory basis in a relatively dense urban area. Since the Steering Committee introduced three strategic core principles for the overall reconstruction the required system had to take those into account for binding various technical, financial and administrative considerations altogether. This operational sociotechnical system linked several organisations and individuals while taking all expectations of design and construction of new homes into account.

The reconstruction group at HFIR, i.e. the ORA, in collaboration with MHUD and guidance from the BAUC designed this required system. Being architects themselves, members of the ORA team were familiar with the existing practice of housing development and urban development in the country. Thus, in order to address the high expectations of the Bam reconstruction, the ORA tried to be purposeful as it approached each core specific principle through direct operational considerations. They also tried to translate the existing and established reconstruction policies and experience that were based on rural reconstruction to the Bam's urban scenario. All the above influenced the formation, implementation and performance of the delivery system of the housing reconstruction programme both in positive and negative ways; and the system at various points of its implementation had to be adjusted accordingly.

These operational considerations were grounded in the realities of housing development practice, its broader urban development and sociopolitical context of the country. For example, the key question around disaster-affected people's participation was in fact in what areas of housing reconstruction those people could have a choice over. In previous reconstruction cases, the level of expected participation was related to the reconstruction context, mostly related to their general perception about the capacity of the local people in that area. The level of participation in rural areas usually was perceived to be higher than in urban areas, as a member of ORA explains:

> The participation of beneficiaries in housing reconstruction is about giving people choices. We should see at what stages people have a choice during a construction operation of their houses in a normal situation. This is an indicator.
>
> Villagers are used to building their houses....but urbanites are teachers, shop-owners and so on; they all have another job. In fact they do not do the labour job to build their houses. They hire a contractor. So we differentiate the level of participation between villages and cities.
>
> (member of ORA/HFIR)

In the case of Bam, the initial perception was that Bami people were not familiar with construction operations. The extent of the destruction even in newer houses was indeed an indicator of the low existing capacity among local people. Thus, beneficiaries of the reconstruction programme could not be fully autonomous for the reconstruction of their houses in a way to be aligned with the two other core principles (Saemian and Erfanian Daneshvar 2011). Beneficiaries initially were not allowed to become a contractor for their own houses. That meant limiting beneficiaries' role compared to housing development in a normal situation. Beneficiaries were seen as applicants and project administrators that were the driving force behind the reconstruction of their houses. But in order to do so, beneficiaries had to be supported technically, financially and administratively all along the way. In an attempt to motivate the traumatised people, debris removal and reconstruction of a house were not started if the beneficiary/applicant did not start the process.

The proposed reconstruction plan by HFIR was approved by the Steering Committee three months after the earthquake, 16/03/2004 (26/12/1382 in Persian Calendar). The main axes of the urban housing reconstruction programme were as follows:

- Dividing the Bam area into districts, and creating district level administration
- Inviting and selecting construction consortia and contractors
- Reaching a broad agreement with the contractors
- Contracts between contractors and beneficiaries
- Creating the Bam Architectural and Urbanism Council for providing, directing and supervising design codes
- Organising local architectural consultancies in Bam
- Establishment of a one-stop-shop compound (construction Bazaar) for provision of technical-professional services, and exhibition of construction materials and techniques
- Supervising the construction operations by HFIR, Engineering Organisation and BHRC
- Financial assistance

- **Dividing the Bam Area to Districts, and Creating District Level Administration**

Dividing the urban area into districts was initially done to manage the search and rescue operations soon after the earthquake. A small group of officials including locals (e.g. the deputy governor) made an ad hoc decision to split the workload. They used their general knowledge of the area (e.g. population density and existing roads), the volume of damage and the capacity of each rescue team. The HFIR used the same division for organising debris removal and reconstruction. In order to cope with the extraordinary workload of debris removal and reconstruction, HFIR called on 14 of its provincial branches as the Assistant Organisations (Setad-e-Moein or in short Setads) in urban and rural areas.

The initial estimation was that 20,000 houses in the urban area were in need of complete reconstruction and 11 Setads were allocated in the urban area which was divided into 11 districts. Figure 5.1 shows the original map for this division, prepared by HFIR. Baravat, the suburban area of Bam, was treated as one district. Each zone was designated to one Setad and one local consultancy to work with the applicants from that district. It was roughly estimated that each district would have around 2,000 applicants for 2,000 houses in the need of reconstruction (Alizamani 2012). Later on, the organisational supervisor of construction operations (Kerman Engineering Organisation) followed the same approach and allocates a task group to each district.

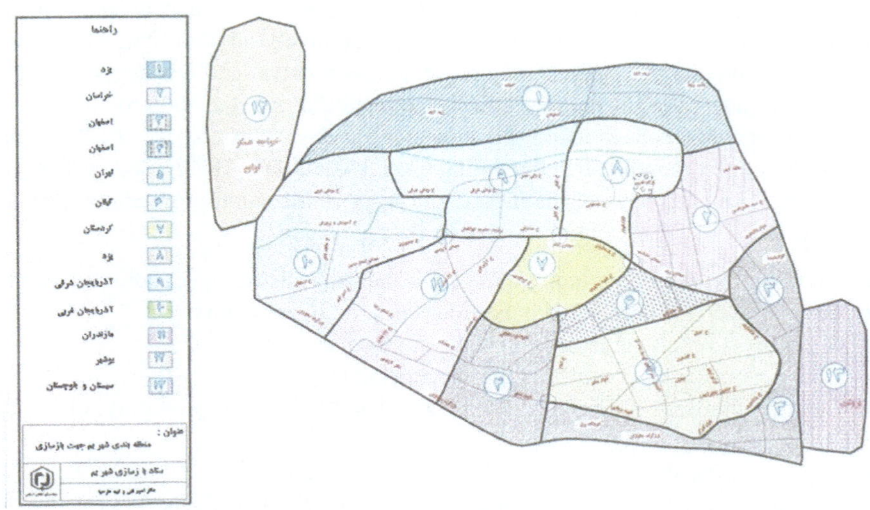

Fig. 5.1 The original map for dividing the Bam areas for reconstruction purposes. *Source* HFIR (2004a, b) (Reprinted with permission)

- **Establishment of a Construction Bazaar for the Provision of Technical-Professional Services, and the Exhibition of Construction Materials and Techniques**

This construction Bazaar was a multifunctional complex within which professional and technical support for people could be comprehensively provided. It acted as a one-stop-shop that included spaces for housing construction service providers, e.g. consultancies; exhibition space appropriate construction materials to people; space for showcasing examples of various construction techniques in real size scale; and an office for contractors' affairs in order to help create jobs for Bami people. The construction Bazaar was among the first building to be completed; until then all engaged professionals were temporarily based in a school building that survived the earthquake (Steering Committee for Reconstruction of Bam 2004).

- **Inviting and Selecting Construction Consortia and Contractors**

The housing reconstruction programme in Bam initially expected to attract consortia and qualified companies which could bring new technologies to the programme. They were expected to bring lightweight materials, speedy construction techniques for earthquake resistant buildings while being considering region's climatic and environmental conditions. The project was advertised in order to invite and select construction consortia and contractors. According to the project advertisement, published on 19/03/2004 (19/12/1382 in Persian calendar), contractors could have an allocated space in the Construction Bazaar for showcasing their technology and build a sample. They also could work throughout the urban area on their choice of property. Priority was given to the companies who additionally provide insurance of their work against earthquake for 10 years. The estimated project size was 500 houses for each contractor company.

- **Reaching a Broad Agreement with the Contractors**

According to the project advertisement, contractors and HFIR would have an overarching agreement. HFIR set a number of conditions for the contractors, for example, contractors' work would be controlled by engineer supervisors and the Engineering Organisation, HFIR or other professional institutions. Contractor works should also comply with the guidance and regulations by regulatory bodies.

- **Contract Between Contractors and Beneficiaries**

Embedded in the general agreement between HFIR and the contractor companies, the construction contract was directly between the beneficiary as the applicant/construction manager and the contractor company. Beneficiaries were the clients. They were in charge of selecting the contractor, administrative follow-ups for providing construction materials and cooperation with the supervising engineers. Reconstruction funds were directly paid to beneficiaries by the banks in a number of construction stages. Beneficiaries/clients would then pay contractors based on the physical progress. In case of any dispute between beneficiaries/clients and contractors, Setads were the reference point for mediation and problem solving (HFIR 2004a, b).

- **Creating the Bam Architectural and Urbanism Council for Providing, Directing and Supervising Design Codes**

One strategic principle of the Bam reconstruction was to safeguard the urban architectural identity that Bam was famous for before the earthquake. The BAUC was established to address this strategic aspect of reconstruction. Before reconstruction in Bam began, the Steering Committee formed the BAUC to ensure that all activities in Bam were design-based and reflected that strategic principle. For the city as a whole, the BAUC had to identify the historic identity of Bam and the way it could be applied during the reconstruction process.

- **Engaging Local Consultancies in Bam**

In order to support disaster-affected people on design and technical matters of reconstruction of their houses, the idea was to establish local consultancies. They would provide free design and technical services for obtaining planning permission as a key stage for housing reconstruction. HFIR advertised the project for identifying and commissioning qualified architectural consultancies and introducing them to beneficiaries. These consultancies from around the country established local branches in Bam. Their work had to comply with the BAUC architectural design codes and guidance, rules and regulations from the Bam urban development plan, and, beneficiaries' individual requirements, as well as technical regulations and seismic codes for earthquake resistant buildings. Before the reconstruction began, consultancies developed conceptual broad typologies based on a variety of locations, direction and sizes of sites. Typologies met design codes and guidance according to Bam's historical and cultural identified by BAUC. They could be personalised later. Another consultancy company was appointed to harmonise and oversee the work of these local companies and link them to the BAUC. It was called the Mother Consultancy (Havaie n.d.; Saemian and Erfanian Daneshvar 2011; Alizamani 2012).

- **Supervising the Construction of Houses by HFIR, the Engineering Organisation and BHRC**

Constructing earthquake resistant buildings was another core strategic objective of the reconstruction. Therefore, seismic regulations and the National Building codes were applied during the construction of new homes. Additionally the quality of construction materials, e.g. cement, brick, etc. had to be standard. The Steering Committee mandated utilising HFIR resources, the Engineering Organisation and BHRC. A multi-layer supervisory network was established to control design and technical maps and construction operations. Engineers of the Kerman Engineering Organisation and at broader level by Setads and HFIR supervised and controlled construction operations. Required laboratories, e.g. a soil mechanic laboratory, were established for quality control of resources at their origins and during the implementation by inspectors (Havaie n.d.; Saemian and Erfanian Daneshvar 2011; Steering Committee for Reconstruction of Bam 2004; Alizamani 2012).

- **Financial Support for Housing Reconstruction**

The Steering Committee allocated $9,500 for the reconstruction of each house. This consisted of $6,000 (60 million Rials) in low-interest loans and $3,500 (3/5 million Rials) in grants. At the time this amount was enough for building a house of 80 m^2 using acceptable quality materials. Additionally, beneficiaries used all technical and administrative support services, e.g. local consultancies and construction supervisions, for free. The Steering Committee paid all associated fees and operational costs.

Following demands by people and follow-ups by the town council on the importance of boundary walls for houses for the safety and security of people, on 29/05/2004 (09/03/1383 in Persian calendar), the Cabinet granted an extra $2,000 (20 million Rials) to each house (The Cabinet, 2004), increasing the total financial support to $11,500. Banks (on behalf the government) paid the loan and grant to applicants/beneficiaries in five instalments: The first was at the start of the construction phase; other instalments were linked to the physical construction progress. In the mid reconstruction period, the overall financial support was increased to total of $16,000 (160 million Rials).

5.2 Putting All Considerations Together

Binding the above operational considerations together, the sociotechnical system for the delivery of urban housing reconstruction was born. The system design treated the housing reconstruction programme as a collection of individual cases. Each case must embed all the above considerations. Implementing the various aforementioned considerations required gathering several principal participants at both individual level (beneficiaries/applicants) and organisation level (all others) as the following: the Housing Foundation of the Islamic Revolution, the Ministry of Housing and Urban Development, Bam Architecture and Urbanism Council, KEO; Setad-e-Moeins; the Mother Consultancy company; the Municipalities of Bam and Baravat; local consultancy companies; Applicants; Contractors; Banks; Laboratories for Quality Control; Construction Material Suppliers; and BHRC. However, HFIR, MHUD and BAUC were not part of the operational activities. They participated during the formation of the operational system and managed it during implementation. Applicants and operational participating organisations worked together in various phases: predesign for entering the system, design and construction. After an applicant's case was verified, the housing reconstruction process started, formed from two separate phases: the design phase for the purpose of obtaining planning permission, and the construction phase on site up until the completion of the construction. Figure 5.2 shows this.

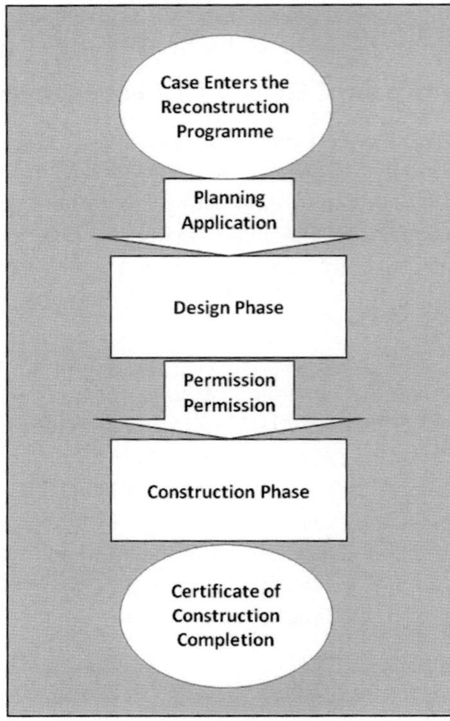

Fig. 5.2 The overall process and stages in the housing reconstruction programme

When a case entered the reconstruction programme it followed a certain workflow process for each phase. The beneficiary applicant was responsible for start and follow-up of her/his case. The reconstruction guidance (HFIR 2005), distributed among the locals, draws the outline of the workflow process as follows: The process started before the design phase when a potential beneficiary submitted his/her request to enter the reconstruction system. Figure 5.3 schematically illustrates the system elements and their connections at this stage which involved beneficiaries as applicants, Setads and municipality, leading to issuing of a planning application. Upon this, the related Setad referred the beneficiary/applicant to the allocated local consultancy to follow the design process in order to receive planning permission, required to start the construction phase. During the design phase, the sociotechnical system gathered Setads, applicants, local consultancies, the Mother Consultancy, KEO and the municipalities worked towards issuing planning permissions. Figure 5.4 illustrates the system elements and their connections for the design phase schematically.

At the construction phase, the sociotechnical system gathered Setads, applicants, KEO, contractors, laboratories for quality control, banks, construction material

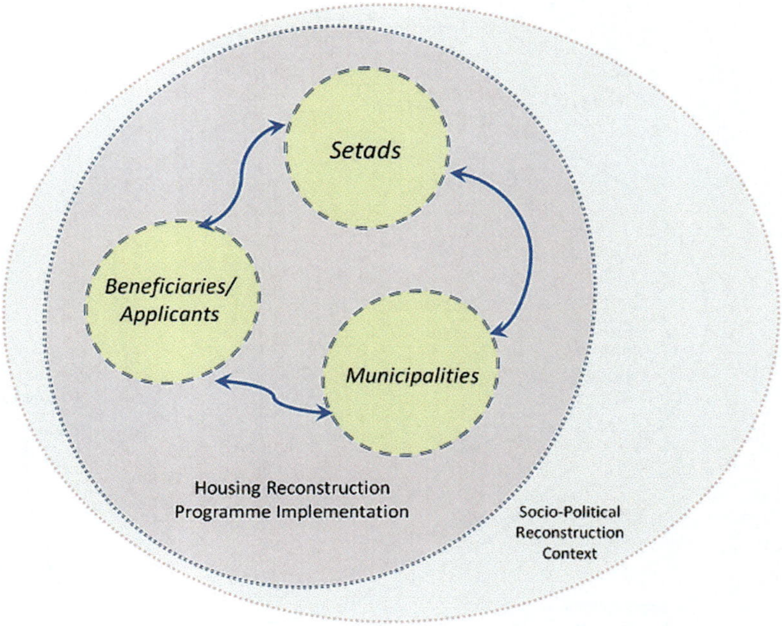

Fig. 5.3 Entering the reconstruction programme

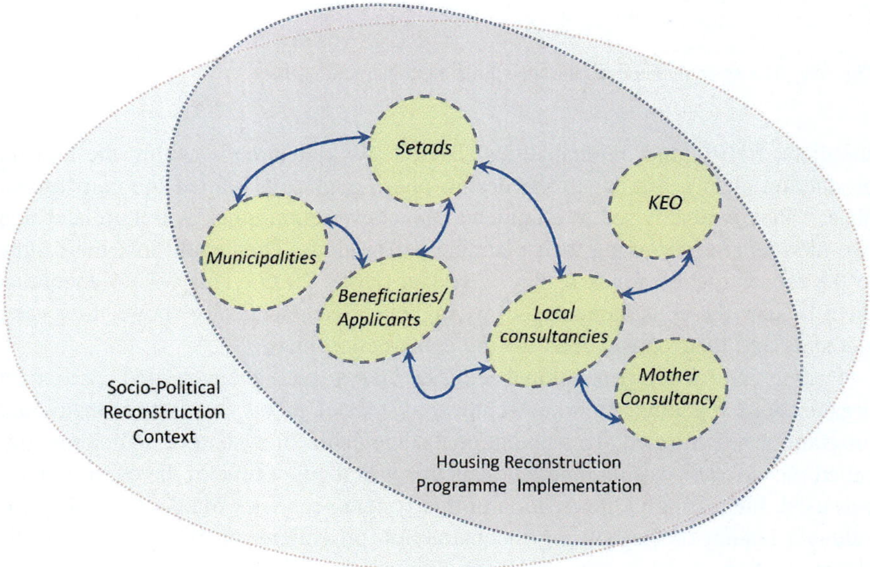

Fig. 5.4 The system elements and their links in design phase

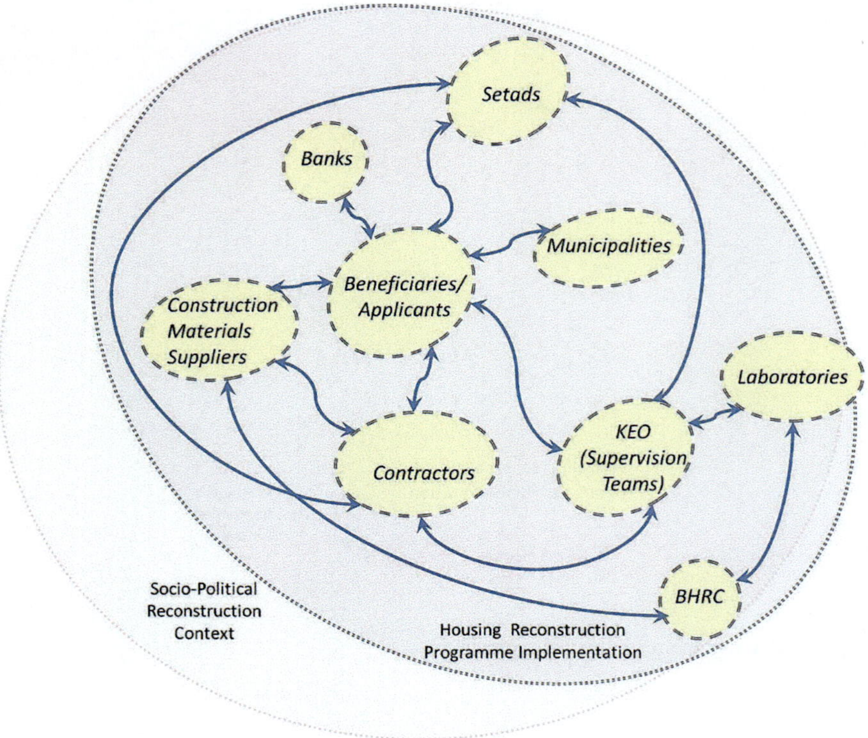

Fig. 5.5. The system elements and their links in construction phase

suppliers, BHRC and municipalities (BHRC was not directly within the housing production chain). Similar to the design phase applicants started the construction phase. They were advised to commence the construction quickly and no later than two months after receiving their planning permissions. They could lose their rights to the reconstruction funds if they were late (HFIR 2005). This was to encourage them to start the construction quickly. Figure 5.5 schematically shows the system elements and their connections for the construction phase.

Putting all above operational considerations and their related participant organisations together, the sociotechnical system of urban housing reconstruction programme was formed. This operational sociotechnical system or as it sometimes called the 'Model' was an innovation as this was the first time in the country that it was used. Interestingly, the creation of this system per se is widely seen as the most valuable feature of the housing reconstruction programme. It is regardless of different positions and opinions about the performance of the system by interviewees for this research. Figure 5.6 shows the sociotechnical system of the housing reconstruction programme.

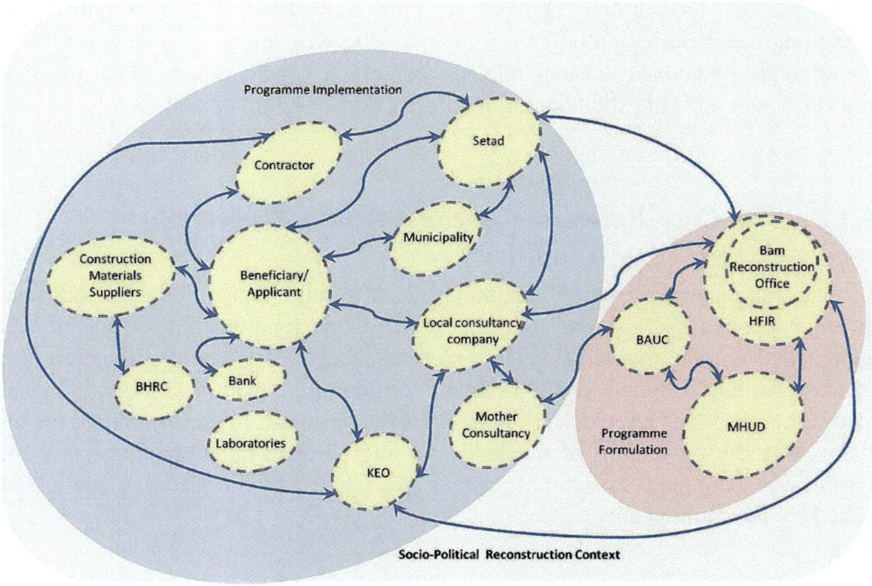

Fig. 5.6 Multi-organisational system of housing reconstruction programme in Bam (Design & Construction)

5.3 The Workflow: The Process of Production

The workflow is the connective process; the manifestation of the functional coordination mechanism that determines the way the necessary activities and tasks are interconnected, their order and their interdependencies. It is the actual basis for bringing the reconstruction programme to life during implementation, linking between various organisational elements, and integrates the system elements to achieve unity of efforts within the organisation system to accomplish the organisation's task. A well-integrated process in a multi-organisational system leads to systemic transition so that one function's output becomes another function's input at the right time; this may result in seamless transferral between activities. A well-designed workflow that creates unity and harmony between actors and subsystems strongly is interrelated with the notions of coordination and communication (Badir 2006; Herber et al. 2000 reproduced in Mintzberg et al. 2003, pp. 234–241; Goold 2002; Mintzberg 1980; Galbraith 1983; Mintzberg 1980).

Designing the interrelationships that bring unity among participant organisations and beneficiaries in post-disaster reconstruction is critically important. The following is to explore how in the Bam's case the connective process of the reconstruction programme delivery system integrated various organisations (including beneficiaries) in order to implement the programme, in a way to achieve the strategic objectives through related operational considerations. This practical workflow was to bring the housing reconstruction programme to life.

5.3.1 The Case Enters the Reconstruction Programme

To enter the reconstruction system, an applicant should approach the related Setad in order to open his/her case and request for debris removal. The applicant then approached to the municipality in order to receive the site specification (i.e. dimensions, land-use and location-related considerations within the town's master plan, for example, if a setback from the road was needed). In the meantime, his/her claim over the ownership of the land and eligibility to enter the reconstruction system was examined. The end of process was marked by the planning application issued by the municipality.

5.3.2 Design Phase Workflow

The workflow of the design phase started upon the applicant's referral to the local consultancy company after the planning application was issued. Once the applicant approached the consultancy, the consultancy personnel conducted a site visit. At the preparation period, before the public commencement of the programme, and at the early stage of its implementation the BAUC requested architectural consultancies to provide their own interpretation of how guidelines on safeguarding urban identity could be realised in housing reconstruction. Other design criteria included: the potential of future extension of houses, keeping the temporary shelter in each plot, keeping trees, reflecting traditional geometry and arrangements in internal layout as well as the bulk of the building. If the applicant's/client's requests and the specification of the site fit into one of the existing typologies, they could personalise it. If not, the consultants and client worked with each other for a new design and produced maps. Maps required approval for architecture and structure, by the Mother Consultancy and KEO as follows:

First, the Mother Consultancy would check architectural design against guidelines and codes provided by BAUC, and design codes and regulations provided by MHUD and the detailed master plan. This control was a formal procedural stage in the workflow. The maps were revised based on the Mother Consultancy's recommendations. After approval of the architectural design, the consultancy would start the structural, mechanic and electronic designs. Second, the KEO would check structural, electronic and mechanical specifications against national building codes, especially seismic regulations, Article 2800. If it was not approved the local company would revise architectural maps which in return needed to receive another approval by the Mother Consultancy. Upon receiving approval from KEO, (a) the case in local consultancies was completed and they submitted it to Setad, (b) KEO assigned an engineering supervisor to the case for the construction phase, introducing him to Setad and the applicant. The Setad would also refer the completed case to the municipality which issued the planning permission. Figure 5.7 presents the workflow at the design phase as documented by HFIR in 2012.

However, it must be noted that this diagram was only produced during the implementation instead of the stage of the system formation. In 2006, I asked my consultancy's team leader for the housing reconstruction project, resident in Bam, to observe and write down what was actually happening in the design process: who does what and when. He did so, and the next time we met we sat together and drew the diagram of the process based on his writings and observations. I gave the diagram to HFIR during an informal meeting. It was later developed and used by HFIR. As will be discussed in the next chapter, despite the hope that the system formation and workflow would facilitate approaching the strategic objectives, realities on the ground meant that it continually evolved.

5.3.3 *Construction Phase Workflow*

The workflow started when applicant hired a contractor. If applicants did not have contractors in mind they would approach the Setads, where they could meet the contractors who had responded to the HFIR tender, choose a contractor, sign the contract and start the work. In the meantime, the application for the construction fund was opened by the Setads and applicants were referred to the banks. The applicant as the project manager and administrator, therefore, had to work with a contractor and follow specific steps for receiving the first instalment of the construction fund. Setads played the role of mediator in cases of dispute between contractor and applicant and also supported applicants if they asked for technical help.

The construction operation was supervised by engineering supervisor who was commissioned by KEO, which held the overall responsibility for construction supervision according to their contract with HFIR. The subject of their contract was the provision of comprehensive supervision over structural, architectural, electronic and mechanical works by qualified engineers and architects, respectively (HFIR and KEO 2004 reprinted in HFIR 2012). However, the existing national Engineering Organisation framework does not make it compulsory to conduct architectural, electronic and mechanical supervision during construction of small houses (similar to the ones that were subject of the contract in the Bam reconstruction); and the only compulsory supervision is structural. Therefore, the control and supervision of the construction phase were focused on structural matters. The number of critical points of the technical control and supervision was introduced more than what was outlined in the construction engineering organisation law. They were as: excavation, laying of concrete, mould and foundation rebar reinforcement, foundation concrete,

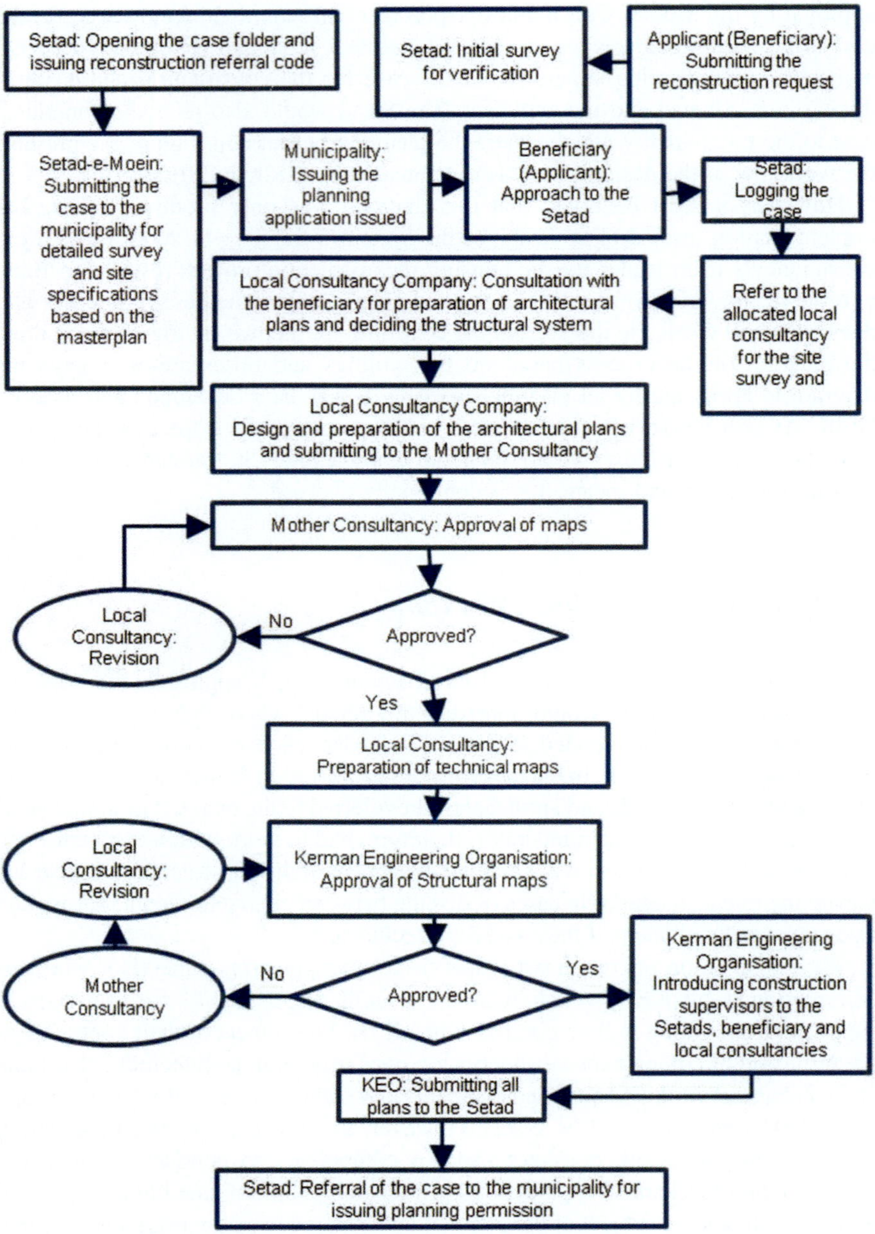

Fig. 5.7 The design workflow in urban housing reconstruction in the Bam area. *Source* HFIR (2012)

erecting structure, roof framing and rebar reinforcement, roof concrete, erection of walls, façade.

The control and supervision mechanism comprised frequent supervision at the above critical points by the engineering supervisor (who was helped by two assistants). Each reconstruction unit had a Building Technical Identification Document (BTID), which included checklists for reporting each of the above construction stages by the engineering supervisor. The construction materials (e.g. cement) also were tested in laboratories. If the supervisor did not approve the quality of the contractor's work the contractor had to fix the issue. Supervisors themselves were supervised by the senior engineer from KEO and the high-level supervision by their related Setads (Saemian and Erfanian Daneshvar 2011). In case of dispute between supervisor and contractor the senior engineers or Setads became involved.

Construction funds were released based on the construction operation progress. A number of the critical control points were introduced by HFIR as milestones for realising the construction funds instalments. At the end of the construction operation, which was documented in the BTID, the applicant applied for the Certificate of Construction Completion from the municipality.

As this research explores, the construction phase (similar to the design phase) saw adjustments and amendments during the implementation such as applicants could become the contractor of their own homes or the way funds were released. Some of these adjustments profoundly changed the dynamics of interactions between the actors involved within construction operation in contrast with what was initially considered. Some of the adjustments of the construction phase are a reflection of the previous adjustments during the design phase. The ways of doing things in the design phase and the construction phase in a number of areas are linked together. For example after applicants could become contractors they tended to quickly choose a design from the design portfolio at the design process because they could change the architectural layout during construction phase as they wished.

5.4 The Sociotechnical Delivery System Elements

As the theoretical discussions showed, reconstruction programmes gather a number of participant organisations that are different in nature, approach, experience and organisational structures. The inherent diversity in housing reconstruction adds to the complexity of organising reconstruction. Similarly, the Bam reconstruction programme brought together 13 organisations as well as the applicants. The organisations, HFIR, MHUD and BAUC participated in setting strategic objectives and formation of the sociotechnical system for housing reconstruction on behalf of the Steering Committee. Additionally, applicants as a group and other participant organisations were the elements of the sociotechnical delivery system. To understand these system elements and the role they played in the formation and implementation of the sociotechnical system the research addresses the following characteristics at the time of the reconstruction programme: type of the organisation and background of the system element; previous reconstruction experience; previous role in housing development practice before the earthquake; role in housing reconstruction programme in Bam; location; links to other elements of the system; and whether they were part of the production chain or not.

- **Housing Foundation of the Islamic Revolution, HFIR**

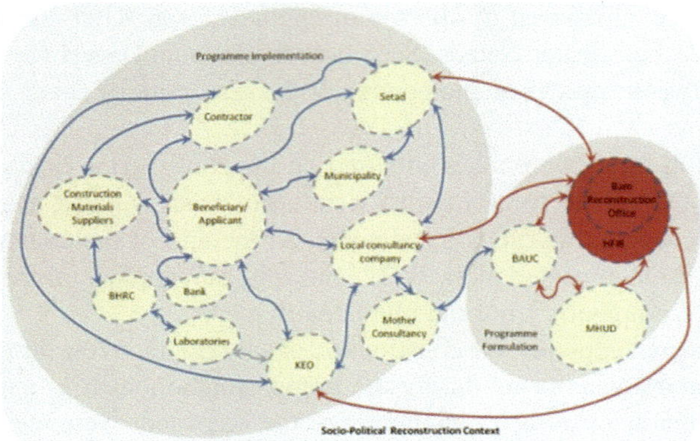

HFIR is a semi-independent organisation in charge of rural development in normal situations. The organisation is also involved in the provision of affordable housing in cities. Originating in the time of the revolution, it was founded by the revolution leader (Imam Khomeini) in 1979 with the mission of providing houses for the poor. The extent of its activities later extended to rural development. It is experienced in fieldwork in rural areas and in difficult conditions. HFIR is well funded as a result of money from Zakat.[1] After the Manjil earthquake, HFIR became the home for the Office of reconstruction Affairs (ORA). HFIR was the executive body in charge of the Bam reconstruction; ORA formed the Bam housing reconstruction programme and its sociotechnical delivery system. HFIR itself is not directly involved in operational activities in Bam, as it performs a coordinating role. At the time of the Bam incident, decision makers at ORA were middle managers, university architecture students who participated in post-war reconstruction, recruited by HFIR and promoted up the career ladder. ORA benefits from organisational stability that allows individual and organisational learning from each earthquake experience, e.g. Manjil and beyond. They have cumulated years of experience and learned the lessons from previous reconstruction cases. However, the Bam case is unprecedented because the previous experiences were mostly based on rural areas or small towns (member of HFIR interview 2013).

[1] An Islamic tradition within which people donate money for good causes to the religious leader (Imam).

- **The Ministry of Housing and Urban Development, MHUD**

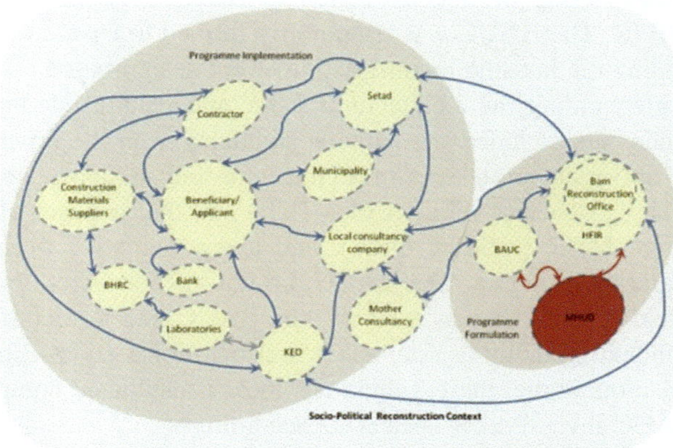

The MHUD is the governmental institution in charge of the country's urban development activities. At the time of the Bam earthquake, decision makers at MHUD were part of the new wave of architects and urban designers who understood the need for a qualitative approach, and the reform of the planning system in the country. Given the history of the Bam, its rich unique urban identity and the ancient monument of Arg-e-Bam, immediately after the incident the MHUD offered help on the qualitative aspects of the reconstruction. The overall project control of the Bam reconstruction in general was controlled by MHUD. Following the establishment of the Bam Architecture and Urbanism Council (BAUC), the ministry hosted the BAUC. The vice minister was the head of the BAUC and the middle managers at the Department of Architecture and Urbanism (DAU) at MHUD organised the BAUC meetings (member of MHUD, interview 2013). The ministry did not have a direct operational role in the housing reconstruction of Bam. Its influence on the programme was through the BAUC.

- **Bam Architectural and Urbanism Council, BAUC**

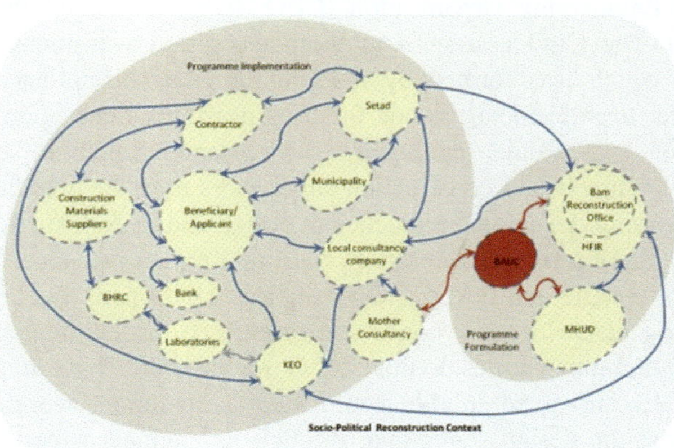

The BAUC was established to help with the qualitative aspects and architectural and urban issues during the reconstruction of Bam. It was a joint initiative by MHUD and HFIR. The BAUC was temporarily formed to ensure that all activities in Bam (including the housing reconstruction) were design-based and reflected the principle of safeguarding the architectural fabric and cultural identity of the city. The BAUC members included two famous architects from the private sector, two distinguished university professors in architecture and urbanism, two representatives from MHUD (the vice minister of MHUD and the manager of the DAU), a representative of HFIR (head of ORA), and a representative from the Cultural Heritage Organisation of Iran. The BAUC did not have a direct operational presence in the field. For the city as a whole, the BAUC had to identify the style of architecture, including defining size, space and traditional styles. For the housing reconstruction programme, these guided the local consultancy companies in their work. The BAUC invited other specialists accordingly.

- **Kerman Engineering Organisation, KEO**

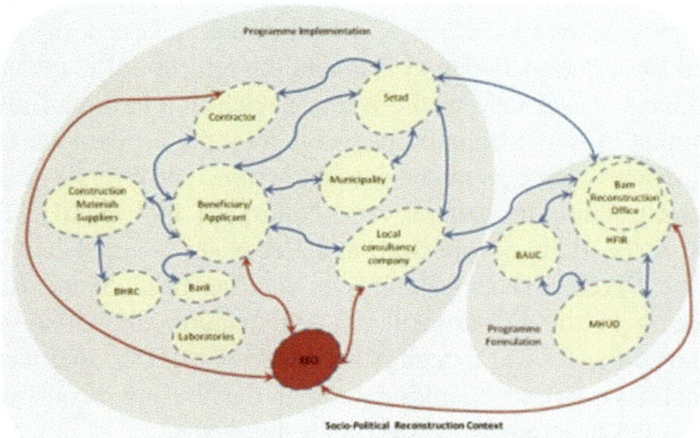

Kerman Engineering Organisation (KEO) is a provincial branch of the country's Construction Engineering Organisation (CEO, it is also called the Engineering Organisation). The CEO is a semi-autonomous, semi-governmental professional organisation, which was formed by merging previous disciplinary qualification bodies in order to implement the law for 'Engineering Organisation and Building Control' after the Manjil earthquake. Engineers (structural, electrical and mechanical), architects and construction supervisors must be qualified by KEO to work in the housing development practice in design and/or construction supervision (Construction Engineering Organisation, n.d.) in Kerman province. Following the suggestion by MHUD and BAUC, KEO was commissioned, by HFIR, to control the engineering plans for each house and to enforce the building regulations (in normal situation municipalities do this). KEO established its base at its Bam branch and was directly involved in the reconstruction operation, working with local consultancies at the design stage, and, through its engineer supervisors, working with applicants and contractors during the construction phase.

- **The Mother Consultancy Company**

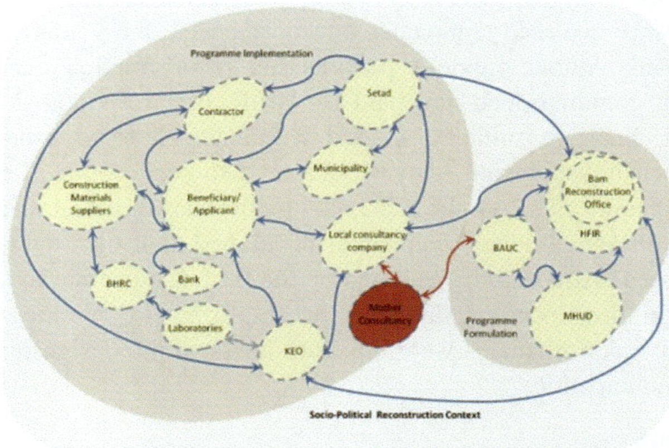

Before the start of the reconstruction, the BAUC commissioned two architectural consultancies to develop the architectural principles to be followed in the housing reconstruction. Later one of these was commissioned as the senior consultancy, called the Mother Consultancy, to make sure that the local consultancy companies working with Setads complying with the BAUC principles. The Mother Consultancy was in charge of controlling and approving the architectural plans and site layouts produced by local consultancies, as the municipalities do in a normal situation. It was a qualified architecture company in the private sector, and was one of a few older, large and high-profile consultancy companies in the field of architecture. The company did not have a previous post-disaster reconstruction experience at this scale. The Mother Consultancy was directly involved in the reconstruction operation in the field, working with local consultancies and Setads at the design phase. It established a local branch in Bam, although the works could be supported from the Tehran branch as well.

- **Local Consultancy Companies**

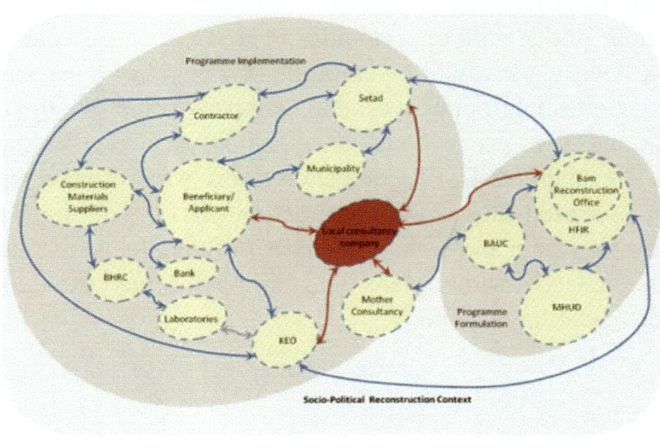

Local consultancy companies technically served applicants to prepare design and technical documents and receive planning permission to start the construction phase. They were qualified companies but mostly were the younger companies in the country and without previous post-disaster reconstruction experience. They were selected, commissioned and paid by HFIR after responding to the project advertisement. Each consultancy worked within an allocated zone and with its Setad. For each house design, they would need to seek approval from both the Mother Consultancy and from the KEO. Before the start of actual reconstruction, at the preparation stage, the local companies developed their own interpretations and ideas and created broad typologies based on the BAUC guidance and the variety of the site possibilities, location, direction, access, size, etc. Working directly with applicants, local consultancies were the front-liners of the design phase similar to Setads and municipalities.

- **Contractors**

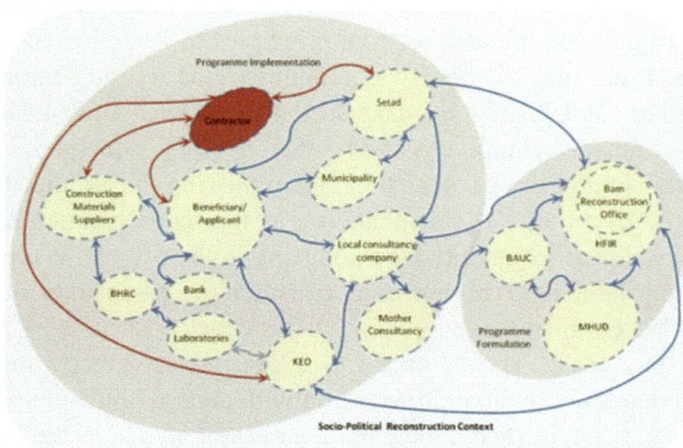

Despite the optimistic expectations reflected in the initial HFIR project advertisement, the construction contractors in Bam were mostly individual contractor teams or small contractor companies who were either Bami or more likely had come to the Bam area because of the work opportunities there. Contractors could work through the urban area; their work would be controlled by engineer supervisors and KEO. The number of controlling points was nine, compare to the five in normal housing development, and the work was documented in the Building Technical Identification Document (BTID). Contractor teams were directly involved in the reconstruction operation during the construction phase, working with applicants, engineer supervision team and Setads. A number of training courses and workshops for local individual contractors were held during the reconstruction in order to update their technical knowledge.

- **Beneficiaries/Applicants**

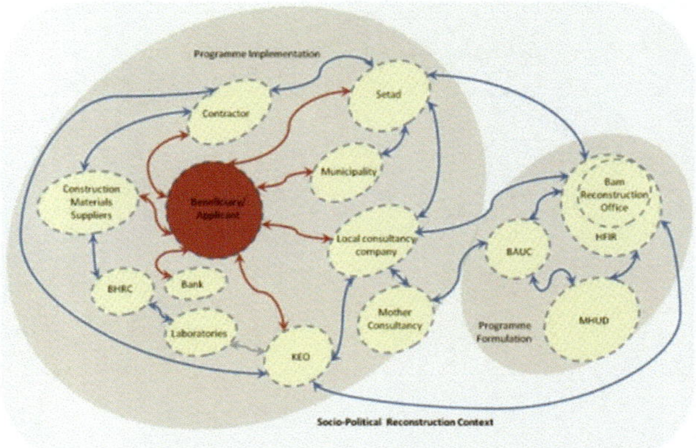

Applicants were individuals representing each beneficiary household. They were at the centre of administrative tasks for the reconstruction of their own houses. They were responsible for starting the reconstruction administrative process for their own case and driving the process forward. They were responsible for managing the reconstruction of their homes, such as submitting the case to the allocated Setad in their neighbourhood and receiving planning application from the municipality, working with the allocated local consultancy for the design of their homes, receiving the planning permission, hiring the contractor and requesting/receiving the construction funds in instalments. They were in direct contact with Setads, municipalities, local consultancies, contractors, engineer supervisors and banks. They had to seek out the support that was available. A number of architectural and technical workshops on architecture and the importance of earthquake resistant buildings, e.g. experimental workshops on shaking tables, were held for local residents.

If the owner had died in the earthquake, the strategy was to allocate the funding and allow the applicant to be the representative of the remainder of the family without taking into account who would finally inherit the house. This was to prevent legal issues on inheritance slowing down the reconstruction process—inheritance law was especially problematic for widows whose houses belonged to their lost husband or if there was some family dispute over the inheritance. Also, applicants who could not follow the work of their house reconstruction were connected to NGOs to act on their behalf and follow the reconstruction process. Moreover, the extended families, where married adult children with their families lived in their parents' houses were allowed to apply to the construction fund and build their houses in the land of their parents' houses.

- **Banks**

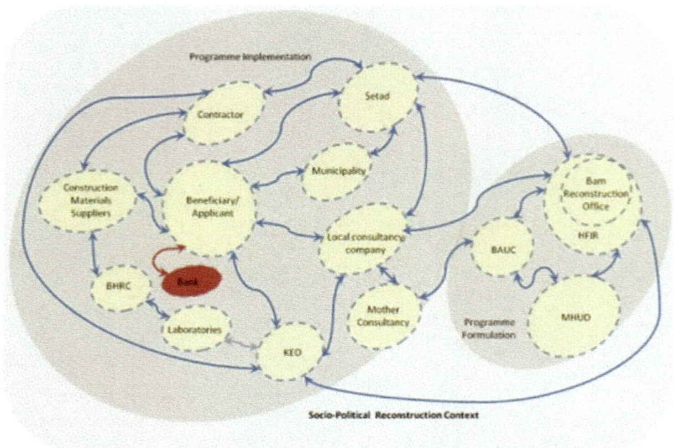

Banks involved in the housing reconstruction in Bam represented the government by providing financial support for each individual housing reconstruction case. The reconstruction fund was partly a grant and partly a 15-year loan with low interest (5%). Local bank branches in Bam and Baravat administrated the individual cases and paid the funds on behalf of the government. Following arrangements with the Steering Committee and HFIR, the Banks were in direct contact with the applicants after the applicant received planning permission and wanted to start construction. The banks paid money to applicants first earlier at the construction phase, followed by instalments based on Setads' recommendation letters (Ghafory-Ashtiany and Hosseini 2007).

- **Municipalities**

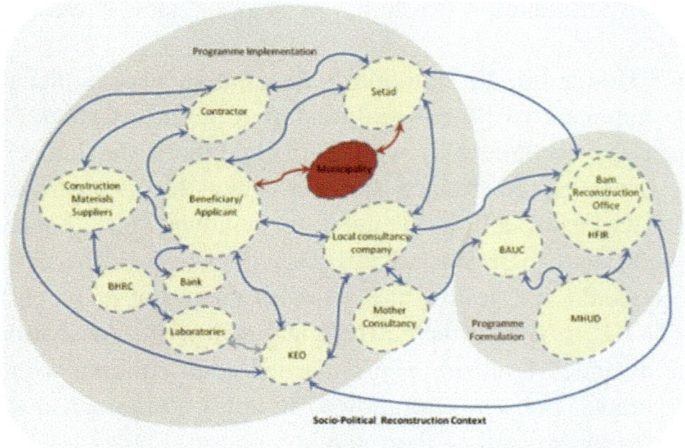

The Bam and Baravat municipalities, similar to other cities, are affiliated with the Ministry of Internal Affairs. Both municipalities were in charge of the housing development process before the earthquake. This included: site visits and examination in accordance with the urban development maps; issuing planning applications; controlling the technical maps and issuing planning permission; applying building controls and enforcing codes and regulations; and issuing certificates of completion. At the time of the earthquake, the Bam and Baravat municipalities ranked grade three and five, respectively, where grade one is the highest after megacities, in the country. The ranking system for municipalities in Iran is based on the city's population, which indicate the capacity of a municipality and the number of planning applications in a certain period the municipality can manage. Decision makers in both municipalities were the mayors at the time of the earthquake, elected by the town councils. The Bam earthquake was followed by debates on the ineffectiveness of municipalities for enforcing the application of existing national seismic regulations, e.g. Regulation 2800.

The scale of the destruction, and therefore housing reconstruction, in Bam covered the whole urban area; therefore municipalities during the reconstruction period temporarily were not in charge of housing development as a whole. Instead, they became participant organisations of the reconstruction programme and in addition to their other responsibilities, for example, roads and public spaces, public services and so on. The municipalities had to help in identifying the land ownership, site examination according to the urban development maps and issuing planning applications for applicants who were introduced by Setads to municipalities, issuing planning permissions for (re)construction of each house after the approval of the technical maps by other participant organisations during the design phase and issuing Certificates of Completion at the end of the construction operation.

- **Setad-E-Moeins**

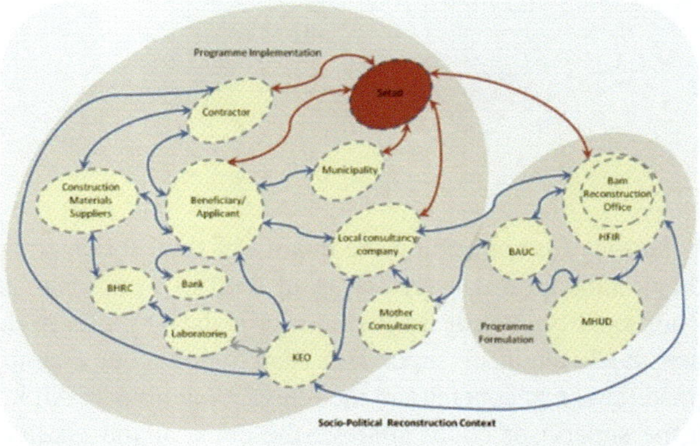

The HFIR has provincial branches throughout the country; they are called Setad-e-Moein (Setad in short) when they are engaged in reconstruction activities in a province other than their own. Each Setad is an independent legal entity that is guided by the umbrella policies of HFIR. Setads are in-house resources for the ORA which is based in the Central HFIR. Every time after a destructive earthquake in country the ORA calls on a number of these Setads in order to cope with the scale of the reconstruction. Usually, the selection of Setads for each reconstruction case is influenced by the proximity to the destroyed location, Setads' previous experience and performance, shared culture and ethnic background.

In the Bam case, 14 Setads were involved in 11 different zones throughout the urban area. They undertook administrative tasks for housing reconstruction. Having local offices, Setads were directly involved in the reconstruction operation, at all stages, working with applicants, municipalities, local consultancies, contractors and engineer supervision teams. They received applicant's requests to enter the reconstruction system, referred them to the municipalities, after the applicant's case was opened they introduced the to the allocated local consultancy, after receiving the approved technical maps they sent them to the municipalities to issue planning permission at the design phase. They also oversaw the contract between contractors and applicants, applied extra technical control to the construction operation and engineer supervisors activities supported applicants' request for construction funds based on the work progress and mediated between applicants, contractors and engineer supervisors in disputes.

- **Quality Control Laboratories**

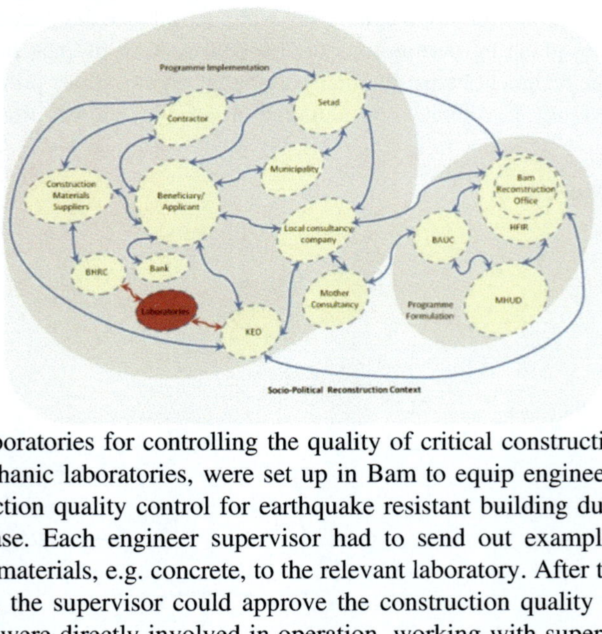

Technical laboratories for controlling the quality of critical construction materials, e.g. soil mechanic laboratories, were set up in Bam to equip engineer supervisors with construction quality control for earthquake resistant building during the construction phase. Each engineer supervisor had to send out examples of critical construction materials, e.g. concrete, to the relevant laboratory. After the laboratory confirmation, the supervisor could approve the construction quality at that stage. Laboratories were directly involved in operation, working with supervision teams. The laboratory reports were attached to the BTID (contractor and engineer supervisor, interviews 2013).

- **Building and Housing Research Centre, BHRC**

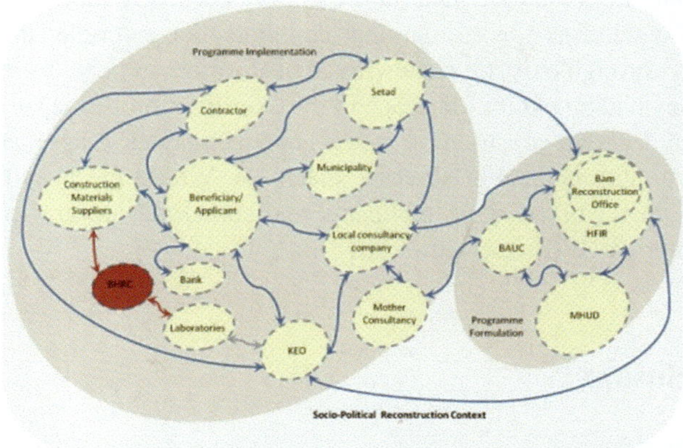

Building and Housing Research Centre, BHRC, is the governmental national research centre for construction methods and materials, the main contributor to introducing and developing National Building Codes in structure, architecture and other specialist disciplines. Following the recommendation by the Steering Committee, BHRC, for the first time, participated in reconstruction. It was commissioned for the project control (overall supervision) for the quality control of construction materials. BHRC was not directly involved in housing reconstruction operation. It worked with and supervised the laboratories and factory producers of strategic construction materials, e.g. brick, steel and cement, preventing non-standard construction materials from entering the construction market in Bam. BHRC teams tested examples of materials, visited factories and laboratories, randomly visited construction operations, and so on. It established a local branch in Bam with the support from its headquarters, disseminating the results and recommendations among the Steering Committee, HFIR, MHUD, KEO and National Standard Institute (BHRC n.d.; Saemian and ErfanianDaneshvar 2011; Steering Committee 2004).

- **Construction Material Suppliers**

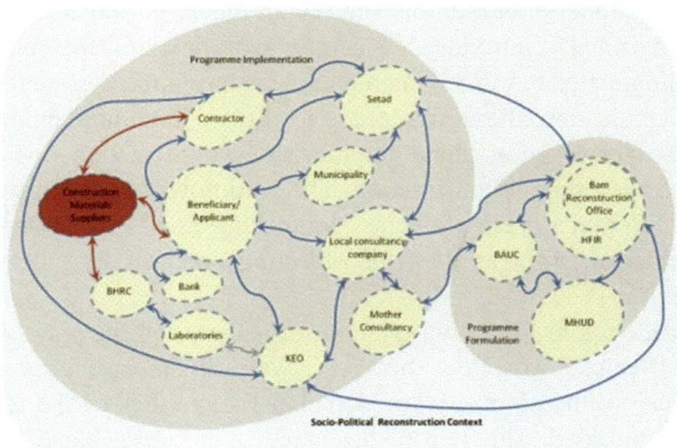

Construction material suppliers provided both strategic materials, e.g. bricks, steel and cement, and non-strategic materials, e.g. doors, kitchen cabinets, etc. HFIR set a number of conditions for the national suppliers, for example distributing their materials only through existing or new local distributors in order to revive the local economy. They had an exhibition space in the Construction Bazaar in order to show the samples of different construction models and techniques to applicants and locals and to allow quality control of strategic materials. Within the housing reconstruction process, they worked with applicants and sometimes contractors (local contractor, member of HFIR interviews 2013).

5.5 Conclusion

This chapter discussed how the housing reconstruction in Bam was initially organised. It was discussed that the programme size which was estimated to include 25,000 cases in a compact urban area triggered the BAUC to request a system in place for the delivery of the housing reconstruction programme. This multi-organisational delivery system created a temporary housing development process across the city. It connected financial, technical, social and practical considerations in the reconstruction programme. Attempts were made to introduce practical considerations for approaching each of the core strategic objectives, which were previously introduced by the Steering Committee for the whole reconstruction, including housing reconstruction. Simultaneously, more practical considerations had to be made to translate the existing reconstruction policies and lessons learned from previous reconstruction cases to the Bam case, the former were mostly rural experiences and the latter was mostly urban. One of these experienced considerations was dividing the Bam urban area into 11 districts. The result was an innovative, yet pragmatic multi-organisational system, which connected applicants and 13 organisational participatory positions together. Each of these positions including applicants was a system element. The diversity of the system elements extended from local individuals to local urban governance, provincial professional bodies, private sector consultancies, national organisations and revolutionary organisations.

The system initially introduced a certain workflow for the production chain within its distinguished but interrelated phases: entering the system towards issuing the planning application, the design phase towards obtaining planning permission, and finally the construction phase for the actual construction operation towards certificate of completion of construction. The workflow introduced the order of the activities by different system elements, controlling points for assuring the works were approaching the primary strategic objectives; and brought the sociotechnical system to life.

These considerations, the initial system formation, and the way of doing things within the introduced workflow, however, were at the initial stage of the programme formation, therefore, yet to encounter the realities of the implementation that would be unfolded. This will be discussed in the next chapter.

References

Alizamani MH (2012) Contemporary approaches in post-disaster housing reconstruction. (Unpublished PowerPoint file presented at various conferences) (in Persian)

Badir Y (2006) The role of network lead company in integrating new product development processes across strategic partners. PhD thesis, EcolePolytechniqueFederale De Lausanne, Lausanne

Davidson C (2009) Multi-actor arrangements and project management. In: Lizarralde G, Johnson C, Davidson C (eds) Rebuilding after disasters: from emergency to sustainability. Spon Press, London, New York

Galbraith JR (1983) Strategy and organization planning. Hum Resour Manage 22(1–2):63–77. https://doi.org/10.1002/hrm.3930220110

Ghafory-Ashtiany M, Hosseini M (2007) Post-Bam earthquake: recovery and reconstruction. Nat Hazards 44:229–241. https://doi.org/10.1007/s11069-007-9108-3

Goold M (2002) Designing effective organizations: how to create structured networks. Jossey-Bass; Wiley (distributor), San Francisco Calif

Havaie MH (n.d.) A brief review on bam reconstruction program (Unpublished Power Point file provided by HFIR)

Herber J, V Singh J Useem M (2000) The design of new organisational forms. In: Wharton and managing technologies, pp 376–392 (Reprinted in Mintzberg H, Ghoshal S, Lampel J, Quinn JB (2003). The strategy process: concepts, contexts, 4th edn. Pearson Education. pp 234–240)

HFIR (2004a) Call for express of interest for consultancies and contractors for participating in housing and commercial units reconstruction in Bam. (Project Advertisement) (A copy of the original appeared in the documentation of the reconstruction of Bam after the earthquake: Design Process in Reconstruction), Tehran (in Persian)

HFIR (2004b) Housing reconstruction programme for damaged regions in the Bam earthquake. (Unpublished report) (in Persian)

HFIR (2005) Guidance for reconstruction, residential, retail and public units of Bam, Baravat and surrounding villages, 2nd edn. Housing Foundation Islamic Revolution, Bam (in Persian)

HFIR (2012) Documenting the reconstruction of Bam: design process in reconstruction (Unpublished report) (in Persian)

HFIR (n.d.) Bonyadmaskan—English page (Website). http://www.bonyadmaskan.com/portal/Home/Default.aspx?Categoryid=e181864e-8d8d-4c9d-991d-b1668e0e9219. Accessed 03 June 2011

HFIR (n.d.) Deputy of reconstruction and rural housing (Website). Housing Foundation of Islamic Revolution. http://www.bonyadmaskan.ir/En/sitepages/Deputy%20of%20Recontruction%20and%20Rural%20Housing.aspx. Accessed 09 Sept 2013

Mintzberg H (1980) Structure in 5's: a synthesis of the research on organization design. Manage Sci 26(3):322–341

Mintzberg H, Ghoshal S, Lampel J, Quinn JB (2003) The strategy process: concepts, contexts, cases, 4th edn. Pearson Education

Saemian S, Erfanian Daneshvar A (2011) The process of technical control and supervision in reconstruction of Bam (Unpublished report) (in Persian)

Steering Committee for Reconstruction of Bam (2004) (1382 in Persian). Decision by the president's special representatives. (A copy of the original appeared in the documentation of the reconstruction of Bam after the earthquake: design process in reconstruction), Tehran. (in Persian)

Chapter 6
Implementation and the Dynamic Evolution of the Bam Housing Reconstruction Organisation

Abstract Plans on how to organise a programme or project are made prior to their actual implementation, when initial assumptions are made while the available information is often limited in the early stages. Corrective adjustments thus are needed over time to reflect the realities of the implementation of the programme. The reconstruction period is already uncertain and dynamic thus upon implementation some initial assumptions be proved problematic and new information or situation regularly emerges. The sociotechnical system for delivery of the strategic objectives of the housing reconstruction programme in Bam provides an illustrative example of a participatory reconstruction programme evolution over the official period of reconstruction, and the variety of the influential drivers of that. This chapter explores the system adjustments of the Bam's urban housing reconstruction programme in response to the emerging issues and forcing drivers either as a result of the initial design of the system or the emerged realities in the field and its broader context. A few examples are adjustments in control mechanisms, introducing creative solutions to an operational bottleneck, changing the role of beneficiaries and the way the programme towards the end of the programme the system was downsized to mark the end of the reconstruction period. The organisational capacity was adjustable in order to address the unknown pace of participatory reconstruction activities. These adjustments also reflected a major political change in the country that influenced the internal dynamics of some participant organisations and introduced a new group of beneficiaries, increasing the size of the programme.

6.1 Changes in Control Mechanisms

As with any other programme or project, plans for organising post-disaster housing reconstruction programmes are made prior to their actual implementation. As Sull (2007) states plans are based on assumptions and the available information/data is often limited in the early stages. He argues that planning for the delivery of the strategy takes place precisely when the least knowledge exists about how events will unfold over time, therefore adaptability and the ability to incorporate new

© Springer International Publishing AG, part of Springer Nature 2018

F. F. Arefian, *Organising Post-Disaster Reconstruction Processes*,
The Urban Book Series, https://doi.org/10.1007/978-3-319-70911-6_6

information and translate it into effective actions during implementation is needed. Hamdi (2010) also advocates adaptability for strategic action plans as corrective adjustments are necessary over time. Within a post-disaster scenario, it is acknowledged that the reconstruction period is uncertain and dynamic (e.g. Coppola 2006; Johnson 2007), therefore upon implementation not only will many of the initial assumptions be proved wrong or problematic but new information regularly emerges. As Lyons et al. (2010) observes several agencies' policies on people's participation changed within a number of international cases, often becoming more liberal and the same agencies' approaches to participation evolved over a short period of 2 years.

In the case of Bam, the reconstruction programme evolution started with adjustments in the control mechanisms in both design and construction phases as the programme was put into practice.

6.1.1 Design Phase

The design phase saw a major challenge in form of a conflict between two of the organisations in charge of the application of architectural and structural controls. This created a bottleneck in the operational workflow.

Based on the programme workflow, in order to obtain planning permission for each house design, they would need to seek approval from both the Mother Consultancy and from the Engineering Body, KEO. However, at early stages of implementing the programme the KEO added more restrictions to the existing national building codes and Article 2800 in the Bam area in a way that many of the architectural features previously developed based on the BAUC guidelines and design types were not acceptable based on the new restrictions. This meant that many of the architectural plans with the approved stamp of the Mother Consultancy could not get the approval from KEO for their structural documents, therefore they had to be adapted to the new more restrict seismic codes and by doing this the architectural plans had to be approved by the Mother Consultancy again. Since both of these bodies, the Mother Consultancy and KEO had only appointed a few people for this controlling task in the field, the waiting time for receiving a reply from both the Mother Consultancy and KEO was very long. The process of obtaining planning permission became unnecessarily lengthy.

This was frustrating for the beneficiary households, who were desperate to get hold of their planning permission and start the construction work, as well as local consultancies and Setads who were directly working with beneficiaries, becoming the face of the reconstruction to them. With the gradual increase in the number of beneficiaries/applicants commencing their houses' reconstruction process the problem significantly increased. In some cases, it could take up to 7 months, and this caused a great level of frustration among participants, especially applicants who were stuck in the middle while construction material prices were rising. Eventually, the problem of the slow processing of planning applications became the most important national issue (Saemian and Erfanian Daneshvar 2011). Efforts for reaching convergence between these two organisations proved to be difficult. KEO demanded a scientific mandate by the authors of the national building code if they wanted KEO to return the Bam regulations to the same national level. But even the distinguished authors of the Article 2800 were reluctant to get involved in this dispute, as a member of ORA recalls that

> ….in order to solve the issue, Housing Foundation approached the KEO and even the authors of the national seismic building codes, but no one accepted the responsibility of returning the seismic requirements for Bam area to the same level as the national building codes. It was Bam (*a special case*) after such (*a severe and destructive*) earthquake!
>
> (member of ORA; italic from author)

The volume of destruction and peoples' desperation to return to their permanent houses were the pressing issues in this dispute, which pushed HFIR to solve the problem and speed up the process. The BAUC took a pragmatic approach reflected in a softer approach by the Mother Consultancy over architectural plans towards accommodating KEO's new restrictions (Arefian 2016). The Mother Consultancy reluctantly started to approve the architectural plans which were based on KEO's new restrictions, which only accepted a simple rectangular pocket volume for houses. The Bam project manager in the Mother Consultancy portrayed the situation as follows:

> ….what is bad is that a structural engineer writes a prescription for an architect…….now that is 8 years passed I don't want to justify; I still feel guilty (*over the compromise*)…..I told our colleagues in Bam (its local branch): "people who are approaching us, without any exaggeration, they definitely lost their belongings and most likely their loved ones. They don't have mood for architectural games. Get on with them, be nice to them".
>
> Once in Bam a lady approached me while crying, I cried together with her, then asked her: what can I do for you? She said: "when my daughter wants to go to toilet she is scared, I must accompany her because I am scared. I want a house". She wanted to protect her Namus (*dignity*).
>
> (member of the Mother Consultancy; italic from author)

Some frustrated beneficiaries directed their frustration towards the consultancies. Consultancies from one side tried to meet the fundamental guidelines, for example the site layout, keeping the temporary house, keeping trees, finishing materials for facades, as well as keeping the houses layouts as a simple cube as directed by KEO. Some consultancies also tend to overdesign the structural elements, e.g. columns and foundations, in fear of receiving KEO's refusal stamp (Arefian 2016). Typical examples of reconstructed houses are presented in Chap. 8.

Although the compromise by the Mother Consultancy prevented an ultimate deadlock in the design process, as this research explores, this operational bottleneck contributed to the overdesigning of the structural elements by some local consultancies that in turn, most likely, contributed to some longer-term implications, such as insufficient reconstruction funds for completing the construction, the system loopholes to be taken advantage of by locals, and the number of abandoned houses in the Bam area. Another solution was to create a design portfolio with over 500 approved design and structural plans. It was an innovative adjustment to speed up the design phase.

6.1.2 Construction Phase

The mechanism for technical control and supervision during the construction phase did not change and it was firmly documented through the checklists in BTID by both the KEO and HFIR (local contractor, interview 2013; a member of Setad, interview 2013; Saemian and Erfanian Daneshvar 2011). Yet, the fluency of the whole process of technical supervision greatly depended on the dynamic of individual interactions between engineer supervisors and contractors (or beneficiaries wherever they became contractors of their own houses) in each individual case.

As the research shows, the structural control mechanism for building earthquake-resistant houses in practice was found to be of greater importance than the control mechanism for safeguarding the urban architectural identity. However, since the reconstruction programme in Bam attracted younger qualified engineers (mostly grade three and a few from grade two[1]) there are examples that suggest an inexperienced resident engineer was not very strict in the supervision, or he was pressurised by the beneficiary, who wanted to receive the supervisor's approval for the checklist in BTID. The latter was the case mostly after beneficiaries were allowed to become contractors of their houses. Training courses were held by KEO for younger engineers, in fact, there is no way of gaining absolute certainty over the accuracy of the control mechanism. Nevertheless, there is a high level of confidence among interviewees (both beneficiary households and among participant organisations) that reconstructed houses are mostly earthquake resistant. Some local public interviewees boasted that Bam now is the strongest city in Iran, and one local explained he does not sleep under a roof other than his own reconstructed house because he is confident his house is safe.

6.2 From Beneficiary-Applicant-Construction Manager to Beneficiary-Applicant-Contractor

The programme int the role of applicants was defined as a construction manager who must initiate and follow-up the process of the housing reconstruction. Within this limited position, people could not build their homes on their own because they

[1]The Country's Engineering Organisation (CEO) qualifies engineers as architects in three grades based on their experiences. It starts with grade three for younger engineers with 3 years work experience after graduation. Grade one is for engineers and architects with 15 years of experience after their graduations.

were not allowed to become the construction contractor for their own houses. However, around a year after the start of the construction phase, a number of cowboy contractors appeared. These contractors deceived applicants and obtained the power of attorney from them to receive directly the funds from the banks, or received money in advance of undertaking construction works. Collecting the first/second payment instalment from a number of applicants scattered all around the city, these cowboy contractors then ran away. A disaster affected local Baravati whose family started construction works in the early stage of reconstruction period in 2004 recalls her family's story as the following:

> … 4 million (*Tomans*)… at the early works (*foundation*) we gave him… He said 'I want to buy steel for construction'. So Seted released (*the money*)… the night he received our money the next morning he escaped and went…..We didn't know him. He was from Kashmar.[2]
>
> (beneficiary in Baravat, italic from author)

Another resident in Bam boasts that he was smart enough not to be tricked by contractors.

> My contractor was from Mashhad[3]….In early days (*a number of*) contractors stole people's money. Our contractor was good. We paid him 1 million or 500,000 Tomans depending on the work progress. We didn't pay him too much at once…..If you were a simple person the contractor had stolen your money….. (*my*) contractor said: Give me the power of attorney so I get money from the Setad and bank. I didn't agree.
>
> (beneficiary in Bam, italic from author)

Despite having an overall agreement with all contractors and receiving guarantees from contractors at the beginning, Setads did not help to chase the cowboy contractors and recover the beneficiaries' money. It appeared that in many cases Setads just had a mobile number and a fake business address somewhere around the country. A member of a Setad describes the situation as the following:

> There were contractors who conned people. For example, (*they*) got 2 million Tomans from 20 applicants and ran away. […] Since contractors just needed to have a (*basic*) contractor permit although we had (overall) agreement and guarantee (*from them*) in Setad (but) in that (extraordinary) situation who could chase them (*contractors*) when they ran away?
>
> In many cases people gave money to contractors in advance without consulting with us. We released funds to people's bank accounts but usually people sought Setads' (*informal*) approval to pay to their contractors at certain construction stage, but in some cases people without consulting with Setads gave the money to the contractor. When we asked why -after the contractor had gone- people said: he looked like a nice guy.
>
> (member of Setad, italic from author)

This damaged reputation of Setads and people's trust in them, and triggered a change as such beneficiaries could become their own contractors if they wanted to. This was an informal change in the way of doing things instead of an official change

[2]Kashmar is a town in Razavi Khorasan Province, northeast Iran.

[3]Mashhad is the capital of the Razavi Khorasan Province, northeast Iran.

in the workflow or the sociotechnical delivery system. This relaxation, in fact, stretched the limited role of beneficiaries within the Bam participatory reconstruction scheme, and influenced the overall reconstruction. One of the results of this change was that quantitative progress rose significantly. Beneficiaries could save money (the contractors' fee) and could spend this money on reconstruction. Another influence was an increase in the number of amendments during the construction operation from the original planning permission documents. However, the effects of this adjustment, in turn, contributed to the emergence of new trends in housing development practice in Bam after the reconstruction period. These will be discussed in detail later.

6.3 Increase of Financial Support and Relaxation of the Way Funds Were Released

In mid-2005, one and half years after the earthquake, the Steering Committee approved an increase of the total $11,500 housing reconstruction fund to $16,500. This was to reflect the ever-rising costs of the construction materials in the Bam area. According to interviewees, the price of construction materials in Bam was even more than other cities in Kerman province, even the capital (Kerman). While at the start of the reconstruction period the reconstruction fund was enough for a house of 80 m^2area, using decent materials, in 2005, the same amount of fund was only enough up until the roof stage.

Housing reconstruction funds (loan and grants) were released by Setads and paid by the bank to the beneficiary's bank account at a number of construction points. The stepped payment of the reconstruction loan and the grant was linked with the progress in the construction stage and the related stage in the checklist; later it was relaxed and the release of funds was decided based on the checklist and judgmental view of Setads. At earlier stages of the construction phase releasing the fund was firmly based on the initial mechanism, which was 5 instalments of $2,000 (20 million Rials equivalent to 2 million Tomans) and releasing the remainder based on the owner's need for construction works. This firm instruction of how to release money was relaxed after the emergence of the following scenarios:

- The financial loss to the beneficiaries who were coned by cowboy contractors was large. If there were more instalments the financial loss was less in each case. Less financial risk was, therefore, involved by increasing the number of instalments.
- Some beneficiaries could not manage their money wisely; they spent it on some other items (e.g. buying a car) rather than reconstructing their homes.

The relaxation of the rules regarding the release of money allowed the Setads to personalise the release of construction funds based on the beneficiary household's construction needs. For example, instead of having 5 instalments of $2,000 (20 million Rials), they could opt for 10 instalments of $1,000 (10 million Rials). Also, if the building was two stories the whole fund could be released soon after the erection of the steel structure.

6.4 Changes in the National Political Scene: Reflections on the Reconstruction

Reconstruction does not occur in a political or state vacuum; as Davis (2007) states the fundamental truth is that all aspects of disaster management, from emergency relief to longer-term recovery and reconstruction, occur within various political contexts. Some observations suggest that the recovery and reconstruction occur within a highly politicised environment. Reconstruction programmes are designed and implemented under certain political circumstances in the aftermath, sometimes filled with optimistic political pledges and slogans, for example as Lyons et al. (2010) reports, following Hurricane Mitch, governments produced reconstruction plans entitled 'Transforming El Salvador to reduce its Vulnerabilities'. In the aftermath of a natural disaster, such political pledges are often declared by heads of state and other high-profile politicians that are never fulfilled (e.g. Hidellage and Usoof 2010; Wilford 2008). Not surprisingly, when the political scene changes it likely affects the programme directly or indirectly, this in return will have short-term or long-term effects on the outcomes of the programme and the city. The Bam case is an example of how the political dynamic in a country becomes the driver for a domino effect on the reconstruction programme and longer-term policies.

The formulation of housing reconstruction programme and the formation of its delivery system happened during Khatami's presidency that allowed greater social and professional new voices. This broader sociopolitical context influenced the way the reconstruction of Bam was strategised. However, the political scene in Iran significantly changed in the middle of the housing reconstruction programme implementation. Upon the end of 8 years presidency of the former president, in summer 2005, after a year of implementing the reconstruction programme, a new president[4] and cabinet took over the office with a new slogan 'Social Justice' and with less political tolerance which made it difficult for professional societies to remain as active as they were before and connect the same minded professionals. This change of cabinet influenced the already existing housing reconstruction programme delivery system in mainly two ways: (a) an administrative change in governmental and public sector that extended to the level of the Bam reconstruction activities, and (b) the new cabinet directly injected a new mandate to the housing reconstruction programme that triggered a change in the way the housing reconstruction programme worked afterwards. The new president gave new political

[4]The new president (Mahmood Ahmadinejad) was elected on 3 August 2005.

pledges to local people that added pressures to the sociotechnical delivery system of the housing reconstruction programme. It was through the introduction of two new qualified groups by the new president during his speech. Although the new promising pledges were popular in the affected areas, but the task of realising those new promises fell onto the reconstruction scheme and its sociotechnical delivery system. As will be discussed later, it not only resulted in diminishing the role of BAUC and pressurising the system, but also made the system itself more vulnerable to abuse. These in turn contributed to further problems the programme encountered and the way it worked afterwards.

6.4.1 Administrative Change

Taking over the office in the second year of reconstruction (summer 2005), the new government created a new administration in the MHUD. The new minister came with his new deputy and middle managers. The new group, who were mostly civil engineers, were placed in charge of organising the BAUC, however, they did not share the same passion for architecture or the need for a qualitative approach in urban planning systems that their predecessors did. HFIR also had a new president but this administrative change did not shake the position of the ORA middle managers who had a more pragmatic approach.

The administrative change affected the BAUC's position within the reconstruction scheme, a fact that was reflected in the frequency of the BAUC's meetings afterwards. The BAUC activity was not officially ended but the frequency of seeking BAUC opinion or approval gradually decreased. The BAUC's position gradually faded as one of its members recalls:

> No one called off the BAUC…they (*the new MHUD team who organised the council meetings*) simply did not organise meetings as they did before… the Council role gradually faded….
>
> (member of the BAUC, italic from author)

Since BAUC was formed to advocate the core principle of safeguarding the architectural fabric of the city, undermining the role of BAUC in practice weakened the importance of this core principle; it should be seen in parallel with the more liberal approach of allowing applicants to become the contractor for their houses. This is despite the protests by HFIR and other officials about the importance of this principle in shaping the sociotechnical system for programme. Given the continuity of management at the Reconstruction Headquarters in HFIR the administrative change strengthened the power of the pragmatic viewpoint (mixed with architectural flavour) of HFIR.

Interestingly, in this situation, the new minister of the MHUD is in fact the former president of HFIR during the period in which the initial arrangements and formation of the sociotechnical system of the housing reconstruction programme were done. This shows the influential role of middle and upper-level managers in both MHUD and HFIR. Their collaboration, which was supported by national and international attention to the architectural qualities of Bam, crafted the core strategic principle on safeguarding architectural fabric in the housing reconstruction programme from the beginning.

6.4.2 New Qualified Groups for Housing Reconstruction Scheme

The new president, Ahmadinejad, visited the Bam area a month after his election and delivered a new promise to locals in 2005. He promised to extend the reconstruction programme to two new qualifying groups: newly wed couples who married in 2005 or 2006, and tenants. These two groups were promised a new conditional scheme whereby they could be qualified for a reconstruction loan (but not the governmental grant) for the sum of $10,000 (100 million Rials) if they could provide a piece of land for building their own house. For the housing reconstruction programme, however, this meant unexpected new entries to its sociotechnical delivery system. According to the work progress reports at the time, produced by HFIR, 6,890 applicants over the period of 2 years entered reconstruction process under this scheme.

Young couples and tenants welcomed the president's pledge, as they now could build and have their own houses. It was also a relief for people who had found the existing system loopholes challenging to their morality, and even threatening their close relationships. It was through removing the temptation of abusing the system. A resident of an opulent neighbourhood in Bam describes how this scheme saved her marriage amid extensive temptation to obtain a reconstruction loan fraudulently:

> Ahmadinejad's promise helped us because we had land. We prayed for him. Back then we wanted to build a house. God bless Ahmadinejad for helping us to become home owners. My brother too had married in August 2005 and used this scheme. My brother in law married in December 2005 (*and used this scheme*). Many with the intention of getting marriage got married after he gave his words.
>
> Earlier (*before his promise*), some people had suggested: "get the letter (*certificate*) of debris removal for the land (*that in fact was empty*), get the reconstruction loan and build your house". But my husband is a conscious man and didn't do it […] We were at an early (*stage of*) marriage; I was worried if (*god forbidden*) my husband accepts to buy the certificate of debris removal. Fortunately he didn't do it, and later we got the marriage loan through Ahmadinejad's pledge.
>
> (beneficiary in Bam, italics from author)

The increase of the entries into the reconstruction scheme influenced the way the system worked by increasing the number of applicants. This forced the system to speed up the process as well as adjust its capacity. At one stage in the middle of programme implementation, more architectural companies were commissioned in order to cope with the volume of work. The introduction of this new scheme not only added pressure to the programme delivery system through the increase in the number of beneficiaries, but also increased the programme vulnerability by creating more loopholes in the system. The latter on its own would have tangible long-term implications on the outcome of the housing reconstruction programme, as will be analysed later.

6.5 Creative Problem-Solving and the Maturity of the System During the Design Phase

At the end of the first year, the programme implementation was mostly at the design phase (before the introduction of the new qualifying groups); and 3,800 planning permissions were issued. At the time it emerged that some applicants were not happy to get the architectural designs from their allocated consultancy company. These applicants preferred the architectural work produced by another company, which was allocated for another district of the city. At the same time, the lengthy process for obtaining approval stamps for the architectural and structural planning permission documents was frustrating. Adding to this was a growing backlog while the number of applicants was growing. As the real-time weekly work progress reports, prepared by Setads, show 9,386 beneficiaries for their reconstruction cases were referred to local consultancies at the end of February 2005, but only 3,834 cases, a mere 40.8%, received approvals for both architectural and structural documents (ORA 2004–2007). Delays in the design phase were so acute, that they were addressed by Parliament (Parliament Website cited in Meskinazarian 2011).

To address this critical backlog and to speed up the process another way of doing things was introduced; the idea was to create a design portfolio shared by all Bam districts. The best approved architectural works with approved structural maps were selected from all local companies. The portfolio included over 520 designs from all companies for various situations and sizes (HFIR 2005). The portfolio was presented to beneficiaries in addition to the plans a consultancy had. Thus, beneficiaries could select a design that was drawn by other companies rather than their allocated company. Since the structural documents for the design were already approved using it could save time for applicants because a beneficiary/applicant could choose from other companies' work instead of requesting his allocated company to do a bespoke job and get trapped in the lengthy design phase. This portfolio was regularly updated by adding new designs and removing less popular designs. If a beneficiary/applicant opted for architectural work from another company in this portfolio, the only work his/her own allocated consultancy had was to design the site layout and administrative works for obtaining planning permission. This company, therefore, had to split the fee for that particular case with the company who originally created the chosen architectural designs (Saemian and Erfanian Daneshvar 2011).

Creating this portfolio was a step forward because it: (a) provided greater choice for applicants; (b) sped up the design phase process; and (c) created a competitive environment among local companies because the company with a greater number of architectural types in the portfolio could enjoy recognition, as well as having more income from HFIR. From then onward, although beneficiaries/applicants could still choose to have a bespoke job from their allocated company in practice they tended to choose one from this portfolio to avoid delays and to avoid rising material prices in Bam. Bespoke jobs were, therefore, usually done for special sites (as a result of their shape, size or location). This new arrangement demonstrates the maturity of the sociotechnical system for the process of the planning permission phase. It was in place until the end of the reconstruction programme. The last version of the portfolio was used for reconstruction by each Setad (and later 'Group 3,000') after consultancy companies left the Bam area, in 2007, before the official end of the reconstruction programme in 2008.

6.6 Downsizing the Delivery System Towards the End of the Programme

The housing reconstruction programme in the Bam area shows a gradual re-adjustment with the workflow that ultimately led to the ending of the programme. Three years after the earthquake, following a reduction in the number of potential beneficiaries/applicants and works in pipeline, it was arranged for all Setads to return to their provinces by November 2007, because it was not worthwhile to keep the previous administrative arrangements in place as most of the cases were completed. It was estimated that in total the remaining 3,000 new beneficiaries/ applicants would be approaching Setads throughout the Bam and Baravat to submit their requests for reconstruction cases. A new arrangement was formed to address this. A number of Setads' employees were selected to form a new group, called 'Group 3,000'. This group was responsible for taking over the uncompleted tasks of all Setads in all districts and finishing the job.

'Group 3,000' was active until May 2008 (03/1387 in Persian calendar), 4 years and 5 months after the earthquake and 3 years and 9 months after the start of housing reconstruction. In May 2008, the Group handed over the remaining cases to the Housing Foundation branch in Bam. This marked the end of the reconstruction programme. The HFIR branch in Bam permanently employed 5 local members of Group 3,000 (they were previously recruited on a contract basis by Setads). HFIR benefited and still benefits from the knowledge of these individuals, as even in 2013, a number of cases were still open, and there were beneficiaries pursuing their reconstruction cases through the HFIR's Bam branch, especially for releasing the reconstruction loans and grants for uncompleted cases which were gradually progressing. From 2008, municipalities took over the tasks relating to housing development activities in Bam and Baravat as normal.

6.7 Adjustable Organisational Capacity

In owner-driven reconstruction, people who lost their houses are given some combination of financial support and technical assistance to repair or rebuild their houses as Jha et al. (2010a, b) state. While the trend of this kind of reconstruction can be more or less directed or influenced it cannot be precisely predicted or managed. The housing reconstruction programme in Bam as an owner-driven reconstruction was not an exception to this.

Although the overall initial estimation was that 25,000 urban homes were in need of reconstruction the pattern of how beneficiaries would start their reconstruction cases was unknown. Thus, the capacity of the delivery system had to be compatible with the number of the applicants entering the system at any time (Joodi 2010). This required flexibility in size of the system elements (Setads, consultancies, contractors, etc.) to cope with the dynamic trend of beneficiaries/applicants entering the reconstruction process. For example, following the preparation of the architectural guidelines by BAUC in summer 2004 (the first year of the reconstruction) the sociotechnical system of housing reconstruction became activated; however, at the beginning the number of applicants submitting their case for reconstruction was low. Given the extent of the loss followed by post-traumatic grief people who had lost their relatives were not in the mood to take action. Additionally, the new post-disaster urban development plan was not approved until autumn 2004, causing uncertainty. At some point incentives, (e.g. large home appliances if owners had started their application for debris removal and reconstruction within a certain period) were introduced to encourage beneficiaries to step forward. It was after the approval of the urban development plan that the programme began to see an increase in the number of beneficiaries applying. Analysis of the real-time weekly reports of the work progress papered based on Setads' reports, to ORA at HFIR is an indication of this trend. Although these reports are not consistent, and the unit of data collection changed during the reconstruction period, the periods which have consistency in data collection offer a quantitative basis for understanding such dynamism (to some extent). For example, Fig. 6.1 presents the trend of applicants entering the reconstruction programme from 15/08/2004, when this data appears in the collection of weekly reports by Setads, to 02/03/2005.

To organise the housing reconstruction programme through the adjustable multi-organisational sociotechnical delivery system, HFIR tried to capture the complexity of coping with unknown trends of applicants by transferring the task of managing the compatibility to each of the organisations involved in implementing the programme. This meant that KEO, Setads and other organisations had to adjust themselves according to the workload through their own procurement strategies. It is reported that at some point 211 contractor teams (Alizamani 2012) and 44 local consultancies (HFIR 2012a, b) were active in the area. Analysis of the real-time

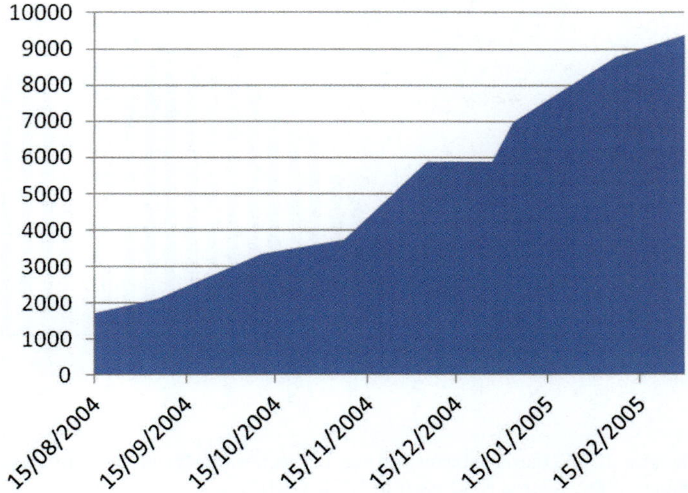

Fig. 6.1 Analysis the number of applicants started design process, from August 2004 to March 2005

weekly reports by Setads, 2004–2007 shows changes in the total Setads' size, in the form of human resources (total urban and rural) up until the formation of Group 3000. This varies from 597 to 1,496 personnel in total for all Setads. It is presented in Fig. 6.2. Another example is KEO, which started its activity with only three people in March 2004 increased to ten people in June 2004 and further grew with the increase of workload to 392 in total in February 2007. Figure 6.3 presents the growing number towards the peak number of 392 (in total for office and site personnel) in the middle of the programme.

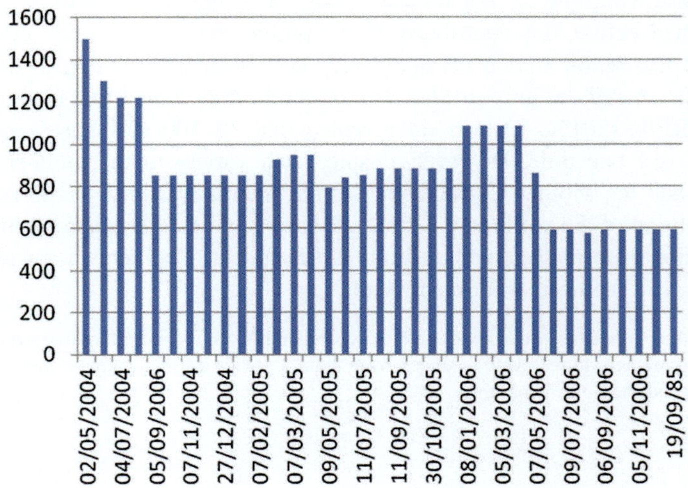

Fig. 6.2 Total number of human resources in Setads from May 2004 to November 2006 (urban and rural in total).

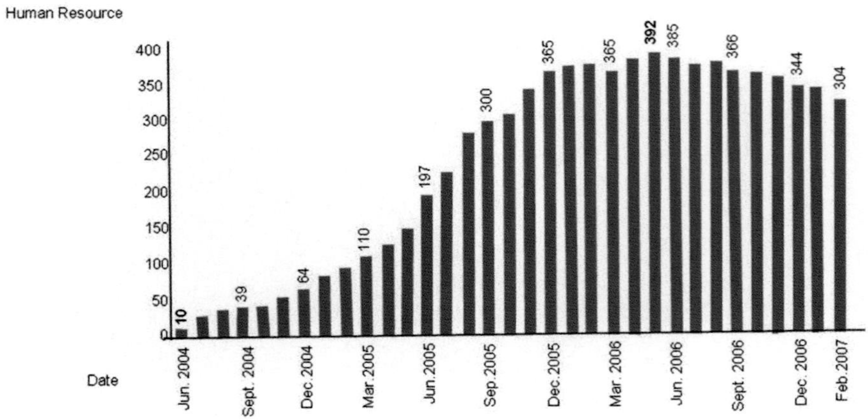

Fig. 6.3 The size of KEO arrangements for housing reconstruction in the Bam area, from May 2004 to February 2007. *Source* KEO cited in HFIR (2012a, b)

This adjustment was combined with efforts to balance the expertise required in organisations to adapt themselves to the demands of the job. For example, in Setads by moving from the stage of debris removal towards reconstruction activities, the number of service based procurement (e.g. truck drivers) decreases from 1200 to 364 and in parallel the number of engineering and technical procurement increases from 166 to 261 people (ORA 2004–2007).

However, despite this crafted flexibility in size and capacity of organisations, such organic growth was mostly during the construction phase in order to cope with the number of construction sites and contractors. During the design phase, KEO and the Mother Consultancy which were both accountable for the critical points of the design phase workflow did not allocate enough personnel to the meet the demands. As explained before, this contributed to the problems caused by their conflict and leading to frustration among the applicants, and eventually leading to the idea of creating the shared design portfolio. This in return contributed to a chain of issues in Bam. HFIR (2012a, b) later documented that 30–100 new cases entered the system at any one time. However, despite such variety in the number of cases the KEO and the Mother Consultancy did not internally adjust themselves to the changing trend of the workload, neither did they appointed an appropriate number of professionals to manage architectural and technical documents compared to the workload, as a member of a Setad recalls:

> The Mother Consultancy had appointed two young architects to control that huge workload. It was taking ages to receive a feedback from them…. Once a friend came to me, his work was delayed there for 4 months.

6.8 Conclusion

This chapter discussed how the sociotechnical delivery system for the housing reconstruction programme in Bam evolved during the implementation through multiple adjustments. The variety of the reasons and motivations for these adjustments was discussed. The system evolved through these adjustments, which in turn contributed to the emergence of some other longer or shorter term issues that in turn affected how the system performed afterwards. From the start, the programme encountered the challenge of cowboy contractors that led to the extension of the previously limited role of people in reconstruction.

The change of approach by one of the controllers and a lack of harmonised activities, or shared understanding of which strategic objective had priority over the others was one of the drivers for creating organisational tensions and lengthy processes. A combination of these with the rising construction material prices affected how other participants (e.g. local consultancies and beneficiaries) performed. This will be analysed further. The system benefited from an inherent flexibility that allowed it to adapt itself according to the workload and through which the participant organisations could also adapt themselves, however, some participant organisations did not adapt themselves accordingly. It was only through an interesting managerial creativity that a severe problem of a slow process and a considerable backlog was solved. Changes in the political scene at the national level not only shook the grounds on which the whole system was initially formed, it also pressurised the system through increasing the size of the programme as a result of new political promises. The extra practical considerations for delivering the new promises within the existing system affected how the system performed against the initial strategic objectives and will be discussed later.

Next chapters will analyse how the emergence of problems and drivers for system adjustments can be linked with the way the system was initially configured for its organisational design and management.

References

Alizamani MH (2012) Contemporary approaches in post disaster housing reconstruction (Unpublished PowerPoint file presented at various conferences) (in Persian)

Arefian F (2016) Getting ready for urban reconstruction: organising housing reconstruction in Bam. In: Arefian FF, Moeini SHI (eds) Urban change in Iran: stories of rooted histories and ever-accelerating developments. The urban book series. Springer International Publishing, Cham, Heidelberg, New York, Dordrecht, London, pp 231–257

Coppola DP (2006) Introduction to international disaster management, 1st edn. Butterworth-Heinemann

Davis I (2007) Learning from disaster recovery; guidance for decision makers. International Recovery Platform (IRP). http://www.undp.org.cu/crmi/docs/irp-decmakersleslearned-td-2006-en.pdf. Accessed 17 Nov 2010

Hamdi N (2010) The placemaker's guide to building community. Earthscan Publications, London, Washington DC
HFIR (2005) Guidance for reconstruction, residential, retail and public units of Bam, Baravat and surrounding villages, 2nd edn. Housing Foundation Islamic Revolution, Bam (in Persian)
HFIR (2012a) Documenting the reconstruction of Bam, role and function of governments, international agencies and NGOs in reconstruction (focusing on Housing Reconstruction) (Unpublished report) (in Persian)
HFIR (2012b). Documenting the reconstruction of Bam: design process in reconstruction (Unpublished report) (in Persian)
HFIR (n.d.) Deputy of reconstruction and rural housing (Website). Housing foundation of Islamic revolution. http://www.bonyadmaskan.ir/En/sitepages/Deputy%20of%20Recontruction%20and%20Rural%20Housing.aspx. Accessed 09 Sept 2013
HFIR (n.d.) Bonyadmaskan—English page (Website). http://www.bonyadmaskan.com/portal/Home/Default.aspx?Categoryid=e181864e-8d8d-4c9d-991d-b1668e0e9219. Accessed 03 June 2011
Hidellage V, Usoof A (2010) Scaling-up people-centred reconstruction: lessons from Sri Lanka's post-tsunami owner-driven programme. In: Lyons M, Schilderman T, Boano C (eds) Building back better: delivering people-centred housing reconstruction at scale. Practical Action Pub, Warwickshire, pp 77–112
Jha AK, Barenstein JD, Phelps PM, Pittat D, Sena S (2010a) Chapter 13, Institutional options for reconstruction management. In: Safer homes, stronger communities: a handbook for reconstructing after natural disasters. World Bank, New York, pp 197–207. http://www.housingreconstruction.org/housing/toc. Accessed 11 Jan 2011
Jha AK, Barenstein JD, Phelps PM, Pittat D, Sena S (2010b) Chapter 6, Reconstruction approaches. In: Safer homes, stronger communities: a handbook for reconstructing after natural disasters. World Bank, New York, pp 93–107. http://www.housingreconstruction.org/housing/toc. Accessed 11 Jan 2011
Johnson C (2007) Strategic planning for post-disaster temporary housing. Disasters 31(4): 435–458. https://doi.org/10.1111/j.1467-7717.2007.01018.x
Joodi M (2010) A brief review on Bam reconstruction program. Presented in International Recovery Forum 2010, organised by International Recovery Platform (IRP), 16/01/2010, Kobe Japan. http://www.recoveryplatform.org/resources/meetings_and_trainings/514/international_recovery_forum_2010. Accessed 23 Dec 2010
Lyons M, Schilderman T, Boano C, D'Urzo S (2010) Conclusions. In: Lyons M, Schilderman T, Boano C (eds) Building back better: delivering people-centred housing reconstruction at scale. Practical Action Pub, Warwickshire UK, pp 345–359
Meskinazarian A (2011) Social resilience of post-earthquake Bam. PhD thesis, Department of Geography, King's College London, London, UK
ORA (Office of Reconstruction Affairs) (2004 to 2007) Weekly physical work progress reports, prepared by Setads from 2004 to 2007 (unpublished reports from HFIR archive)
Saemian S, Erfanian Daneshvar A (2011) The process of technical control and supervision in reconstruction of Bam. (Unpublished report). (in Persian)
Sull DN (2007). Closing the gap between strategy and execution. MIT Sloan Manage Rev 48(4). http://sloanreview.mit.edu/the-magazine/articles/2007/summer/48412/closing-the-gap-between-strategy-and-execution/?type=x&reprint=48412. Accessed 19 Jan 2011
Wilford J (2008) Out of rubble: natural disaster and the materiality of the house. Environ Plann D 26(4):647–662. https://doi.org/10.1068/d4207

Part III
Analyses

Chapter 7
Analysis of Organisational Configuration of the Delivery System: Consistencies and Inconsistencies

Abstract Organisation design deals with the manipulation of a series of organisational attributes that determine the division of the workforce and achieving coordination. The organisational configuration is shaped by bringing those and situational factors towards creating a fit between them. A synthesis of theoretical discussions in Part I suggested an insight for understanding and examining the way reconstruction programmes are organised is the contingency, fit for organisational attributes and their consistency with each other, regardless of their form. This chapter analyses the initial organisational configuration, the actual implementation and evolution of the sociotechnical delivery system of the housing reconstruction programme in Bam. It identifies the most influential organisational attributes of the initial configuration and analyses their surrounding issues such as their interconnections and how they contributed to the way the programme worked in practice. Dynamics of those led to the need for further evolutions, which had domino effects. Unit grouping, control mechanisms, delegating accountability, (de)centralisation and the workflow process, were interlinked; and some of them had considerable influences on other parameters. The innovative approach to the formation of the system including the workflow was based on the modification of previous housing development. The fluency of the workflow was highly influenced by other attributes. Building on the contingency view, the Bam case presents a mixture of consistencies and inconsistencies in its organisational attributes that influenced the way the programme performed and its outcomes. Graphical presentations of those main influential organisational attributes and the final composition of the whole are explanatory, offering a visual analytical method of examination for reconstruction programme organisations.

7.1 Introduction

The essence of organisation design in organisation theory is the manipulation of a series of parameters that determine the division of labour (the workforce) and the achievement of coordination. Some of these parameters are concerned with the

© Springer International Publishing AG, part of Springer Nature 2018
F. F. Arefian, *Organising Post-Disaster Reconstruction Processes*,
The Urban Book Series, https://doi.org/10.1007/978-3-319-70911-6_7

design of individual positions, others the design of the superstructure (the overall network of subunits, reflected in the organisational chart), some of the design of lateral linkages and workflows to flesh out that superstructure (network) and then finally with the decision-making system of the organisation. In parallel to design parameters, there are a number of contingencies (or situational factors) which influence the choice of these design parameters and vice versa. They include the age and size of the organisation, its technical system of production, various characteristics of its environment, such as stability and complexity, and its power system (e.g. Galbraith 1983; Goold 2002; Mintzberg et al. 2003). The organisational configuration is, therefore, shaped by bringing these together. The general conclusion on organisational configuration, based on the contingency view, the manifestation of systems theory in organisational studies, is that the best organisation design is the contingent one. In other words, effective configuration requires consistency among design parameters and contingency factors. Such organisations create a fit between the organisation and its environment. This follows the assumption in organisation theory by Mintzberg (1980) that a limited number of configurations can help to explain much of what is observed in organisations. Regardless of what the organisational configurations are they should create consistency and fit with the design parameters (e.g. Herber et al. 2000; Mintzberg 1980; Mintzberg et al. 2003).

Theoretical discussions in the field of reconstruction showed some scholars such as Zetter and Boano (2010), Quzai (2010) and Maskrey (1994, 2011) highlight the importance of the design and delivery process for housing reconstruction and its outcomes from an architectural perspective. From the reconstruction management perspective, Davidson et al. (2007) links the concept to that of organisation design by stating that there is a need to design the relations between various participant organisations, and this should be integral to the procurement. These acknowledgements indicate the need to examine this subject in more detail and link the discussions in this section to the literature on reconstruction. This section examines the organisational configuration of the sociotechnical delivery system of the housing reconstruction programme and its organisational attributes, and discusses how consistency or inconsistency with other organisational attributes and or with major contingency factors (size and environment) influenced the way in which the programme achieved its strategic objectives and its longer-term influences. We saw the sociotechnical system in Bam was a temporary and innovative organisation within the dynamic and complex context of extreme trauma and the devastating loss of 30,000 human lives and the destruction of 85% of the buildings.This corresponds with adhocracy structural configurations (innovative organisations) based on contingency view. It tried to capture the emerging opportunities and to approach the strategic objectives without knowing how the dynamic reality would unfold during implementation. This innovative system tried to blend previous successfully considerations with the new demands, expectations and the specific situational factors in the Bam case.

However, for the purpose of this research, I go beyond the categorisation of the reconstruction system as an innovative organisation because the final destination of any organisation is to provide internal and external consistency among organisational attributes and their situational factors, in order to serve the purpose of creating the system. The research identifies the following organisational attributes and design parameters as being critical to the success or failure of an organisation. These attributes were interlinked; and some of them had a considerable influence on other parameters.

7.2 Unit Grouping

Unit grouping is one of the main organisation design parameters and refers to the bases by which positions are grouped together to create a unit (Mintzberg 2003). The Bam case shows that application of the unit grouping is directly related to the size of the reconstruction programme, and the way it is done has widespread influences on the programme performance and its outcomes. One of the main considerations for the initial formation of sociotechnical system of the programme was to divide the Bam area 11 zones. Each zone had its own allocated Setad, consultancy and later engineer supervisor teams to serve around 2,000 households (applicants) all from that zone. This initial zoning and allocation of Setads and consultancies, in addition to dividing the workload, provided a combination of geographic-client-size bases for grouping a number of reconstruction participants into an operational unit. Unit grouping, in essence, relies on two fundamental approaches: function-based that refers to the means and single links in the production chain (workflow); and customer-served that refers to the whole chain for specific end products (Mintzberg 1980, 2003). The application of operational zoning in Bam was an example of the customer-served basis for unit grouping that also had taken their geographic proximity and the unit size into account. As Mintzberg clarifies, unit grouping encourages coordination by requiring them to share common resources and achieve common measures of performance and uses proximity to achieve mutual adjustment. This smaller scale of the operational unit was formed to serve applicants in specific geographical zones. Each zone was an operational unit, which potentially allowed Setads, consultancies and later engineer-supervisors, and ultimately residents of a zone, to form the whole production chain for housing reconstruction. They could know each other, the geography of the area in detail, and also build a personal relationship. All of these could potentially help to achieve mutual adjustments for doing things in the workflow about interdependencies. This operational unit zoning was also an application of operational vertical decentralisation and delegation of accountability because Setads were in charge of the operated reconstruction of each zone. Figure 7.1 schematically shows this.

Such arrangement for geographic, customer service, size based and end-focused operational zoning was a valuable lesson in managing the scale of the reconstruction from previous reconstruction programmes, mainly Manjil and post-war reconstruction. What was different in Bam compared to other reconstructions was that in previous experiences operational unit zones were at some distance from each

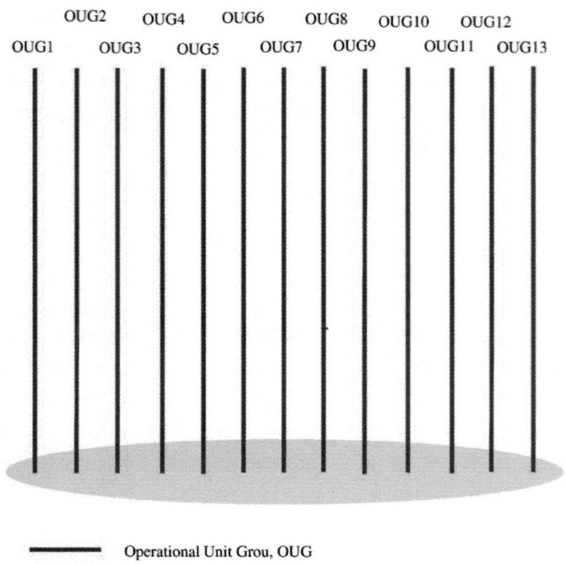

Fig. 7.1 Size-geography-end-focused operational unit grouping in Bam (schematic illustration)

other. Dispersed villages or small cities created a natural geographic division for the workload, for instance, Manjil town itself had 2,500 homes in need of reconstruction, whereas, the number in Bam was 25,000. Thus, in Bam the zones for operational unit grouping were artificially defined; they were adjacent to each other and covered a relatively dense urban area (member of ORA, interview 2013; member of local NGO, personal communication 2013). It was a creative way of applying the old lessons in a new urban context.

The Bam case illustrates the importance of the combination of size/scale of the reconstruction programme and the location characteristics/density of the earthquake-affected area for decisions on unit grouping. If the overall housing reconstruction programme is on a large scale, it necessitates managing the scale by dividing the workload into manageable packages. Deciding on the scale of a manageable workload (the overall size and the unit size) depends on the availability of human resources, technical capacity and the managerial experience available in that field. For the Bam case, the perceived manageable unit size for each of the provisional Setads started at 2,000 households, and increased as a result of the increase in the number of qualified applicants. Forming Group 3,000 for the whole urban area towards the end of the programme also indicates the manageable size for each operational unit. The combination of the two characteristics of size and location/density characteristics influence the need for unit grouping. If the volume of destruction is high but the locations are dispersed, for example villages in an area then these naturally provide unit groups because of the lower population, the number of households and thus workload. Their dispersion makes it easier to dedicate task groups to each village, and divide the workload in a way that they do not interfere with each other. This was the main difference between Manjil and Bam. When a large volume of destruction occurs within a relatively dense urban area, which was not divided before, the task of unit grouping is the fundamental design parameter which influences organisational attributes.

However, the Bam reconstruction programme could not fully benefit from such a creative customer-served, end-focused, size-related and geographical unit grouping. Although the geographic-customer-served basis for unit grouping (zoning) was a creative way for managing the scale, delegating accountability and facilitating mutual adjustment among participants, HFIR did not reflect it on all positions within the production chain (workflow). Although consultancies, Setads, applicants and later engineer-supervisors, all were tied to a specific zone, but the operational unit grouping was not complete because contractors were not part of this unit grouping. This was due to contractors' job specifications that allowed them to be active in whatever zone they wanted.

This inconsistency potentially contributed to the situation that housing reconstruction in Bam could not attract bigger contractor companies. This specification, although primarily intended to make the Bam project more attractive by offering greater freedom to contracting companies, in practice proved counterproductive. By detaching contractors from the geographical operational unit grouping, the job description for contractor companies became a collection of single small projects (up to 500 units with an average size of 80 m^2 each), directly contracted with clients, and dispersed throughout the city. The nature of work in construction industry involves taking construction equipment and materials to the construction sites, hence, the proximity of the construction sites for a large number of small projects is a considerable issue. Consequently, the Bam Housing reconstruction lost its attraction for the bigger companies. In the absence of big or medium size qualified companies HFIR had to reduce its initial expectations compared to what was specified in their reconstruction plan and the project advertisement we saw in Chap. 5. From then on, the housing reconstruction relied on unknown contractor companies mainly individual contractor teams, increasing the system vulnerability and opening the door for cowboy contractors to enter the housing reconstruction programme. The unforeseen problem of cowboy contractors had a damaging domino effect on the reputation of Setads and contributed to a number of uncompleted structures across the area; all led to the system adjustment from applicant-administrator to applicant-contractor, which in turn had its own reflections on the programme performance as will be discussed later. Figure 7.2 schematically presents this.

Talking about numbers, as later documented, 211 contractor teams were active in the housing reconstruction programme (HFIR 2012a, b). This should be seen in parallel to the initial number of 28 consultancies, 11 Setads, 1 KEO (with its 11 task groups during the construction phase). This inconsistency led to implications for delegating accountability, managing scale and decentralisation. In the absence of any unit grouping and/or delegating accountability, dealing with the 211 contractor teams was a difficult task, even if we do not take the possibility of corruption into account. Whilst the reputation of Setads was damaged as a result of the problem of

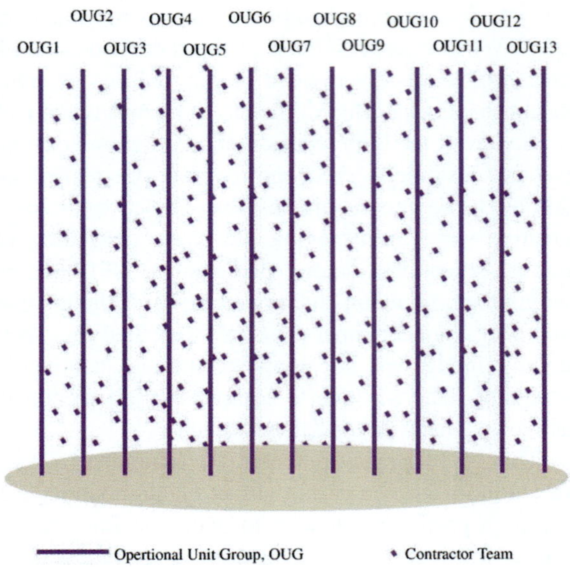

Fig. 7.2 Inconsistency in unit grouping that linked to job description in Bam (schematic illustration)

dodgy contractors, a member of Setad linked the problem with the initial arrangements for job description and operational unit zoning.

> There were (*some*) contractors who conned people. For example, (*they*) got 2,000,000 Tomans ($*2,000)* from 20 applicants and ran away. The contractors who came to Bam were not good and professional, (*they*) had a contractor permit; (*but*) Grade (*professional qualification*) was not considered. Of course at the beginning Grade was important but for good contractors it was not feasible to operate in scattered way (*throughout the city*). If they (*contractors*) were allocated to a certain locality it could have been feasible.

Even if the application of zoning for contractors would not have helped to attract larger contractor companies to the housing reconstruction, it could have helped locals to recognise dodgy contractors sooner. The city-wide operational area was an opportunity for dodgy contractors to remain unidentified during the earlier stages of the construction phase. This was not difficult within the traumatic, devastating and chaotic situation in the aftermath of the earthquake.

The case also shows another inconsistency between organisational design parameters and the size of the programme that is function-based unit grouping for the control mechanism, and poor internal arrangements in the responsible organisations. Parallel to the end-focused size-geography-customer-based unit group zoning; the case also presents function-based units for the control mechanisms that covered the whole urban area. The initial estimation for each of the operational unit zones was around 2,000 households; but the operational unit for institutions in charge of the control mechanisms was, 20,000 initially (25,000 individual residential mixed-use units in total). HFIR decentralised the decision-making power to achieve a multilayer control mechanism and held these controller organisations accountable for the tasks, creating a matrix organisational configuration, which as

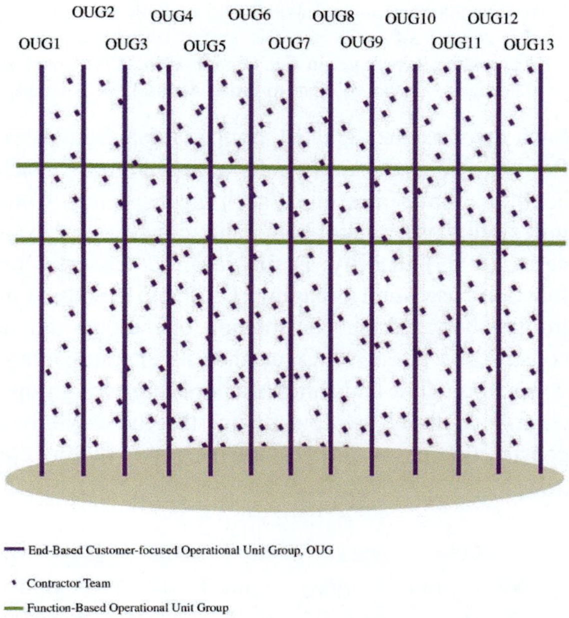

Fig. 7.3 End-focused unit grouping and function-focused unit grouping, which linked to the control mechanism

Minzberg states can be perceived as a basic form for network organisations (e.g. Morgan 2006; Mintzberg 1980). Figure 7.3 schematically presents the initial organisation configuration for the unit group zoning and the control mechanisms.

To achieve flexibility HFIR transferred the task of adapting to the workload to participant organisations. Despite this intended flexibility in size, and being responsible for the whole urban area, the controlling organisations did not accordingly adapt their internal procurement arrangements. During the design phase, KEO and Mother Consultancy did not fit their internal arrangements to the unit grouping, neither had they appointed sufficient number of personnel compatible with the total variable size of the workload. The consequences of such inconsistencies were exposed during the implementation in form of lengthy waiting time for receiving feedback from the controllers. This negatively contributed to aforementioned backlog at the end of the first year, frustration among front-liners, planning permissions to become a national concern, leading to the creative solution of creating shared design portfolio, as well as applicants' reluctances towards personalising their work, and overdesigning structural elements by some consultancies, all in turn contributing to even longer-term issues in Bam. We saw that 30–100 cases entered the reconstruction programme for the design phase but, the number of controlling personnel in KEO and Mother Consultancy remained the same and there were insufficient staffing. A member of local consultancy company recalls:

> The initial estimation was 2,000 cases over 3 years for the consultancies. Mathematically, this meant an average of almost 2 cases a day for each of us (*local consultancies*), to complete and send to the Mother Consultancy or the Engineering Organisation (*KEO*) to

> receive approval stamps. We should multiple these by the number of consultancies to see the volume of work for the senior consultancy and the Engineering Organisation. Appointing 4 persons in the Organisation (*KEO*), and appointing 2 persons in Mother Consultancy, who all want to prove themselves, is simply not enough.

Such inconsistency created an operational bottleneck in the workflow that was far from an effective matrix configuration. Figure 7.4 schematically presents the actual configuration as a result of such inconsistency. Controllers saw such self-oriented inflexibility as a sign of better and tougher controls; an example of what Mintzberg et al. (2003) identify as the intrinsic tendency towards narrow-mindedness in function-based unit grouping. Learning lessons, it appears that KEO, as the controlling body of the construction operations, did not repeat this mistake in the construction phase. KEO matched its internal arrangements to the operational unit zones, through establishing a supervisory task group in each zone. This might have contributed to the previous finding that the workflow and control mechanism for the construction phase during implementation did not change, unlike the design phase.

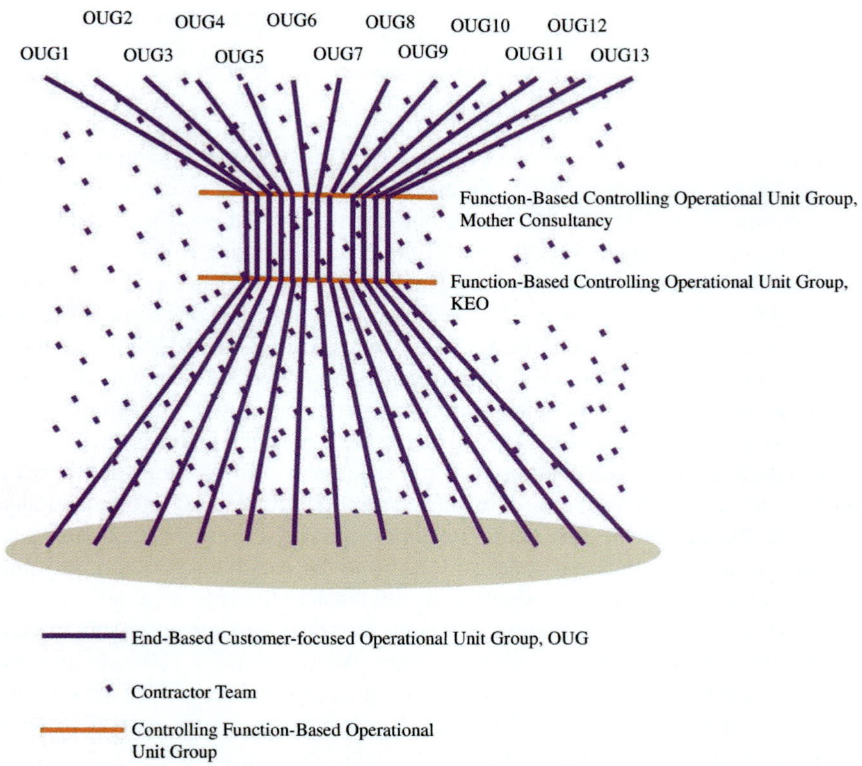

Fig. 7.4 Inconsistency in internal operational arrangements in function-based units (schematic illustration)

Reaching mutual adjustments and agreements on the way forward was easier during the construction phase as a result of consistent unit grouping in size and capabilities. This was despite the great amount of documentation in that phase.

7.3 Control Mechanism

Control mechanisms are design parameters used to standardise the output by specifying the desired results of the whole chain of actions (Mintzberg et al. 2003; Mintzberg 1980). A sign of decentralisation in the Bam case was indeed to delegate the task of controlling mechanisms with others, so that the multilayered controls, were performed by various organisations, including municipalities, Mother Consultancy, KEO and BHRC (without operational power). This was also a way of decentralising HFIR's power to reduce the risk of potential corruption, as HFIR documented later (HFIR 2012a, b). However, as previously analysed, the housing reconstruction programme suffered from a conflicting duality and inconsistency of the human resources in controller organisations against the actual workload in its control mechanisms at the design stage. The case clearly illustrates the extent of the damage created by a major bottleneck in the workflow that in turn had domino effects, such as the lengthy waiting time for the approval of the designs, people's frustration who encountered the ever-increasing construction material prices, who then abused the system in order to get more funds to finish the building, and so on. The consequences, therefore, are not only physical, but also financial and social. The situation clearly demonstrated narrow-mindedness in both controller organisations towards their own subject of control. From the organisational perspective, this narrow-mindedness and rigidity to the assumed accountability by the controllers can be linked to the organisational configuration and consequence of discipline-focused unit grouping.

The control mechanisms were discipline based, which is a manifestation of function-focused unit grouping; another way of unit grouping against end-based grouping that was the basis for zoning in Bam. As Mintzberg (2003) states, grouping by function narrows perspectives, and focuses on the means instead of the ends, potentially hindering the big picture. The unforeseen conflict between controller organisations in the Bam case also showed a diverse understanding of the priorities in the whole reconstruction programme that was undefined at the beginning. This contributed to the divergent approach in the control mechanism. Each of the controllers assumed its related disciplinary strategic objective had priority over the other.

Since, the executive body for the reconstruction, HFIR, linked the control mechanisms to the strategic objectives, implicitly it delegated the accountability for achieving two strategic objectives with two participant organisations. Creating such intertwined linkages between each of the strategic objectives and their related control mechanisms also showed (to all interested individuals and organisations) that the reconstruction of Bam addressed all primary concerns, which were materialised in the strategic objectives. Such arrangements added more weight to the responsibility that controller organisations assumed. Simultaneously, the responsibility of any achievements or underachievement towards two of the strategic objectives was transferred to their related controller organisations. Thus, KEO as the technical controller became accountable for making reconstructed homes resistant against earthquakes, while the BAUC, Mother Consultancy and to some degree local consultancies were responsible for addressing qualitative concerns over the unique architectural characteristics of the Bam area. This potentially contributed to the rigidity and narrow-mindedness in KEO and Mother Consultancy's approaches.

Given the above, this research stresses the importance of a collaborative and integrated approach in designing the control mechanism in the organisational configuration of the reconstruction process. There is a hidden threat in implementing multiple parallel activities that is growing divergent directions, especially if mutual understandings of the priorities do not exist. If dividing the control mechanism by function between multiple participant organisations is needed, then it is crucial to be aware of such a hidden threat and the possibility of narrow-mindedness in those organisations. Therefore, further efforts for harmonising various function-based controllers and reaching an integrated approach are needed, for example, prioritising objectives, and facilitating mutual understanding of what is the best for the programme, not for each function.

7.4 Accountability

Previously, I discussed how the links between the strategic objectives, control mechanisms and accountability in forming the programme delivery system contributed to narrow-mindedness of the controller organisations. The case also presents an example of an inconsistent approach in delegating accountability for other participant organisations that contributed to the emergence of unforeseen problems, mostly related to the emergence of dodgy contractor teams, and its entailed effects on the programme implementation. The Bam case exemplifies the necessity of having a clear accountability mechanism in place for dealing with various teams, especially when the scale of housing reconstruction increases.

By appointing consultancies, Setads, and KEO, HFIR delegated accountability for managing numerous working teams to their respective organisations as legal responsible entities. For example, through commissioning KEO to be responsible for construction supervision KEO was made accountable for managing the supervision teams, which engaged 392 engineers and their assistants. Managing this number of working supervisory personnel would not have been possible (or at least not easy) if HFIR had not commissioned KEO for this. However, there was a lack of management accountability for the construction teams. There was no legal entity and institution to take accountability for dealing with the construction teams in the field while 211 contractor teams worked in the Bam housing reconstruction programme (HFIR 2012a, b).

It appears that HFIR's initial intention, reflected in the reconstruction plan and public project advertisement in early 2004, was to rely on construction companies to be accountable for the contractor teams, similar to consultancies' accountabilities for their local design teams. But, by losing contractor companies—linked to the inconsistency in unit grouping and job specification—the Bam reconstruction programme lost an important layer of accountability. Therefore, the delivery system required further adjustments to reflect this unfolding reality. When HFIR understood that it would not be possible to manage accountability for the contractor teams in the same way that they managed accountability for the design teams, they should have tried a different approach. This situation necessitated a shift in operational considerations for the accountability and management of a large number of small construction teams in Bam. It did not happen.

In addition to the problem of cowboy contractors who could not be traced, another complication the lack of an accountable organisation for contractors in Bam was that contractors did not have a voice within the overall reconstruction programme. Each of the system elements, i.e. consultancies, supervisors and Setads, and applicants, had a point of reference and communication for raising issues surrounding their activities and keep information; contractors did not.

The need for clear delegation of accountability increases when the size of the reconstruction programme increases, because the programme demands more participant individuals and working teams. In the Bam case, adding an accountable organisation, e.g. a legal professional institution, however, might have added a layer of hierarchy to contractors' activities. Within the field of organisation studies it is acknowledged that with the growing size of an organisation, it tends to be more bureaucratic (Mintzberg 1980). Therefore, any potential adjustment requires an examination of its own possible effects on the way a delivery system works towards its strategic objectives.

7.5 (De)Centralisation

Centralisation or decentralisation refers to the diffusion of decision-making power and the degree that it is in the hand of one organisational element (usually the top manager) in an organisation, or if the power is dispersed among many elements (Moenaert et al. cited in Badir 2006; Mintzberg 1980; Mintzberg et al. 2003). In disaster and reconstruction field researchers such as Lyons et al. (2010) promote decentralisation, and this is echoed in the draft guidance for Disaster Recovery Framework (2014) by GFDRR. The Bam case however, shows that not all decentralisations are functionally positive.

The decision-making power in the Bam housing reconstruction programme was relatively decentralised at a higher level through the formation the BAUC. At an operational level decentralisation was achieved by transferring the power to Setads,

Mother Consultancy, municipalities and KEO; and by transferring the decision-making for self-adaptation to each participant organisation. Two main approaches to decentralisation influenced the programme: (a) functioning vertical operational decentralisation that corresponded with end-focused unit grouping; and (b) problematic horizontal decentralisation that corresponded with control mechanism and function-based unit grouping. Here, I examine the underlying differences between these two kinds of decentralisations in the Bam case and draw on organisational literature, mostly by Mintzberg to explain them.

The operational decentralisation through zoning and the appointment of Setads was vertical as the literature on organisation design suggests. The overall workload required dividing it into manageable operational units. Setads were responsible for the reconstruction of their allocated zones. Interconnections between these Setads were minimal; they performed in parallel and there was no interdependency between them, and they were not competing with each other. There was a standard working process in place but Setads had decision-making power for handling relevant affairs, e.g. administration, controlling the fluency of the work process and progress, mediating disputes and overall supervision of all stages. Notably, Setads are provincial Housing Foundation branches. Although they are independent legal entities, and are not subordered institutions of the ORA, they all part of the national HFIR and share the same organisational culture. They meet each other at various national events. Also, they probably have the experience of working on other reconstruction activities, as their previous performance is considered when allocating Setads to a new reconstruction case. The Setads' reflections on the field situation were influential in adjusting the system. These Setads members had weekly harmonising meetings with HFIR to raise issues, share experiences and seek solutions for emerging problems during implementation. For instance, the necessity of the system adjustment towards a judgemental approach to the payment of fund instalments at construction stage was raised by the Setads front-liners after they had noticed a number of applicants could not manage their funds wisely.

Such effective functional vertical operational decentralisation for managing the scale of reconstruction is traced back to post-war reconstruction. The UNDP National Programme Analyst in charge of the Bam reconstruction programme at UNDP Iran elaborated on this decentralisation for housing reconstruction in Bam. The interview was conducted in Persian and was translated into English by the me:

> The government acted in decentralised way in the reconstruction of Bam, by bringing Setads from all around the country. No one gave this idea to the government; it had learnt this from other reconstruction experiences and from post-war reconstruction. The nation was mobilised to assist in the reconstruction. I learned that the Government of Japan invited the Government of Iran to share its experience with Setads from different provinces who took responsibility for reconstruction of zones in Bam.
>
> Given that 40% of the civil servants lost their lives due to the earthquake, the government was required to mobilise expertise to support the reconstruction. For example, the Ministry

> of Health arranged deployments to the Bam area by its personnel throughout the country, to work in Bam. This was helpful because it brought experience and expertise to Bam.
>
> (Dr. Victoria Kianpour)

Contrarily, the case also presents dysfunctional horizontal decentralisation with regards to the application of control mechanisms. The horizontal decentralisation of control power equally to both Mother Consultancy and KEO created competing parallel priorities. Additionally, KEO and Mother Consultancy did not have any previous experience of working together on that scale; they did not have the experience of working in a post-disaster context or an understanding of other important post-disaster factors, such as the level of urgency, people's despair, public emotions; and they did not share an organisational culture. Mother Consultancy was a nationally famous private sector company; and the KEO was a semi-autonomous provincial professional organisation. One advocated the architectural quality, which was the concern of the professional community and public; the other one took the blame for the poor construction quality of the destroyed buildings in Bam, and was, therefore, determined to preserve their reputation for technical reliability. Their formal joint meetings were held at the BAUC but were not effective. HFIR, the BAUC and MHUD did not enable KEO and Mother Consultancy to understand the post-disaster situation in general, work together and reach a mutual understanding of priorities for managing their interconnected activities. A member of the Mother Consultancy reflected that:

> The important factor for the Foundation (*HFIR*) was speed. They sought speed. ….Each of these three focuses (*safeguarding architectural fabric, earthquake resistant buildings, and speed*) were important. KEO was worried this incident would happen again; Bam is located on a fault.
>
> We had built homes; we hadn't built Bami homes (on *such a scale*). None of these three organisations had been faced with such a volume before. If someone thought wisely he could put all three on track and realise them.
>
> (member of Mother Consultancy, italic from author)

Interestingly, although HFIR decentralised the power, but HFIR itself could not mediate and solve the conflict between these two bodies. It appears that even the creative problem-solving solution at the design stage, in the form of creating shared a design portfolio, was a smart managerial tactic for avoiding the problematic control mechanisms by both KEO and Mother Consultancy without decommissioning them. As literature review showed, other international empirical examples of competing priorities, lack of previous experience in working with each other, and differences in organisational cultures as sources of problems in organising reconstruction were provided by scholars. However, whilst the Bam case adds another example to the list, this research analyses further and identifies organisational weaknesses in dual and equal horizontal decentralisation.

Not all types of decentralisation are effective or efficient in a housing reconstruction programme. Vertical decentralisation, linked with end-focused unit grouping and accountability, corresponds to managing the scale for large-scale reconstruction programmes, towards manageable size, flexibility and is adjustable to the specific situations of households in each unit. Contrarily, parallel dual horizontal decentralisation, in the presence of competing operational tasks and duties for non-prioritised objectives creates dysfunctional decentralisation, increasing the problems. Because such dysfunctional decentralisation eliminates the ultimate decision-making position that can have the last say. In the chaotic situation of a post-disaster reconstruction context, with its pressing issues, quick problem-solving is required. While HFIR's intention in creating horizontal decentralisation was valid in order to avoid concentrating powers in Setads by creating a multi-layer control mechanism (Saemian and Erfanian Daneshvar 2011), the application of such decentralisation required perquisites, such as shared understanding of the priorities, non-competing agendas and harmonised activities, as the vertical decentralising in Bam presented.

The point, however, is not that all organisational participants need to be affiliated with the executive body of the reconstruction programme. The point is that the hidden problematic situations in organisational configuration have to be exposed and dealt with in advance and accordingly.

7.6 The Workflow: The System Integrating Process

7.6.1 An Improved Version of the Previous Housing Development Process

At the start of the programme formation, the BAUC requested a clear system and mechanism in order to minimise the risk of chaos, given the estimated demand. It appears that to create that system, HFIR modified and tried to improve the existing housing development process to integrate measures and practical considerations for approaching the strategic objectives. Since the strategic principles of the reconstruction were reflections of the need for these improvements, the organisational arrangements for the workflow were towards modifying the previous housing development system to address its shortfalls and to support the existing organisational system. The workflow standardised the application of improvements (through guidance, regulations and control mechanisms) among all groups of applicants and in all operational unit zones which covered the whole urban area. This, in fact, created a temporary process for housing development for the urban area.

The workflow of this proposed system, similar to the normal housing development in the country, consisted of three main phases: predesign for issuing planning application (the qualification phase to enter the system), the design phase for issuing planning permission, and the construction phase. The workflow arrangements in each phase corresponded to the extraordinary post-disaster situation and improvements for approaching the strategic objectives. For example, the Mother Consultancy and KEO took the responsibility for controlling the architectural, structural and technical maps for the design phase; this would normally be the task of municipalities. KEO and Setads also took the responsibility for supervising the construction phase that in normal situations would be done by individual engineer-supervisors and municipalities. Commissioning KEO—as the qualifying body for construction professionals (including structural engineering supervisors) and responsible for the implementation of the Engineering Organisation Law—also addressed the inherent conflict of interests in normal situations, where engineer-supervisors were commissioned and paid by developers. Although the municipalities remained as participant organisations but they had fewer tasks than in normal situations. Nevertheless, they issued official planning applications, planning permissions and certificates of construction completion, as they did in normal situations.

Additionally, the system standardised the workflow for different groups of applicants. Despite the great human loss that left a considerable number of vulnerable groups with a variety of needs, the system did not provide a variety of workflows for different groups. Instead, at the beginning of dividing the tasks, NGOs were appointed to represent poor and vulnerable groups in the housing reconstruction process and undertake all tasks on behalf of the applicants (HFIR 2012a, b). This simplified making arrangements for the workflow. Even the adjustments during the implementation were adjustments to that standard workflow and applicable for all zones and all groups.

Moreover, given the scale of the destruction in Bam, in which 85% of the buildings collapsed, the reconstruction programme covered the whole urban area. This modified version was indeed a temporary process for housing development, from 2004 (1383 in Persian calendar) until after the end of the programme. A member of ORA at HFIR explains the overall organisational approach on this:

> The plan that we have for ourselves (in *HFIR)* is established. After the earthquake in Gilan and Zanjan*(known as Manjil earthquake)*, a plan was defined: (*For example*) not to change the technology too much; respect what people have; we must respect the architecture, what they had (*before*), don't say whatever was before was wrong and now an incident happened so we must erase them, and bring something new from scratch.
>
> Our policy for local organisations is that we don't want to go into Bam to create organisations parallel to what they have. We say, in Bam there is a line of local organisations. It has a governor; it has a mayor; it has an Engineering Organisation; producers for supplying construction materials; a town council. We don't want to disturb these; I mean create a copy for each to do the job and leave after the reconstruction completes. Our policy for fundamental operational organisations is that we empower them.
>
> Think of the concept of (*utilising*) Setads. We called it Setad-e-Moein- in Kerman (*province*) we have a Housing Foundation. This Housing Foundation has a branch in the city of Bam in which the earthquake happened. We don't create a new organisation called the housing foundation for reconstruction there. Moein means assistant. Based on the scale of the incident, in order for the Bam Housing Foundation to be able to do its job, it needs help, because, the job exceeds the capacity of the branch.
>
> These are our main policies; however, when an incident happens in a region, depending on the regional situation, social condition, and cultural situation these policies may change. They are flexible and are not holy verses not to be adjusted.

> It is in our policies not to shuffle internal organisational arrangements but to support them. When a city does have a mayor we should not bring another system to the city to issue planning permission. (*We*) must use the same mayor and town council. The city had an Engineering Organisation. We didn't want to change all existing arrangements and enter new arrangements.
>
> The mayor of Bam might have issued 100 planning permissions in a year (*before the earthquake*) but here he wanted to issue 30,000 planning permissions in a year. Was the (*previous*) system capable of that (*volume*)? We supported the system, injected human resources.

Creating the workflow as a modified, temporary process for housing development can be linked to a number of sustained improvements of the normal housing process after the reconstruction period, as will be discussed in Chap. 8.

The Bam case shows that modification of the existing housing development process makes it easier to connect the reconstruction programme to the normal housing development practice afterwards. Treating the reconstruction workflow as the modified/improved version of the housing development process benefits both decision makers and people. It allows decision makers to deal with critical areas for improvements based on available human and technical resources and capacities instead of working from scratch. It is advantageous to have quicker decision-making in the aftermath of a disaster when time is a crucial matter. The new reconstruction workflow is more likely to fit the broader sociopolitical reconstruction context. It is potentially less challenging because such improvements are applied to an existing system; and they are supported by the momentum after a disaster for improving the status quo. Also, a modified version presents elements of familiarity for people in participatory reconstruction. The workflow is not a completely strange procedure to people and other participants in the programme, thus, reducing confusion.

The above in fact pave the way towards sustaining improvements in the normal housing development practice after the reconstruction period. This is also aligned with the desire to improve the previous status quo through recovery and its strategic agent reconstruction. The theoretical discussion showed the need for mainstreaming measures and strategies for disaster risk reduction in development activities and delivering improvements during reconstruction frequently expressed by scholars (e.g. Wisner 2004; Alexander 2002; Benson and Twig 2007). However, some such as Christopolos et al. (2006) state the window of opportunity opened in post-disaster reconstruction to mainstream disaster risk reduction is illusive. I argue that too many expectations for fundamental changes might be illusive but there is always room for improvement and space for progress in an existing system. Specific improvements, such as new regulations, supervision and enforcing mechanisms, can start from the reconstruction phase, by modifying and improving the previous practices. The level of sustained improvements and progress depends on how well the improvements are crafted and combined with local capabilities and attitudes.

7.6.2 Fluency in Workflow

The above discussion, in fact, re-emphasises the necessity of fluency and effective organisational configuration and workflow because it has potential influences on future normal development practice. Any bottleneck within this organisational workflow reduces the chances of success in not only achieving the strategic objectives but also influencing the future normal development practices. The Bam case is an exemplary case to highlight the various multidimensional effects that are traced back to the bottleneck within the housing reconstruction programme's workflow. The emergence of this problematic bottleneck was linked to the combination of three major syndromes: first, the divergent approaches of two parallel controllers, with equal decision making approval power; second, the lack of attention to operational interdependencies of these controllers in the workflow which left those interdependencies unexamined; and third, the lack of appropriate self-adaptation of human resources in the two controller organisations, whose their operational unit size was the whole urban area. Applying two horizontal control mechanisms throughout the urban area created a quasimatrix configuration. The capacity of the system, therefore, depended on the capacity of each of these controllers. The fluency of each operational unit zones' activities depended to the outcomes of these two nodal points. While insufficient human resource in these two controllers narrowed down the capacity of the system, their divergent approaches tightly knotted the whole system at these two points. The bottleneck in the workflow was a manifestation of a complex combination of various inconsistencies in actual organisational configuration. Figure 7.5 presents how the bottleneck changed the actual organisational configuration for the sociotechnical system of housing reconstruction.

The real workflow in practice was, therefore, far away from the initial expectations and potentially intended quasimatrix configuration. The domino effect of negative consequences of the poorly designed workflow and creating such a bottleneck can be summarised as: the applicants were waiting a long time for their designs to be approved, while the cost of materials continues to rise; it created the tendency among consultancies to overdesign the structural elements of the buildings in the hope of gaining quicker approval, and this also added to the costs; thus the allocated funds were insufficient to build the houses required; and this, in turn, contributed to further issues as will be discussed in the next chapter.

The Bam case shows that due to the importance of the workflow with potential longer-term effects after the reconstruction period, it is not enough to put all the practical considerations together within an overall process of the previous housing development—as it was the case in Bam. It is crucial to recognise a workflow as an organisational attribute is a product by itself part of the organisational configuration. For each set of considerations for a sociotechnical system, there are some potential alternatives for the workflow. For example, in Bam's case by having the same delivery system the order of seeking approval from KEO and Mother Consultancy could be reversed, or they could create a working group together.

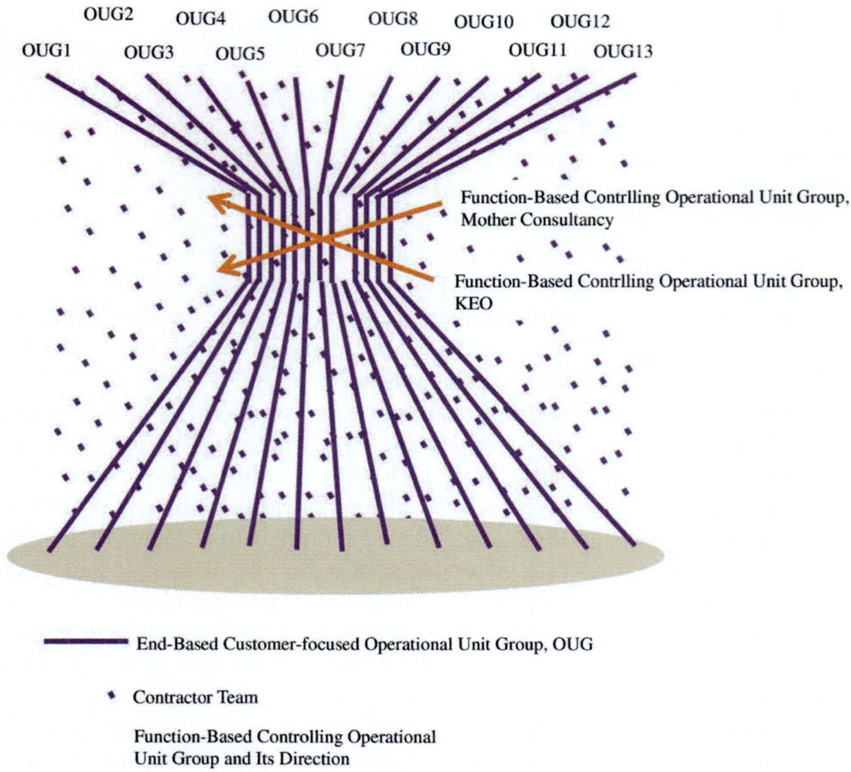

Fig. 7.5 A schematic presentation of the actual organisational configuration in Bam shows how the inconsistency in organisational attributes and diverse duality created a bottleneck in the workflow

The importance of the actual graphically drawing and examining of possible alternatives for the workflow in the reconstruction scene finds its immense importance in post-disaster situations if organisations: do not have previous work experience together on that scale and context; and have different organisational cultures, e.g. private sector, professional institutions and governmental ministries, as it was the case in Bam, and also in other cases. As will be discussed later, the workflow, therefore, is not only a production chain but also a standard coordination mechanism as Mintzberg (2003) states. By graphically drawing the organisational configuration and actual workflow as production chain the potential areas of interdependencies between organisational tasks, their orders, and practical actions for approaching all objectives with together are exposed. It is especially the case if the sociotechnical system addresses more than one strategic objective, as was the case for Bam. If HFIR drew its proposed working process in advance it could identify the potential conflicting areas and address them, especially if HFIR discussed the workflow with a multidisciplinary team from various participant organisations, including both KEO and consultancy companies. This could have provided a platform for starting their collaborations towards a shared understanding of priorities and mutual adjustments.

7.7 Conclusion

The sociotechnical system of the Bam housing programme was an innovative organisation, trying to garner the established practical considerations of previous reconstruction and demands for improvements, in form of strategic objectives, within a complex situation of a large-scale devastating destruction in a historic urban fabric. Within the contingency view, that is the manifestation of systems theory in organisational studies, there is a need for consistency in all forms of organisations. This chapter examined this for the intended quasimatrix organisational configuration of the delivery system of the housing reconstruction programme in Bam. It identified the most influential organisational attributes, namely, unit grouping, control mechanisms, accountability, decentralisation mechanisms and workflow, and examined their consistencies and inconsistencies which influenced the way the programme performed and/or longer-term impacts.

Graphical drawings of influential organisational attributes and the whole initial composition of the organisational configuration exposed hidden bottlenecks and areas for attention which could have been addressed before putting the programme into practice, if the organisational design was illustrated in advance. The chapter introduces this as a visual analytical method of understanding organisational configuration and their examinations.

Unit grouping is a major influential organisational attribute for forming and managing a sociotechnical system for a reconstruction programme. The case showed that the end-focused basis for unit grouping includes the whole production chain for the operational unit and corresponds with vertical decentralisation. Therefore, is adaptable to various sizes of the reconstruction programmes, depends on the available human, technical and organisational capacity. On the contrary, means-focused basis for unit grouping, corresponding with horizontal decentralisation, is likely to lead towards competing priorities. Such narrow-mindedness in the control mechanisms overshadowed the Bam reconstruction programme and greatly contributed to problems which can be linked to a lack of shared understanding of the reconstruction programme priorities, function-based unit grouping and lack of self-adaptation in controller organisations, as well as linking strategic objectives and control mechanisms.

Accountability and decentralisation became important factors given the size of the reconstruction programme. This is mainly because an increase in the size of the programme also increases the number of engaged teams and individuals. The lack of accountability for managing the great number of contractor teams negatively influenced the system implementation and the programme performance with longer-term damaging effects, such as financial loss of people, damaged trust between Setads and locals, and meant that contractors' voice was missing from the programme. While post-disaster reconstruction literature mostly advocate decentralisation of the reconstruction programmes (Lyons et al. 2010), in the Bam case, the housing reconstruction programme was relatively decentralised, compare to the highly centralised attitude of the government. However, the research identified and differentiated two types of decentralisation, vertical and horizontal decentralisation, linked to the bases for unit grouping and managing accountability. Decentralisation without adequate examination of other organisational attributes and overall configuration and mutual understanding of priorities creates more harm than good; because it eliminates the ultimate decision-making power, which is crucial in that post-traumatic situation.

All the above organisational attributes, which are linked and shape the organisational configuration, come to life through the workflow. As the manifestation of the coordination mechanism among the system elements, the workflow is an important organisational attribute that connects all participants, defines the order of activities, and their interconnections. While treating the workflow in Bam as a modified version of existing housing development practice paved the way to mainstream some improvements after the end of the reconstruction period, a major bottleneck prevented the programme from fully benefiting from it. The impacts of this approach exceeded the reconstruction period, and influenced the housing development practice after the reconstruction period.

References

Alexander DE (2002) Principles of emergency planning and management, Illustrated, ed. Oxford University Press, England

Badir Y (2006) The role of network lead company in integrating new product development processes across strategic partners. PhD thesis, Ecole Polytechnique Federale De Lausanne, Lausanne

Benson C, Twigg J (2007) Tools for mainstreaming disaster risk reduction: guidance notes for development organisations. http://www.preventionweb.net/english/professional/publications/v.php?id=1066. Accessed 22 Nov 2010

Christopolos I, ALNAP, Glemminge Development Research (2006) Provention consortium forum 2006. Strengthening global collaboration in disaster risk reduction. The elusive 'window of opportunity' for risk reduction in post-disaster recovery. ipcc-wg2.gov/njlite_download.php?id=5282. Accessed 20 Dec 2010

Davidson C, Johnson C, Lizarralde G, Dikmen N, Sliwinski A (2007) Truths and myths about community participation in post-disaster housing projects. Habitat Int 31(1):100–115

Galbraith JR (1983) Strategy and organization planning. Hum Resour Manage 22(1–2):63–77. https://doi.org/10.1002/hrm.3930220110

Global Facility for Disaster Reduction and Facility, GFDRR (2014) Guide to developing disaster recovery framework (World Reconstruction Conference Version), GFDRR, Washington, DC

Goold M (2002) Designing effective organizations: how to create structured networks. Jossey-Bass , Wiley (distributor), San Francisco Calif

Herber J, Singh JV, Useem M (2000) The design of new organisational forms. In: Wharton and managing technologies, pp 376–392 (Reprinted in Mintzberg H, Ghoshal S, Lampel J, Quinn JB (2003) The strategy process: concepts, contexts, cases, 4th edn. Pearson Education, pp 234–240)

HFIR (2012a) Documenting the reconstruction of Bam, role and function of governments, international agencies and NGOs in reconstruction, (focusing on Housing Reconstruction) (Unpublished report) (in Persian)

HFIR (2012b) Documenting the reconstruction of Bam: design process in reconstruction, (unpublished report) (in Persian)

Lyons M, Schilderman T, Boano C, D'Urzo S (2010) Conclusions. In: Lyons M, Schilderman T, Boano C (eds) Building back better: delivering people-centred housing reconstruction at scale. Pract Action Pub, Warwickshire, UK, pp 345–359

Maskrey A (1994) Disaster mitigation as a crisis of paradigms: reconstructing after the Alto Mayo earthquake, Per. In: Varley A, Wiley J (eds) Disasters, development and environment. Chichester, New York

Maskrey A (2011) Revisiting community-based disaster risk management. Environ Hazards 10 (1):42–52. https://doi.org/10.3763/ehaz.2011.0005

Mintzberg H (1980) Structure in 5's: a synthesis of the research on organization design. Manage Sci 26(3):322–341

Mintzberg H, Ghoshal S, Lampel J, Quinn JB (2003) The strategy process: concepts, contexts, cases, 4th edn. Pearson Education

Morgan G (2006) Images of organization, updated edition (first edition 1997). Sage, Thousand Oaks

Quzai U (2010) Pakistan: implementing people-centred reconstruction in urban and rural areas. In: Lyons M, Schilderman T, Boano C (eds) Building back better: delivering people-centred housing reconstruction at scale. Pract Action Pub, Warwickshire UK, pp 113–134

Saemian S, Erfanian Daneshvar A (2011) The process of technical control and supervision in reconstruction of Bam (Unpublished report) (in Persian)

Wisner B (2004) Chapter 9, Towards a safer environment. At risk: natural hazards, People's Vulnerability and Disasters. Routledge, London, pp 321–376

Zetter R, Boano C (2010) Space and place after natural disasters and forced displacement. In: Lizarralde G, Johnson C, Davidson C (eds) Rebuilding after disasters: from emergency to sustainability. Spon Press, London, New York, pp 206–230

Chapter 8
Analysis of the Strategic Nature: The Bam Housing Reconstruction Organisation

Abstract Theoretical discussions in Part I characterised reconstruction programmes as the strategic physical agent for facilitating a multidimensional developmental recovery that pursues their objectives. Organisation theory suggests that approaching objectives requires strategic thinking on 'how' to approach objectives in addition to 'what' should be achieved. In essence vertical cascading and horizontal convergence, as well as objective setting are crucial. This chapter analyses how the intrinsic strategic nature of the programme, in form of introduction of the three objectives and ways of approaching them, influenced the way the programme was implemented; how the performed against each objective; and its related potential contributors to that performance, as well as potential longer-term effects. Performance towards the objective of safeguarding the historical urban identity relatively shows underachievement with a compromise on architectural criteria during the implementation. Building earthquake-resistant buildings included tougher control mechanisms and improved supervisory services that also became part of the mainstream after the post-reconstruction housing development process. Interestingly, the case shows the local knowledge on structural aspects of safe construction improved. As an owner-driven programme, the beneficiaries' role ultimately equated people's role in a housing development procedure in normal situations, overarched by the broader administrative and sociopolitical context of the country. The case demonstrates some interesting signs of objective-oriented organisational configuration and creative problem-solving. However, there were also gaps in the full application of strategic thinking. Learning opportunities draw on both strengths and lessons learned and graphical illustrations analyse the essence strategic thinking for reconstruction activities.

8.1 Introduction

Theoretical discussions in Chap. 2 have characterised reconstruction programmes as the strategic physical agent for a multidimensional developmental recovery that pursue their strategic objectives. Decisions made in early days after a disaster have

© Springer International Publishing AG, part of Springer Nature 2018
F. F. Arefian, *Organising Post-Disaster Reconstruction Processes*,
The Urban Book Series, https://doi.org/10.1007/978-3-319-70911-6_8

longer-term effects (Davis 2007). Therefore, the question of organising reconstruction is not only about constructing the physical assets but also how to constructing the physical assets in a way that they contribute towards achieving the strategic objectives. Objectives define the 'what' and the way of approaching them defines the 'how'. Literature on organisation theory suggests that approaching strategic objectives requires strategic thinking on 'how' to approach objectives in addition to 'what' should be achieved. Discussions on 'how' to approach objectives highlight the importance of both vertical and horizontal convergence (Hannagan 2002). This 'how' includes both the formation of the system in order to facilitate approaching a specific purpose and also its potential evolution through the adjustments during its implementation.

The Bam reconstruction programme explicitly introduced itself as a strategic agent for three objectives: safeguarding the historical urban identity, building earthquake-resistant houses and mobilising the disaster-affected population. As the Chartered Management Institute (n.d.) states objectives set out 'what' an organisation (sociotechnical systems) is trying to achieve. This chapter aims to unpack how the strategic nature of the programme, in form of introduction of the three objectives and ways of approaching them, influenced the way the programme was implemented. This chapter examines the strategic nature of the formation and adjustments of the sociotechnical multi-organisational delivery system for the Bam reconstruction programme. It starts by examining how the sociotechnical system for the delivery of the housing reconstruction programme performed regarding each strategic objective; it will be followed with discussions of the potential contributors. In search of the longer-term effects of the programme, the research provides an overview of the local situation with regards to the strategic objectives and broader national-scale influences of the Bam reconstruction activities, 10 years after the earthquake, 2013.

I then analyse the strategic thinking behind the formation and adjustments of the sociotechnical multi-organisational delivery system of the programme against the essence of the strategic thinking as the literature on organisation theory suggests: vertical cascading and horizontal harmony, as well as setting the objectives in the first place.

8.2 Strategic Objectives and the Programme Performance

8.2.1 Safeguarding the Historical Urban Architectural Identity

The programme saw a change of approach towards the strategic objective, of safeguarding the historical urban identity of Bam from the initial expectations towards a more pragmatic approach. It was linked to the following: the dynamics of the conflict with the newly introduced regulation for structural aspects, the power balance between the organisations and the urge to speed up the process and people's despair and frustration (Arefian 2016). The new structural regulations for the objective of building earthquake-resistant buildings in practice undermined some of the initial guidelines for this objective. They greatly shaped the architectural fabric of the city by mono-typing the outer design of the houses, as they all became simple cubes. From the long list of the guidelines from the initial BAUC recommendations the most frequently applied ones were: the guidelines on respecting the garden-city characteristics of the city by locating the new houses at exactly the same locations as the previous ones, and using various textures of brick for the appearance of houses. The reason for the latter was that traditional materials (adobe brick structures) were not allowed to use because there were no seismic regulations for traditional construction techniques at the time) (ibid). Also, wherever possible, those temporary accommodations were kept to compensate for the relatively small size of the new houses. Figure 8.1 present images of typical reconstructed houses in Bam and Baravat.

The number of unfinished abandoned structures at various stages from foundation to structural elements is also noticeable, especially in the less advantaged neighbourhoods. The existence and extent of uncompleted and abandoned structures across the Bam area can be seen as a physical materialisation of inconsistencies in the initial organisational configuration, as well as the system adjustments. There is no data on the number of uncompleted buildings and it is impossible to know their exact number because although the number of planning permissions can be found but there is no reliable data on the number of completed buildings. Many people did not apply for the Certificate of Building Completion after they finished the construction work and moved into their homes. Locals see that applying for the Certificate of Building Completion might force them to repay the loan they received for the reconstruction of their homes and they do not want to do that.

A number of issues and conditions have contributed to the appearance and accumulation of such uncompleted structures. Some contributing issues were associated with different stages of the reconstruction programme within its official period. They can be listed as follows. First, the problem of dodgy contractors: during the early stages of reconstruction deceitful contractors came to the Bam area, deceived local people and took their money (the reconstruction loan). While some owners could manage to secure money from some other resources many of them were unable to continue their construction activities. Some waited for even 7 years in order to resume the reconstruction of their homes, while some could not find another source of finance.

Second, inflation hit people: As time passed in the reconstruction period, inflation (which is high in Iran) devalued the money and the prices went up. People who delayed the reconstruction of their homes encountered the constant increases of expenses on material, e.g. cement and steel. Thus, they short of cash for completing the construction operation of their homes. While some of them secured other resources and completed their homes; others had to stop their construction operations; some also solved this problem (shortage of money) by double-crossing the system, applying for the second time and receiving the second or third main/

Fig. 8.1 Typical examples of reconstructed houses in Bam and Baravat, *Photos* Arefian 2009 and 2013

subsidiary financial package/loan. Thus, they erected one or two more uncompleted structures in order to use the money released for them to finish the construction operation of their first project.

Third, unrealistic choose of housing typology: Beneficiaries/Applicants underestimated the expenses and did not realistically choose the plans and the area of their desired home based on their own individual financial situation. For example, when they approached the local companies they requested 120 m^2 designs to apply for the planning application, but the reconstruction fund was considered for 80–90 m^2. According to a member of Setad, such people were usually optimistic that they would be able to find extra funds to construct their houses, however, when the time arrived—and it arrived with inflation and more expenses—they got trapped and could not find other resources. Some of these people followed the temptation of faking new applications and receiving the second loan and grant by taking advantage of the system.

Logically these uncompleted structures will gradually be completed as the applicants are able to find financial resources and complete their houses. Other uncompleted structures, which are there for other reasons will also be completed upon the growth of demand and when the housing development market and housing prices in the Bam area justify it. Until then, these uncompleted buildings will be left on their own. Figure 8.2a, b, c provide examples of uncompleted abandoned structures.

The way the delivery system approached safeguarding the historical urban fabric was also influenced by the consequences of the system adjustments and evolution. In practice, this objective lost its importance over time, especially during the later stages. A manifestation of this is reflected in emerging new trends of architectural changes in both layout and architectural elements. My observations in 2004, 2005, 2006, 2007, 2009, 2011 and 2013 show an increase in architectural changes, e.g. unusual motifs,[1] from the later stages of construction phase until the present. This phenomenon can be linked with the system adjustment for allowing people to become contractors without relating the possible effects on the whole strategic dimensions of the programme; this will be discussed later. According to some members of Setads and KEO, the number of cases in which the owners did not build the approved architectural style but changed it to something else, increased after beneficiaries could become their own contractors. Before the adjustments, beneficiaries were in direct contact with contractors. Genuine contractors acted as a medium between people, builders, and engineers. They were directing people tastes towards the initial design. Genuine contractors had already worked on other projects and overall were knowledgeable and up-to-date with a more urban style of building. However, when people became contractors that medium was removed.

[1]Motifs include very big eagles, Roman columns, ancient Persian figures outside and the romantic figures—from Persian poems, ancient kings, and religious Imams in decorative tiles inside their homes.

Fig. 8.2 Typical examples of uncompleted abandoned structures, *Photos* Arefian 2013

Thus, people could change the approved designs during construction more easily than before.

Additionally, there was no architectural supervision in place because according to the National Building Codes and Engineering Organisation architectural supervision for such small dwellings is not compulsory. After the system adjustment on beneficiaries' participation, they not only came into direct contact with engineers but also into direct contact with builders and masons, who entered the Bam area from all around Iran and Afghanistan to benefit from its construction boom. Through direct informal social interactions at the construction site, where people also lived, the taste of these labourers about what a good building is (especially in terms of architectural taste) influenced people. Some of local people's interviewees related the use of some decorative motifs and architectural style of their homes to the advice and suggestions of their builders and masons. This expanded and became fashionable since beneficiaries copied each other.

Also, the political promises by the new president and the introduction of new criteria for qualifying applicants directly influenced the programme performance towards this strategic objective. In addition to increasing the system vulnerability and the number of uncompleted structures, the practical considerations for this new scheme created a trend of splitting lands by tenants and newlyweds. Renters and newlyweds required land in order to meet the second condition for being qualified for the reconstruction fund. This damaged the unique garden city characteristics which, as a member of ORA/HFIR clarified, was the essence of safeguarding the architectural fabric. Research observation and interviews present examples in which the garden at the edge of a street was divided to provide a number of small building plots for reconstruction purposes. This is schematically illustrated in Fig. 8.3. Each plot has either a completed house or uncompleted structure. Figure 8.4 presents an example of this.

After the reconstruction period (at the time of research) the architectural considerations for housing development practice in urban areas address the preservation of the architectural fabric of the city including advice on finishing materials. The municipalities of Bam and Baravat are in charge of such matters, similar to the pre-disaster practice. However, there is lenient trend in the municipalities about meeting the architectural features/codes required in planning permission. The planning manager in Bam municipality explains the situation as the following:

> If architectural codes and regulations are breached during construction the owner will be called to correct the issue, if the issue is not corrected, the case will go for the Commission Article 5 to decide what to do. This usually ends up with financial penalty. Even if the case goes to the court it is unlikely that the court orders the demolition of the building.

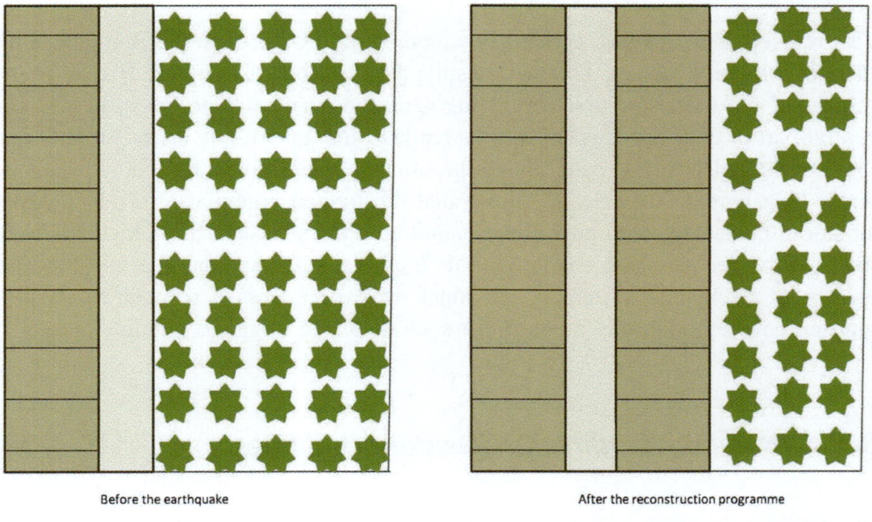

Fig. 8.3 A schematic illustration of how in some cases a garden was divided into small plots at the edge of a road to provide building sites residential for the purpose of reconstruction

Fig. 8.4 According to the neighbours, these building sites were all part of the big garden behind before the earthquake. Part of the garden, adjacent to the road, was divided to provide land for the reconstruction purposes. *Photo* Arefian 2013

A member of Baravat municipality provides a justification, delivered in a quasi-spiritual tone, on this lenient approach that reflects the general social attitude observed during the research.

> Why should we bother people when an earthquake will come (*and might kill us all*)? Life is short!

This relaxed approach also appears in widening narrow roads in the Bam area after the reconstruction period. This was despite the clear mandate by the BAUC in the middle of the reconstruction period that stated no road was to be widened. The central part of the Bam area still has its gardens and garden-city characteristics and the morphological urban patterns are the same as the old ones but many roads are wider than before. This clearly shows that the limited potential of a post-disaster situation for dealing with prevailing quantitative planning system, which has been in place for decades, and challenging it. Improvements are possible but a radical change of the whole system is too much to expect from a reconstruction programme, even though the intention for such a change might be genuine.

8.2.2 Building Earthquake-Resistant Buildings

In practice, the structural control mechanism for building earthquake-resistant houses was prioritised over the mechanisms to safeguard the architectural fabric.

Construction supervision was applied and material quality control was implemented at nine checkpoints instead of the five compulsory ones in normal situations and had to be documented in the BTID. There was a high level of confidence among interviewees of this research that reconstructed houses were earthquake resistant. In fact, 57 interviewees out of 659 ones from various organisations including locals were positive that the reconstructed houses were earthquake resistant. Still, since the reconstruction programme in Bam attracted younger qualified engineers, there were examples of inexperienced engineer supervisors who were not very strict in the supervision or were pressurised by the homeowner to receive approval for the checklist in BTID. The latter was the case mostly after homeowners were allowed to become contractors for their homes. To address this, KEO organised training courses for younger engineers. It is impossible to gain absolute quantitative certainty over the accuracy of the control mechanisms.

Astonishingly, there is a high level of local knowledge about construction operations and specific measures for seismic regulations, for example, structural knowledge on bracing or enforced bars in the foundations. A Ph.D. study on knowledge transfer during the reconstruction of Bam, conducted by Gharaati Kopaei (2009), also elaborates on this newly gained local knowledge among both builders and local people. He and local engineers point out that knowledge was transferred during the directed reconstruction phase and knowledge on safe construction was mainly internalised by builders who were in frequent formal and informal contact with a large number of engineers in the area. Interestingly this local knowledge extends to the group of people who usually are not expected to know much about construction operations and structural matters, for example, housewives. Observing the basic general knowledge of structural matters among less-educated ordinary women during the fieldwork was very interesting. For example, a local housewife approached me and sought reconfirmation on what she did:

> Are you an architect? Do you know if it is good to add water to the concrete for foundation? I was told that adding water to the foundation concrete makes it stronger. I took a water hose and watered the concrete as much as I could.

Such awareness among locals who would not traditionally be expected to know about construction operations, can be linked back to the practical considerations for temporary accommodation. The amended approach by HFIR, of erecting individual temporary accommodation in each plot, positively contributed the local knowledge of the construction operation and earthquake-resistant buildings. Knowledge was transferred through informal social interaction facilitated by their shared location of activities and living, a manifestation of social learning that will be discussed in Chap. 9. However, this knowledge about construction operation and building is not an informed expert engineering knowledge and depends on the whom the locals talked with more and were influenced by, whether it was an engineer, a contractor if the case was applicant-administrator or builders and engineers in case of applicant-contractors.

Also, as a result of the overall turn of events with regards to this objective (e.g. KEO winning the conflict and local consultancies overdesigning the structural elements), the concept of safe construction was misinterpreted as having bigger and more columns and more bars in re-enforced concrete for the foundation. By doing this, it potentially contributed to the shortage of the already-limited fund for other parts of the construction operation and this might have contributed to the number of uncompleted structures. This might be seen as a reaction to a bottleneck in the workflow, the conflict between KEO and the Mother Consultancy, and the fear of refusal by KEO among local consultancies. An extensive phenomenon within reconstructed houses is the imbalance between the small size of houses (typically 80–100 m^2), and the number of columns inside the house that in some examples reached 16 columns. This issue of imbalance between structural design of a building and the size of the house was even raised by the president of housing reconstruction during a field visit as a local resident recalls:

> Our house was the one of the first completed houses, and it was visited by Eng Saidi Kia when he came to Bam. He asked his entourage why this one storey building has a structure strong enough for a five storey building.

The imbalance also can be linked to the locals' fear of another earthquake after their traumatic experience that motivated 'the more columns the safer house' idea, even when they knew about the imbalance. A local builder boasted about the safety of his house although he knew that in technical terms such an extreme cautionary approach was not needed.

> My house is 80 m^2, it has 16 columns, I know they are more than enough…I spent extra 1,000,000 Tomans (*$1,000*) on them. I wanted to have a strong house.

Interestingly, the structural control mechanism after the reconstruction period did not return to the pre-disaster situation. Those strict seismic codes have since been in place. The owner forms a contract with the Bam branch of KEO, and the construction supervisor is selected and paid by the Bam branch of KEO unlike other parts of the country at that time. With this approach, they reduce the chance of conflicting interests for construction supervisors. Moreover, according to the Bam branch of KEO, the number of construction/structural engineers in the Bam area has significantly increased from three persons in 2004 to 70 persons in 2013. The Bam branch of KEO acts as the construction police on structural matters within the national Engineering Organisation framework. However, similar to the reconstruction period, the accuracy of the construction supervision cannot be guaranteed and there are issues that happen after the completion of the construction and the receipt of the last certificate. For example, a member of the Bam branch of KEO pointed out that people attach dangerous elements to their buildings or even add another story to their building after they obtained the last certificate of the Construction Completion (interview 2013).

The above scenario clearly shows that the existing public local knowledge on earthquake-resistant building houses lacks an understanding of the underlying concepts and principles of safe construction. Safe construction is translated to mean

Fig. 8.5 Attaching additional elements to reconstructed houses. *Photo* Arefian 2013

a few of the main structural elements without paying attention to other elements of safe construction. For example, while local people boasted about their buildings' earthquake-resistant characteristic and the number of columns their houses have, at the same time they had added attachments to their building that were not safe. Figure 8.5 presents an example of this. Gharaati Kopaei (2009) concluded that while locals (builders and public) know what to do; they do not know why they do it. In the absence of such understanding of the principles of safe construction little can be done to assure the further knowledge transfer in the longer-term. However, in my view, even a little knowledge is better than no knowledge since it creates a ground upon which those improvements can be applied.

8.2.3 Mobilisation and Participation of People

The strategic objective of mobilising beneficiaries' participation called for mobilising people to take part in the reconstruction of their own houses, and follow the procedure for urban housing reconstruction plan in Bam. The system performance on this can be understood in both quantitative and qualitative ways. Quantitatively, according to the HFIR reports and the World Bank evaluation report (2010) the rate of beneficiaries' participation in the housing reconstruction programme in Bam is 138%, exceeding from the initial estimation. The total number of reconstruction cases reached 36,225 (urban housing and commercial units) compare to the initial estimation of 26,284 cases. This final figure includes 6,890 cases for new qualified group (renters and newlyweds), and 2,304 cases of secondary households living with extended families before the earthquake. Therefore, the housing reconstruction programme in Bam succeeded to mobilise people in the housing reconstruction programme, which was owner-driven.

In owner-driven reconstruction (ODR), as Jha et al. (2010, p. 95) state, people who lost their houses are given some combination of cash, vouchers, and in kind and technical assistance to repair or rebuild their houses. ODR is seen as the most empowering and dignified approach for households, and it should be used whenever the conditions are right for it. However, ODR requires good oversight and governance, that is, a government capable of establishing and enforcing standards, and some agency (governmental or non-governmental) to ensure the quality of construction. According to them, when engineered building technologies are being used, or multifamily housing is being rebuilt, using ODR is more challenging, but not impossible.

This owner-driven approach in Bam meant it was the beneficiary who could start reconstruction of his/her house not the executive body of reconstruction, HFIR. Based on the collection of weekly real-time progress reports provided by Setads to HFIR, from 2004 to 2007, it can be concluded that people welcomed the reconstruction programme. Table 8.1 presents a conclusive report on the physical progress of the reconstruction programme in urban areas.

On a qualitative understanding of system performance on this objective, the research also finds that there is a disagreement on whether the Bam reconstruction was or was not a participatory reconstruction at all. The reconstruction programme initially opted for a limiting approach by defining people as applicant-construction managers. Encountering problems previously linked with various organisational inconsistencies, the system was adjusted and extended the role of people to be able to become contractors as they could in a normal situation. Officials boast about the participatory nature of the reconstruction of Bam that ‘Bam is built by its people’. However, a number of researchers and interviewees, e.g. Meskinazarian (2011) and post-war academic veterans reject that and dispute claims that the Bam reconstruction was a true participatory programme.

Positioning the research within the above spectrum, I argue that the Bam reconstruction, despite its limiting approach at the beginning, later became as much

Table 8.1 Reconstruction physical progress in urban areas. Ref: HFIR (2004–2007)

Description	Number of units
Initial estimation of the total number of houses and private commercial in need of reconstruction	26,284
Number of houses before the earthquake	19,224
Number of cases as main households	22,983
Number of cases as extended household	23,04
Number of cases as renters and newlyweds	6,890
Number of cases for private commercial units	4,048

participatory as the normal housing development practice in Iran. Each country and locality has its own pre-disaster housing development procedure which defines the role of various stakeholders in a housing development practice. People (applicants) in a normal urban housing development practice follow a procedure that is formed of local authorities/municipalities in charge of issuing planning applications, planning permissions, and certificates and so on. People do not design and form the actual procedure, and even if they enjoy a certain level of power it is not usually the ultimate decision making step in the ladder of participation that was discussed in literature review chapter. When configuring the workflow process for housing development activities, it is not people (in their totality) who shape the procedure. People might be informed or consulted when an operational system for housing development procedure is formed. Although it might be desirable to stretch people's manoeuvring/decision-making power for reconstruction purposes as a trigger to influence the future, in more practical matters the role of people (in fact, the step they are on in the participation ladder) has a ceiling. It is mostly informed and limited by people's role in the pre-disaster housing development condition which does not challenge the tolerance of the established sociopolitical system in a country. Beneficiaries' participation and its practical consideration for the system formation and implementation should be seen in the light of pre-disaster sociopolitical context.

The theoretical discussion in Part II concluded that it is the purpose of beneficiaries' participation that needs to be understood. What was interesting in the Bam case was that it saw participation as means for psychological healing, to develop a house that would be acceptable and to return to normal life as soon as possible. This is aligned with what Davis (2007) identifies on the multidimensional advantages of participatory reconstruction and the link between reconstruction and recovery. In Bam, the deeper underlying purpose of beneficiaries' participation was away from avoiding architectural standardisation or even reflecting the local lifestyle in new houses as it is suggested in by scholars such as Lizarralde et al. (2009). Nevertheless, a number of local interviewees (local people, local NGOs, contractors) pointed out the benefit of engaging people as a mechanism for avoiding the creation of boring similarity throughout the city. Members of HFIR, which formed the system clarify that in addition to learning lessons from previous reconstruction experiences, the idea in Bam was that the participation of disaster-affected families had a healing value for locals. Given the scale of the despair that the whole city was in, engaging disaster-affected people could help them to overcome their grief and return to taking responsibility for their lives and move on.

> What we mean by reconstruction is the reconstruction of society….In very early days we told people anyone who helps with debris removal of his own house could keep all recyclable materials, doors, bricks, etc.…and use them for the (*re*)construction (*in addition to the available funds*). Many questioned us: 'how much money would you like to save after this?' But we didn't think about the money. The point was not about saving money. Every single brick they (*people*) moved was a step forward to move on.
>
> (member of ORA/HFIR, Italic from author)

Moreover, beneficiaries' participation in the Bam was seen as a means for making beneficiaries active on an individual/household basis. It was not intended to politically empower communities and so on. As discussed previously, the workflow was an improved version of the previous housing development process with regards to clear objectives. If additional expectations, such as community empowerment, had been on the agenda, it would have had its related objective or policy that, in turn, they would also add more layers of complexity.

8.2.4 National Level Influences

Similar to the Manjil earthquake the Bam earthquake led to improvements at the national level. They can be identified in the three following aspects. First, the Bam earthquake became another turning point for the institutional development of the Engineering Organisation by exposing that having compulsory seismic codes is not enough if they are not applied in practice. We saw the Manjil earthquake created a momentum for further development of National Building Codes and seeking ways for their application through proposing and endorsing the Engineering Organisation Law, which became responsible for developing, upgrading and applying the National Building Codes. The Bam earthquake also created the momentum to address the application of National Building Codes (e.g. Regulation 2800), which was on agenda from the beginning that they were not properly applied. Thus, the first detailed operational framework for implementing the application of National Building Codes was prepared and approved 6 months after the Bam earthquake. According to the framework, an engineer supervisor is not hired by the owner/developer. This engineer supervisor is introduced and paid by the Engineering Organisation, as was already happening in the Bam. However, according to a manager in the Tehran municipality despite having this operational framework it has not been put in practice in Tehran.

Second, the Bam earthquake and reconstruction contributed to the extension of BHRC research areas for the National Building Codes on earthquake seismic regulations and activities to improve the construction industry. A number of National Building Codes and regulations were introduced to cover a greater variety of construction techniques in the country. For example, the seismic codes for traditional construction techniques, which at the time of the Bam earthquake did not exist. Also, according to members of BHRC and HFIR and independent architects in Tehran, another longer-term improvement in the construction industry is the standardisation of construction materials. It was suggested by BHRC during its participation in the Bam reconstruction that quality control for construction material should be imposed. The Bam case played a facilitative role. The facilitation of such longer-term impacts was linked with the engagement of influential organisations. This will be discussed further.

Third and very interestingly, the Bam case contributed to a strategic initiative on reconstruction activities in the country that links the built environment to disaster studies. The housing reconstruction programme in Bam has been criticised by post-war reconstruction veterans and academics over its insufficient attention to social aspects and beneficiaries' participation; this will be discussed further. This criticism led to the establishment of a specialised reconstruction department in the Faculty of Architecture and Urban Planning in SBU,[2] proposed and established by post-war university veterans and academics, who became concerned about the future of the reconstruction trend in the country. According to a post-war reconstruction academic veteran, there were concerns about the possibility of losing all previous valuable reconstruction experiences. It can be expected that the establishment of this department will greatly contribute to how the future reconstruction programmes are undertaken by HFIR and other organisations. Alumni will be gradually employed by ORA and other stakeholders, thus they will be the next generation of professionals and managers in the field of reconstruction (both post-disaster and post-war).

8.3 Vertical Cascading and Horizontal Alignment Towards Core Directions

Some of the problems that the programme encountered and the way the system performed can be linked with insufficient attention to the strategic nature of the programme, despite its purposeful formation and implementation of the housing reconstruction programme and its sociotechnical delivery system. The ORA group addressed the strategic nature of the reconstruction through the introduction of three core objectives, corresponding to strategic principles set by the Steering Committee. They then introduced objective oriented practical considerations. However, what was missing was to examine the interrelations between the practical considerations of different strategic objectives.

The literature on strategic management in Chap. 2 showed that despite the variety of prescriptive texts on strategic management, keeping the core direction between various levels of strategy and cascading them is a crucial element in strategic decision making. Strategy needs to be translated correctly and accordingly into lower level strategic objectives and then lower levels of practical considerations, which must be put in place effectively in order to implement the strategy. Cascading various levels from highest levels to the operational ones must be supported by horizontal alignment or harmony of the considerations at each level. This horizontal alignment directs all considerations towards the core direction as the main purpose of the activities. Higher level policies and lower level operational practical considerations must contribute to the strategic objectives and corporate strategy and they are aligned with other considerations for doing so. These

[2]SBU was one of the universities, which took part in post-war reconstruction.

considerations exist to serve the strategic objectives not themselves per se (e.g. Armstrong 2009; Morgan 2006; Mintzberg et al 2003; Hannagan 2002; Ansoff 1990). In the Bam case, each core principle of reconstruction as a whole became an objective and created one core direction on its own that was a reference direction for vertical cascading and horizontal harmony accordingly, for example, building earthquake-resistant buildings. Therefore, each of these strategic objectives needed its specifically targeted practical considerations for the initial system formation and organisational configuration. Figure 8.6 shows this.

The need for vertical convergence was accordingly noted and practical considerations to approach each objective were made by ORA at the system formation stage; however, what was missing was to create horizontal harmony by identifying interconnections of practical considerations towards different objectives. This was related to how organisations at each operational level act in consistence with the objectives. The three strategic objectives in the Bam reconstruction programme did not exist in isolation from each other, and the organisations which were brought into approach an objective did not work in isolation. The participation of beneficiaries, building earthquake-resistant houses and safeguarding the architectural fabric of the city, at the design stage of the delivery system were treated as separate

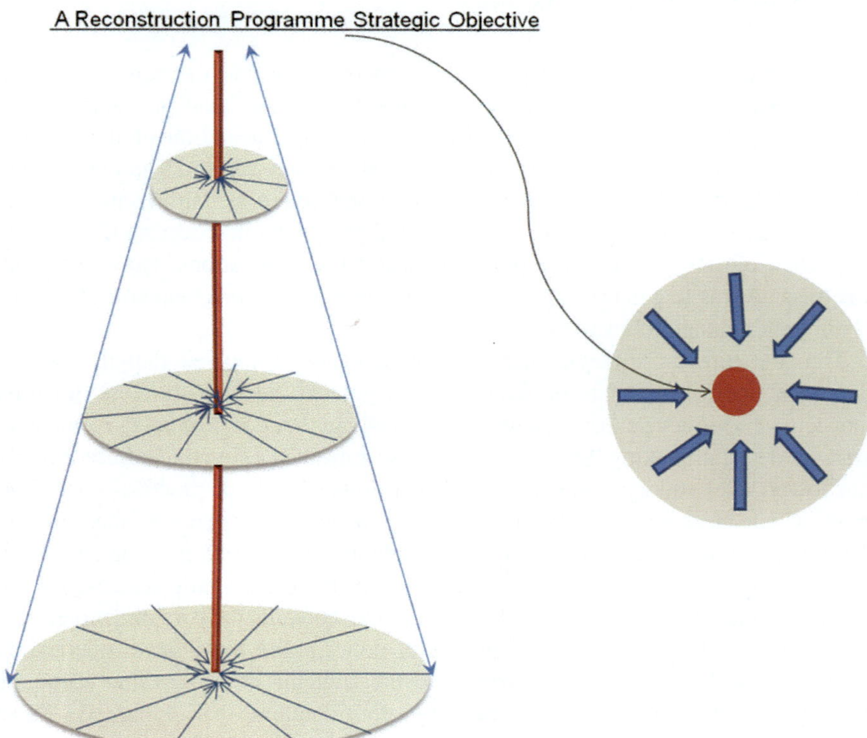

Fig. 8.6 Each objective created a strategic direction (Red Thread) in Bam

and isolated objectives. The practical considerations for each objective were not examined against its possible effects on other objectives or how they were interconnected with practical considerations for achieving other objectives. All these three objectives were related to the same product—houses. For example, the technical interrelations between architectural and structural considerations were not identified and therefore possibilities for how things might go wrong were not examined. But in reality, there were overlapping areas and interdependencies between the considerations for two or more objectives.

This argument should be seen in the light that having more objectives in reconstruction multiplies the complexity of the operational scene because: the number of practical considerations increases; actions for achieving one objectives might affect achieving another objective as a result of interrelated activities; and having more objectives probably requires more organisational elements, which means more participant organisations and individuals. Therefore, the alignment of lateral actors' internal arrangements with the overall strategy and strategic objectives is more difficult as the system created is more complex, increasing the chance of bottlenecks and unforeseen situation in combination with the chaotic context in an aftermath, as exemplified in the Bam case. Having more objectives increases the chances of contradictions in operational policies and frameworks. Approaching one strategic objective in practice is easier than two strategic objectives in one programme.

The Bam case shows that the lack of defining overlapping interconnections horizontally and harmonising participant organisations before the implementation of the system increases the chance of divergent approach or change of direction that affect the system workflow, which in turn affects the programme performance, both in the short and long term. A greater number of organisational objectives lead to a greater number of horizontal interrelations, for example, if the Bam housing programme had two objectives, e.g. beneficiaries' participation and earthquake-resistant buildings, the number of interrelations would have been reduced and the operational bottleneck of the reconstruction programme would not have existed. On the other hand, if the number of objectives was increased to four, more practical considerations would have existed that would have led to more interdependencies, more chance of omitting the big picture during decision making over the considerations and ultimately increasing the chance of operational conflicts/bottleneck during operation. As Bonabeau (2007) states with a growing complexity the risk of failure for a system increases. As a system becomes more complex the system itself becomes vulnerable to internal risks simply because the system incorporates an increasing number of interrelated activities and elements that can contribute to loosing the big picture along the way. Figure 8.7 schematically shows a show a delivery system for a multi-objective housing reconstruction programme becomes more complex.

Logically, paying attention to vertical and horizontal alignments and harmonisation (keeping the red thread in mind) for a sociotechnical delivery system is easier during the initial system formation phase compared to the real-time implementation. This is because the pressing issues in the field require quick decisions for the system

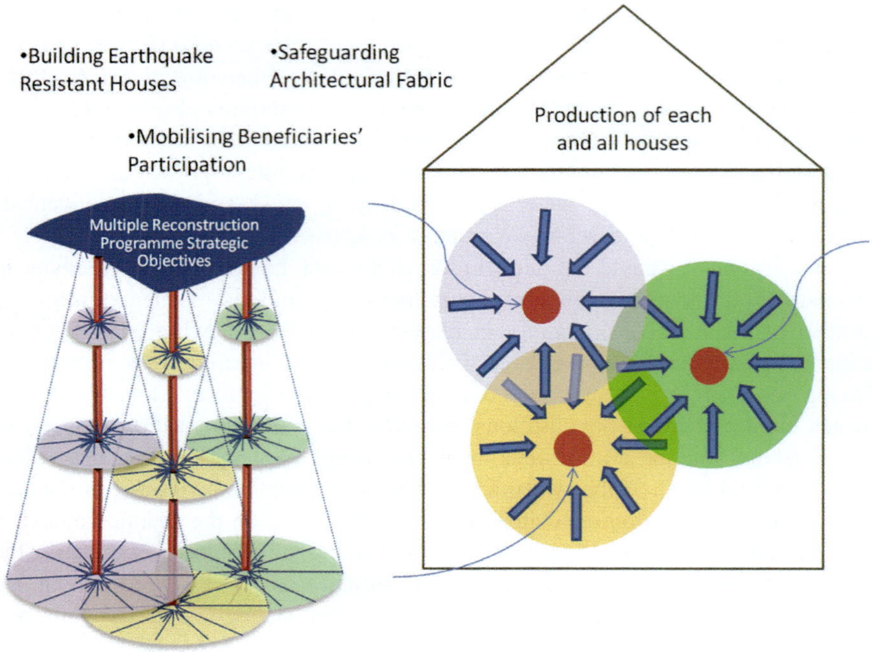

Fig. 8.7 Vertical cascading and horizontal alignment in multi-objective housing reconstruction programme (schematic illustration)

to be adjusted. It appears that the system formation in Bam paid attention to its strategic nature (at least on vertical cascading) through its purposeful approach but during the implementation and adjustments such attention was missing. Targeting strategic objectives was the way to form the delivery system but the implementation and adjustments of the system showed evidences of forgetting the red-thread (main objective) when adjusting the system to the realities of the context. For example, the system adjustment extended beneficiaries' role from applicant-administrator to applicant-contractor but this was not examined against the possible influences on the other two objectives. The issue around having contractors as a separate element of the system or allowing beneficiaries to become contractors during programme implementation is not only about which one is better and more workable in general. The system was initially designed based on the separate contractor and changed in nature after that adjustment but the other practical considerations did not reflect that change. Such reflections could have been more practical considerations (e.g. extra supports in the form of training workshops for applicant-contractors could have been introduced). Similarly, the shift from bigger qualified contractor companies towards individual contractor teams should have led to an examination of what it meant regarding the strategic objectives. When there is a need for adjustment the new way of working in the system might be in conflict or inconsistent with the reason the system was introduced in the first place. This change should have been seen against other objectives as Davidson (2009) points out and also the possible effects and results. This change, in fact, shaped system in another way that was not thought of at the beginning and therefore the consequences and the functionalities of the process in this new system were not examined.

As previously mentioned, the Bam case supports the current literature (e.g. Quzai 2010; Davis 2007; Wisner 2004) on the influential role of the political context on reconstruction and recovery programme. Keeping the red-thread in mind during design and implementation greatly depends on the broader political alignment with the programme. The political intervention in the middle of the Bam reconstruction programme not only increased the system vulnerability from social perspective (this will be discussed later), it also increased the number of practical considerations in the system to meet the political promise. Through this, it increased the complexity and distorted the big picture at decision-making points. Thus, the vertical cascading suggested in the strategic approach should stretch up to the highest political position. The practicalities of implementing political promises should be (at least in an ideal situation) examined against the reconstruction programme strategic objectives and their related multilayer considerations. At the operational level again the new practical considerations are needed for realising the political promises, so similar to other adjustments, they should be examined against the strategic objectives. The practical consideration in the Bam case that linked reconstruction fund for the new group with provision of land damaged the most prominent feature of safeguarding the historical urban identity of the area that was a garden-city.

While creativity is seen as crucial in strategic management, including project and programme management (Hannagan 2002; Morris and Jamieson 2005), the Bam case provides an interesting example of how managerial creativity solved the problems which emerged as a result of inconsistent organisational configurations and other inherited inconsistencies. Arguably, this creativity saved the whole Bam case. The case reconfirms the need for creativity in finding solutions and exploring out-of-the-box solutions based on the available resources, specific conditions while approaching multiple objectives and finally the fit between the system and its environment stressed in the literature (Markides 2008; Mintzberg 1980).

8.4 Strategic Overview of Objectives

This discussion has highlighted the need for vertical cascading and horizontal alignments towards strategic core directions for the formation and implementation of the programme delivery system. But the very beginning and yet critical activity is to set objectives that provide the core directions. According to the Chartered Management Institute (n.d.) objectives set out what an organisation, business or programme is trying to achieve, through a statement which describes what an individual, team or organisation is hoping to achieve. It is important for leaders and managers to get the process of setting objectives right, as inadequately formulated objectives could guide an organisation in the wrong direction. While specific and measurable objectives provide a definition of the success of a project or initiative, achievable and realistic objectives engage and motivate individuals, and time-bound objectives ensure that all stakeholders agree when it is to be achieved. Objectives can be stretching but not unachievable. They should be realistic even if they are challenging and difficult. Objectives can be set which are demanding but not so much that the chance of success is small. Realistic objectives take into account the available resources such as skills, funding, and equipment (e.g. Adair 2008; Chartered Management Institute n.d.). In case of multi-objective systems there might be a potential conflict between high-level strategic objectives (Hamdi 2010) that create conflicting directions and dysfunctional organisations.

Returning to the Bam case, the objectives did not naturally conflict because the structural and architectural technicalities in the construction industry always work with together and reach a mutual understanding of what should be achieved and how it is done. Also, the objective of beneficiaries' participation did not stretch the overarching role of people beyond the previous housing development procedure. In principle these three strategic objectives were compatible. However, the lack of clear prioritisation of the objectives and a lack of shared understanding among the controller organisations negatively contributed to conflict and a bottleneck in the workflow. In the absence of stating priorities, reaching a mutual understanding or agreement, participant organisations were left on their own to interpret the priorities, which naturally lean towards their own professional disciplines. KEO prioritised conservative structural design thus the architecture must follow the structure; in contrast, Mother Consultancy and local consultancies prioritised architectural features thus structure must follow architectural design. Here the third objective of beneficiaries' participation brought a sudden increase in demand, adding another crucial and decisive factor of speed and timing to this dispute. The inevitable consequence of such a loose approach towards objective-setting at the beginning in post-disaster reconstruction programmes is that priorities will be determined during implementation, as it was the case in Bam. When during the system implementation a conflict occurs the winner is determined through power balance, the emotional mood after a destructive disaster and the willingness of participant organisations to compromise.

Moreover, the strategic objective of safeguarding the historical urban identity in Bam was neither specific nor realistic as an objective. It was initially defined idealistically and approached broadly. This objective was missing a brief specific answer to the question of what is meant by safeguarding architectural urban fabric in a housing reconstruction programme. The answer to that single question in architecture and urban design fields can be very vague and broad. The BAUC initially examined traditional local architecture that existed in the city centre (and not in the newer urban areas); a broad spectrum of architectural guidelines and codes were extracted (categorised as elements related to function, physical characteristics, identity, climate) from the location of reconstructed houses (to protect the garden-city), to architectural elements, extendibility of the house, climatic consideration, room dimensions and proportions and so on (Armanshahr 2004; Golpayegani and Einifar 2004). While many public buildings and projects in Bam present interesting examples of contemporary interpretations of traditional architectural features, the pragmatic shift during implementation specified such broad expectations within the scope of an owner-driven individually built housing reconstruction. It focused on a few important achievable features, including protecting the garden-city, use of materials and future extendibility. However, and arguably more features could have been achievable if this objective was dealt with in advance accordingly.

Reminding that such broad expectations from the Bam housing reconstruction was a result of the desire (and seize the opportunity) to correct the existing failure of the planning system to protect architectural and urban identity in cities. This was a great challenge on the national scale and its scope was greater than the scope of a housing reconstruction programme to be fully addressed in the aftermath of a disaster, although specific improvements are achievable. The initial guidance and expectations covered a wide range of architectural and urbanism elements for the housing reconstruction. In addition to protecting the garden-city characteristics the BAUC requested all architectural typologies should address the climatic and traditional architectural elements such as shading, material, semi-open spaces and so on. A member of MHUD recalls that all saw the Bam reconstruction as an opportunity to influence the comprehensive planning system at national level and the idea was to establish architectural urbanism councils (similar to the BAUC) for other cities as well. Surely those fundamental changes need to be addressed in the country's quantitative planning system but within the post-disaster situation, such idealism raised the level of expectations. The initial guidelines were extrapolated from the older urban fabrics and houses and were set as guidance for reconstruction without examining them against the existing influential realities at the national level. For example, traditional materials could not be used because at the time BHRC had not developed seismic regulations on traditional construction techniques; and the post-earthquake disastrous situation and the importance of timing and finance were missing. This objective was treated as if it was happening in an ordinary situation as a result of a non-catastrophic sudden demand (Arefian 2016). According to Armanshahr (2004), the consultancy which conducted the previous and new urban development plans for the city before the earthquake, the average size of Bami houses was over 200 m^2 whereas the reconstruction finance was calculated for 80 m^2. Given this big difference, the logical preferences of applicants were to spend their limited finance on more space. Whereas architectural workshops were held both in Tehran and Bam, they were held to allow consultancies to share ideas and understand local people's requirements from architectural perspective instead of extracting the priorities against time and costs.

It appears that the important reality of the post-disaster context was absent in the initial system formation and organisational configuration stage. The BAUC, as the reference reference point for this objective, was biased towards a single disciplinary view of architecture. It was formed of distinguished architects, university professors and related public sector officials who themselves did not carry the experience of working in chaotic situations, instead they were mostly used to idealistic urban projects. Although HFIR/ORA was a member of BAUC and had the experiences of

Fig. 8.8 Example of initial architectural ideas on safeguarding the architectural fabric of the city, *Photo* Arefian 2005

post-disaster reconstruction (including Manjil) it appears that HFIR members did not accordingly communicate the non-architectural aspects of working in post-disaster situation with BAUC. A related exhibition of initial architectural ideas for the housing reconstruction of Bam (organised by MHUD and HFIR, held in Tehran, 03/01/2005) provides examples of architectural elements and motifs would cost money that was already not sufficient for building new houses as large as they were before the earthquake. Figure 8.8 presents an example of the initial ideas on this objective presented at that exhibition.

8.5 Conclusion

This chapter examined the strategic nature of the housing reconstruction programme in Bam during the system formation and implementation. The Bam case presents elements of strategic thinking in form purposefulness of vertical cascading.

However, despite paying attention to form a purposeful system, directed by strategic objectives what was missing was to identify horizontal interconnections of different groups of practical considerations that aimed to address each strategic objective. The system lacked horizontal harmonisations since different objectives were treated in isolation from each other and had equal priorities on paper. But during the implementations, the emergence of problematic realities that were linked to the inconsistent organisational configuration influenced the programme performance towards each strategic objective. As a result, in practice, some strategic objectives were prioritised over the other ones.

As it was elaborated in the case, and presented in graphic illustrations as analytical tools, the greater number of objectives and core expectations in reconstruction increases the complexities of the delivery system in two main ways: firstly, it increases the potential interconnections, and that makes it more difficult to harmonise between operational tasks. Second, it potentially increases the number of participant organisations that must perform those tasks and increases the chance of organisational rivalry or conflicts. This will be discussed in Chap. 10. The above implications can easily distort the strategic nature of the reconstruction programme by forcing the system performance to become imbalanced in favour of one strategic objective and against the other one, especially during the implementation. The system also did not prioritise the three strategic objectives from the outset. Such lack of clarity and mutual understanding or agreement opened the doors for individual interpretations by participant organisations. Their interpretations were biased towards their own professional disciplines and contributed to the emergent problems and bottleneck in the workflow.

As an owner-driven programme, the beneficiaries' role initially was limited to applicant-project manager ultimately became similar to people's role in a housing development procedure in normal situations, overarched by the broader existing administrative and sociopolitical context of the country. Building earthquake-resistant buildings in practice were prioritised and included tougher control mechanisms and improved supervisory services. Those improvements and control mechanisms became part of the mainstream within the post-reconstruction housing development process. Additionally, local knowledge on structural aspects of safe construction improved. The existence of such public knowledge is linked to the early decisions that locals lived in the same location as their construction site. This facilitated an informal knowledge transfer and social learning which will be analysed in Chap. 9. However, performance towards the objective of safeguarding the historical urban identity that on its own was idealistic and tried to capture a broader opportunity at the national level was not as initially expected. The Bam earthquake also created a momentum for institutional improvements in different construction industry related aspects that were introduced by the national level participant organisations such as BHRC and Engineering Organisation.

References

Adair J (2008) The best of John Adair on leadership and management. Thorogood Publishing, London

Ansoff H (1990) Implanting strategic management, 2nd ed. Prentice Hall, New York

Arefian FF (2016) Getting ready for urban reconstruction: organising housing reconstruction in Bam. In: Arefian FF, Moeini SHI (eds) Urban change in Iran: stories of rooted histories and ever-accelerating developments. The urban book series. Springer International Publishing, Cham, Heidelberg, New York, Dordrecht, London, pp 231–257

Armanshahr Consultancy Company (2004) Special structural and strategic plan for Bam (Unpublished urban development project document). MHUD, Tehran (in Persian)

Armstrong M (2009) Armstrong's handbook of management and leadership: a guide to managing for results, 2nd edn. Kogan

Bonabeau E (2007) Understanding and managing complexity risk. MIT Sloan Manage Rev 48

Chartered Management Institute (CMI) (n.d.) Setting SMART objectives, checklist 231. http://www.managers.org.uk for CMI members. Accessed 14 Feb 2011

Davidson C (2009) Multi-actor arrangements and project management. In: Lizarralde G, Johnson C, Davidson C (eds) Rebuilding after disasters: from emergency to sustainability. Spon Press, London, New York

Davis I (2007) Learning from disaster recovery; guidance for decision makers. International Recovery Platform (IRP). http://www.undp.org.cu/crmi/docs/irp-decmakersleslearned-td-2006-en.pdf. Accessed 17 Nov 2010

Gharaati Kopaei M (2009) Knowledge transfer in post-disaster reconstruction; the problem of post-post-disaster reconstruction. PhD thesis, McGill University, Montreal. http://digitool.library.mcgill.ca/webclient/StreamGate?folder_id=0&dvs=1420020480363~426. Accessed 21 Aug 2014

Golpayegani A, Einifar A (2004) Typology and design guide for housing in Bam… ministry of housing and urban development-office of architecture and urban planning, Tehran (in Persian)

Hamdi N (2010) The placemaker's guide to building community. Earthscan Publications, London, Washington, DC

Hannagan T (2002) Mastering strategic management. Palgrave, Basingstoke Hampshire [u.a.]

HFIR (2004–2007) Collection of work progress weekly reports by Setads (Unpublished report) (in Persian)

Jha AK, Barenstein JD, Phelps PM, Pittat D, Sena S (2010) Chapter 6, Reconstruction Approaches. In: Safer homes, stronger communities: a handbook for reconstructing after natural disasters. World Bank, New York, pp 93–107. http://www.housingreconstruction.org/housing/toc. Accessed 11 Jan 2011

Lizarralde G, Johnson C, Davidson C (2009) From complexity to strategic planning for sustainable reconstruction. In: Lizarralde G, Johnson C, Davidson C (eds) Rebuilding after disasters: from emergency to sustainability. Spon Press, London, New York

Markides C (2008) Strategy as balance. Bus Strategy Rev (Special report: strategy classics) 19(3):50–91. https://doi.org/10.1111/j.1467-8616.2008.00553.x

Meskinazarian A (2011) Social resilience of post-earthquake Bam. PhD thesis, Department of Geography, King's College London, London, UK

Mintzberg H (1980) Structure in 5's: a synthesis of the research on organization design. Manage Sci 26(3):322–341

Mintzberg H, Ghoshal S, Lampel J, Quinn JB (2003) The strategy process: concepts, contexts, cases, 4th edn. Pearson Education

Morgan G (2006) Images of organization, updated edition (first edition 1997). Sage, Thousand Oaks

Morris PWG, Jamieson A (2005) Moving from corporate strategy to project strategy. Project Manage J 36(4):5–18

Quzai U (2010) Pakistan: implementing people-centred reconstruction in urban and rural areas. In: Lyons M, Schilderman T, Boano C (eds) Building back better: delivering people-centred housing reconstruction at scale. Practical Action Pub, Warwickshire, UK, pp 113–134

The World Bank (2010) Islamic Republic of Iran—Bam Earthquake emergency reconstruction project (Implementation completion and results report no. ICR1207). http://www-wds.worldbank.org/external/default/WDSContentServer/WDSP/IB/2010/09/28/000356161_20100928235953/Rendered/PDF/ICR12070P088061C0disclosed091271101.pdf. Accessed 09 Sep 2013

Wisner B (2004) Chapter 9, Towards a safer environment. At risk: natural hazards, people's vulnerability and disasters. Routledge, London, pp 321–376

Chapter 9
Analysis of the Social Nature: The Bam Housing Reconstruction Organisation

Abstract A participatory reconstruction programme blurs boundaries between the programme's organisational society and society as a whole. Therefore, the dynamics of society as a whole or its representatives mixes with the way of doing things and the culture of all other participant organisations towards influencing the way the sociotechnical delivery system of the reconstruction programme works. This identifies social dynamics as an area for examination of the sociotechnical delivery system of the reconstruction programme. This chapter analyses the major social dynamics of the housing reconstruction programme in Bam that influenced the programme approaching its objectives through influences on system formation and implementation. The case of Bam interestingly presents an unintentional yet strong social learning in housing reconstruction programme and dynamisms behind that are examined. Informal social interactions facilitated by a common physical place contributed to longer-term improvements after reconstruction activities. This was despite a technocratic approach to the housing reconstruction programme system formation. The sociotechnical delivery system of reconstruction programme itself was vulnerable. Such vulnerability had physical implications and was linked to operational loophole that also was linked to inconsistencies in the initial system's formation and implementation, and broader economic and social contexts.

9.1 Introduction

Epistemological advances in organisational theory see organisations as social systems as much as they are a technical system, and programmes/projects are temporary organisations, hence they are temporary sociotechnical systems. This understanding of organisations is linked to the open system perspective of organisations. Elements in a system make up the organisational society. Dynamic interactions (both formal and informal) among the organisational society influence the way an organisation performs, and through this it influences the outcomes of the implementation of that system. The ultimate interpretation of such understanding is

© Springer International Publishing AG, part of Springer Nature 2018
F. F. Arefian, *Organising Post-Disaster Reconstruction Processes*,
The Urban Book Series, https://doi.org/10.1007/978-3-319-70911-6_9

that organisations are socially constructed (Morgan 1989, 2006). Within reconstruction, while organisational society in general influence the formation and performance of the sociotechnical delivery system of a reconstruction programme, in a participatory reconstruction this organisational society includes the beneficiaries, who pursue reconstruction of their houses. Thus, this system includes representatives of the broader society. A participatory reconstruction programme intensively blurs the boundaries between an organisational society and society as a whole.

9.2 The Informal Way People Learn (Social Learning in Bam)

The housing reconstruction programme implementation in Bam showed two distinct phenomena that although are different but linked by the informal social interactions at individual levels and informal social learning. First is the increased public knowledge about construction operations and some elements of safe construction, even among locals, for example, ordinary non-educated housewives. Earthquake-safe building techniques were transferred to the local builders and the beneficiaries'/applicants' households. Local people now have more knowledge about construction industry technicalities than before the earthquake, and structural regulations have become their priority (Gharaati Kopaei 2009). Although their knowledge underlying those technicalities is limited, any knowledge is better than no knowledge. This informal knowledge transfer and observations of construction activities also gave people the confidence about the city resilience as local residents boasted that 'Bam is the strongest city in Iran'. This perception is so robust that some locals avoid travelling other cities and spending the nights in other places.

Second is the emergence of new trends for changes to the already approved plans that accelerated after the system adjustment that allowed beneficiaries to become the contractor of their houses if they wanted to. In those cases, by eliminating the contractors, the intermediary elements, residents were in direct contact with labourers from all around the country and even Afghanistan, and were therefore influenced by them about the way their house should be. According to engineer supervisors and members of Setads this gradually influenced people's taste, especially in the later stages of reconstruction.

Both of the above phenomena were linked to the fact that beneficiaries lived at the construction site while their own house was under construction, allowing them to observe, interact and learn from engineers, contractors and labourers. Within post-disaster recovery literature Wisner (2004) stresses the importance of local knowledge, including local technical knowledge as the precondition of beneficiaries' participation in reconstruction. Both phenomena identify the housing construction sites as the common space for facilitating such informal social interactions. Applicants (as the programme participants) and their families therefore supervised and observed the construction operations, asked questions about it and learned about the construction operation. They lived in a living construction workshop. The Bam case exemplifies the importance of a common social space for facilitating individual learning, although it was created by accident. It was linked to a decision made during the temporary housing stage that changed the approach from large-scale temporary camp compounds on the outskirts of the city to erecting individual temporary houses in each site for affected families.

Such individual learning which is conditioned by its social environment is an interpretation of social learning (Pelling et al. 2008). Individuals and group learning within organisational context as Miller (cited in Curado 2006, p. 28) states is 'the knowledge acquisition made by actors (individuals and groups) when these can and are available to apply it in the decision-making process, or use it to influence others within the organisation'. Jervis et al. (cited in Pelling et al. 2008) reports of a longstanding interest in the extent to which such learning is determined by culture and socialisation in a social system. Despite debates on the importance of local knowledge, the social context and social position for information flow, the notion of informal social learning seems to be under-explored within the post-disaster recovery literature (Wisner et al. 2004). However, there are a few emerging empirical studies, e.g. on disaster management in Malaysia (Roosli and O'Brien 2011) that highlight the need for more attention to social learning in complying with regulations in disaster management. Although a social system is a complex set of human relationships interacting in many ways, the importance of organisational structures, such as hierarchy, working groups, the geographic locations, and the work space, structure of roles, interests and power create dynamics influence the learning process. Thus, organisations as sociotechnical systems are seen as environments that enable or inhibit individual learning (Curado 2006; Roosli and O'Brien 2011; Wang and Ahmed, cited in Pelling et al. 2008).

This research identified two missed opportunities to attend the social aspects of the reconstruction despite the existing potentials: First was the opportunity of using this common social space (where household members could actively and constantly engage with construction and supervision teams) that existed from the outset of the system formation for the housing reconstruction programme. It was indeed an existing socio-operational context, providing an opportunity to inform and influence social attitude through informal social learning towards the strategic objectives of the programme. This opportunity was missed. Therefore, it left the influences of such constant social learning processes undirected and accidental. Second was a grassroots innovative social initiative on participation that created a bottom-up channel of communication and coordination among locals. Immediately after the earthquake a social initiative was tested by a university sociology lecturer, with the support of the Bam and Baravat Councils and Kerman University. Testing a new local community approach in the country, local clusters were formed, and 30–60 neighbouring households elected two representatives, a man and a woman, called Shora-yar (Town Council's Friend). There were a community coordination office and a weekly local newsletter, *Citizens and Participation*, which shared information (e.g. informative experiences on the reconstruction process), reports of follow-ups and results. This initiative was intended to identify problems and raise issues to be solved at a higher level. Representatives met people, formed task groups to categorise issues and organised group meetings. The newsletters of this bottom-up approach were published from 2004 to 2007—1383 to 1385 in Persian calendar—(Citizens and Participation Newsletters 2004–2007; The Society of Children's Friends 2009). Again HFIR did not recognise a potential for using this related social initiative as a two-ways communication channel for promoting strategic objectives of the reconstruction. Given the existence of this bottom-up initiative that had a very simple social organisation, the reconstruction programme could have benefited from it to informally educate and influence people through the same social information channel, but this did not happen.

The above missed opportunities are representative of the insufficient attention paid to the social nature of the housing reconstruction programme in Bam that goes beyond debates on people's participation, which discussed in Chap. 8. The lack of such attention hindered the need for understanding underlying social dynamics among beneficiaries and other individual participants. HFIR declared the reconstruction of Bam as a social task rather than merely a physical operation but the attention to the social dimension was only seen as creating a formal participatory mechanism and owner-driven reconstruction process (Fallahi 2007). Paying less attention to the social dimension of the sociotechnical system, the delivery system was formed around technical formalisations. Therefore, and not surprisingly the programme lacked a social learning perspective. Although dividing Bam to operational zones might have supported informal interactions between applicants, design teams, supervision teams and Setad members, the common space for such interactions did not exist, and constant deeper interaction only happened at the construction sites.

Conversely, as echoed by post-war reconstruction veterans, the cumulated practical know-how from previous reconstruction experiences highlighted the importance of social learning through people's informal interactions and trusted social organisations, e.g. mosque groups, trusted people and neighbourhood communities. Such experiences confirmed the literature on social learning that suggests that collaborative learning among peers is believed to be faster and deeper than learning received from an instructor (Elwyn et al. cited in Pelling et al. 2008). Such lessons were not used in the Bam case, stimulating criticism and leading to the establishment of a reconstruction university department as we saw. Among those who criticise the Bam housing reconstruction programme for being a mere technical programme and ignoring its underpinning social dimension is also the former president of HFIR, who at the time of the Manjil reconstruction developed overall reconstruction policies and linked the know-how gained from post-war experiences to post-disaster situations. Interestingly, the need for addressing the social dimension is now acknowledged by an influential independent architect member of BAUC. A post-war reconstruction veteran and university lecturer, also active in Bam for non-residential buildings, calls the Bam case a setback. He implicitly emphasises the importance of paying attention to social learning in organising reconstruction.

> Understanding what happens in people's decision-making process is not a technical matter; it is a cultural matter, rooted in, and dependant on the history of their habitat.
>
> Different to formal relations which one (*might easily*) sees (*in a city*), some (*informal*) relations exist that one must understand otherwise makes a mistake. These are cultural things, not technical things to take the people of Bam to a number of class sessions, teach them the importance of architecture; organise a number of classes, teaching them the importance of structure. If we believe this we can better begin (*the reconstruction*). This experience did not happen in Bam. I believe Bam was a setback.
>
> This returns to the problem in conceptualising people's participation. When we talk about involving people in something, they directly refer to those whose houses were destroyed. But that is not people's participation (*as a whole*); that is one side of people's participation.
>
> In Bam 100,000 inhabitants lived (*before the earthquake*). A mechanism will be being created in the city, for example, someone from Aligoodarz[1] comes and builds something for them. The relationships that exist under the skin of the city, these should be discovered. If one recognises this, there is no difference in Bam. In Roodbar and Manjil too they (*beneficiaries*) did not build their houses by themselves (*labouring work*); there were carpenters who came and built for people, like Bam.
>
> What is important and could have happened in Bam too is that one finds these (*underlying*) mechanisms and supports them. Because what was important in Bam was that one know whatever one does during the reconstruction period contributes to improving people's knowledge, understanding and capability in (*as a result of*) reconstruction. But if we built strong houses during the reconstruction (period) but (after that) people are again building the previous rubbish, we did nothing indeed; we did the contracted job; (*but*) we did not do reconstruction programming….
>
> We should know what to put where (*in the system configuration for programming*). If for example people in Bam rely to you (*an expert, builder, influential person,* etc.) for the plans of their houses, we should come to you. Do not cut short the pathways. Believe that there are (*social*) relations underneath. If we believe in them (then) we work on them….We did not do so (*in Bam*).
>
> (italic from author)

Acknowledging the above argument, in this research, however, I found that social learning still happened in the Bam case, but it was unintentional, non-examined and unsupported, and that therefore its influences were undirected and unpredictable. Since beneficiaries and their households lived in a living workshop, there was a constant flow of social learning in all directions among the household, contractors, supervisors and builders.

This technocratic approach in Bam can be linked to: (a) an attempt to implement previous experiences in urban areas, reduce complexities in organising reconstruction through a modification of the existing housing development practice in order to form the delivery system; and (b) individual characteristics at the highest level of decision-making at HFIR. It is worth remembering that the organisational continuity at ORA and HFIR meant that the group formulating the programme

---

[1]A city in the West of Iran.

understood lessons from other national experiences. However, the programme delivery system was formed through the modification of the previous housing development practice in an urban area, which is individualistic and does not deal with social aspects including social learning during its formal procedure. Creating a temporary housing development practice on its own meant that the ad hoc advisory methods could not enter the formal housing process because it would make managing the reconstruction even more complex. As a member of ORA stated all good national examples of previous experiences (both post-war and post-disaster) had been in the context of rural reconstruction or very small towns, e.g. Manjil, Rafieieh and Sarieh. This is also a case on the international scale, for example, the successful village reconstruction in Kashmir in Pakistan or the parallel urban reconstruction in Muzaffarabad after the earthquake in Pakistan in 2005 (Quzai 2010). Previous discussions, showing the uniqueness of the Bam case in form of a historic urban area, as well as the high level of expectations on multiple strategic objectives, explain the complexities in applying those lessons. Adding other expectations or practical considerations to support and direct social learning could have increased the complexity of the programme and its delivery system, however, it was never explored.

Such technocratic bias, in fact, started from the stage of the objective-setting at the first place. Strategic objectives lacked attention to social learning. For example, the objective relating to beneficiaries' participation focused on mobilising beneficiaries to participate in reconstruction was concerned with encouraging locals to move on. The objective of earthquake resistant building was a technical interpretation of safe construction. Recognising the importance of deeper social dynamism and seeking practical considerations to enable, support or direct social learning is also linked to characteristics of the highest level decision makers at HFIR. The president of HFIR, at the time of the Bam earthquake, was known within the high-level public sector managers and professional community as a technocrat engineer who believed in engineering solutions and did not believe in social approaches.

Nevertheless, paying attention to social nature of reconstruction programmes and facilitating strategic social learning is crucial. The recognition of existing and emerging social interactions and opportunities to informally educate people might not even require a change in formal technicalities and individualist approach to the reconstruction workflow. Therefore, it might not require changing the delivery system by itself. Instead, it could be carried out through supportive practical considerations for educating representatives, disseminating practical knowledge on safe construction, urban identity, providing advices, and likewise. It is about creating links between bottom-up approaches and formal top-bottom procedures and formalities. By supporting informal interactions they can be directed to strengthen public knowledge and desire towards the strategic objectives by developing local knowledge. The sociologist who initiated and directed the Shora-Yar initiative stresses the importance of supportive mechanisms for such initiatives in order for them to be fully effective.

9.3 A Vulnerable System

As the research previously illustrated, the sociotechnical programme delivery system in the Bam case was to some extent vulnerable as non-genuine reconstruction cases and dodgy contractor teams entered the system. Such vulnerability, directly and indirectly, influenced the way that the delivery system worked and had some tangible implications, e.g. splitting the gardens at the edge of the residential roads with an abandoned structure. There is no exact measurement to understand the extent of system vulnerability, but indirectly it can be presented by the extent of uncompleted abandoned building structures. While corruption can always be a problem, a deeper analysis identifies there were underlying contributors to taking advantage of the Bam reconstruction programme: (a) issues and inconsistencies organisational configuration of the delivery system as analysis in Chaps. 7 and 8 elaborated; (b) specific and general socio-economic context; and (c) degree of human loss in local organisations and the emotional sympathy of non-local personnel.

From system vulnerability perspective, more number of system elements (participant organisations and individuals in reconstruction programme and process) increases system vulnerability as the number of practical considerations and interactions increase and so does the likelihood of the distortion of the big picture during decision-making (Sull 2007). The Bam reconstruction programme with three strategic objectives and various local and national organisations had a big delivery system with more chances of emergent of system loopholes. The aforementioned problems were resulted from inconsistencies within both initial system formation and evolution triggered or made it possible to take advantage of the system. While there were severe delays during the design phase, prices of construction material rose significantly and there was also high inflation during the programme period. Although the reconstruction fund was increased by 63%, from 9,500,000 Tomans ($9,500 in total for the grant and loan) to 15,500,000 Tomans ($15,500 in total), prices rose constantly. Whereas the reconstruction fund at the beginning was sufficient to complete constructing a house of 80 m^2, at the end of the programme, even the increased fund could only cover the costs of construction up to the stage of erecting the walls, as members of Setads, KEO and local residents recall. Moreover, the tendency of some local consultancies towards over-designing the structure of houses (linked to the conflict between controllers) meant that more money was spent on structural features, therefore, less money was left for the remainders, adding to the problem of money shortfalls.

We saw those inconsistencies discouraged the bigger qualified contractors from engaging in housing reconstruction, which removed a layer of accountability from construction operations and opened doors for dodgy contractors that led a number of applicants losing their reconstruction fund. Each of the above scenarios in return caused more vulnerability so that the applicant re-applied, for the same land or another piece of land, in order to be able to cover the cost of completing the main house. Research observations showed more abandoned structures in disadvantaged neighbourhoods. Taking advantage of the system became easier after the political interventions by the new president because it eased the way people could enter the programme. Within this scheme, tenants could enter the reconstruction process if they could provide land. In small traditional cities like Bam, tenancy agreements are not official deeds; and hand written tenancy agreements are acceptable, both before and after the earthquake. Thus providing fake tenancy agreements was not difficult. This new scheme created more opportunity to enter the reconstruction programme fraudulently, and it also created a tendency of splitting lands in order to meet the second requirement of the scheme.

Local interviewees mentioned that a number of non-locals also entered the reconstruction system, claiming to be disaster-affected Bamis. In order to enter the reconstruction process, and receive a planning application, the applicant's claim to the property right had to be verified through supporting evidence. As the informative brochure on the reconstruction process for locals explained, acceptable evidence were a utility bill, title deed (if there was an official one), previous planning permission (if it existed), and the birth certificate of the householder and the survivor registration card (issued by the Red Crescent during the relief stage). However, the Bam case saw an influx of non-locals to the area immediately after the earthquake. Reports later showed that the Red Crescent issued 130,000 survivors cards, giving them access to emergency relief aid and temporary accommodation. This means that almost 30,000 people, probably impoverished, came to the Bam area and claimed to be Bami in order to benefit from the national and international aid. This is almost the same as the number of people who died in the earthquake. Recognising who was a genuine survivor in that devastating situation was difficult. For example, a sociologist who resided in Bam immediately after the earthquake, later reported that it was only after six months of living in a neighbourhood that she could recognise who was Bami and who was not Bami there. Some of these non-locals could enter the system through corruption and fake documents, which were sold on the black market (Ekhlaspour 2009; HFIR 2005).

On the issue of non-local applicants, it should also be noted that the Bam municipality had lost more than 58 personnel (The Municipality of Bam 2007). Surviving personnel had themselves lost relatives in addition to their colleagues; they were not different from other locals in suffering from post-traumatic grief. As members of the Bam municipality recalled, many of them could not return to work until 6 months after the earthquake. This affected the capacity of the municipality to work accordingly. The municipality (similar to some other local organisations) was supported by temporary non-local personnel deployed from other cities to cope with the extraordinary workload. Non-local support teams did not have local knowledge; and they had a great deal of sympathy with the locals who had lost everything. The humanitarian nature in an aftermath emotionally encourages outsiders to give the benefit of the doubt to the ones who claim to be a disaster victim. Later on, when the surviving local employees returned to work they shared the same experience of living in temporary houses and problems related to the rising costs of construction materials.

The above scenarios were also influenced by a broader social phenomenon, which is a general attitude of mistrust among people towards their governments in Iran. This mistrust is rooted in the history of feudalism in Iranian society that continues in contemporary Iran, linked to the government's reliance on the oil-based economy. This in return contributes to the government's elitism and adds to the mistrust in the society (Ekhlaspour 2009). Given this historical background, the reconstruction fund, provided by the government, was seen by some disaster-affected people as their 'right' because it came from the 'Oil Money'.

9.4 Conclusion

This chapter examined the social dynamics that influenced the way the delivery system of the programme worked and its longer-term effects. The Bam case provided an explicit example of the existence of social learning and knowledge transfer through informal social interactions that are facilitated by a common place or location. This was despite the fact that the programme delivery system of the programme was formed around the technicalities of housing development; and the social dimension of the programme was reduced to formal, similar to the normal urban housing development procedure. Despite the existing and emergent opportunities, the housing reconstruction programme in Bam missed opportunities to benefit from informal social interactions and ignored spaces for social learning which would have enabled the programme to enhance its effectiveness. The overall technocratic approach and lack of attention to the programme's social nature existed during both periods of formation and implementation of its delivery system.

The extent of the uncompleted structures is an approximate indication of the system's vulnerability for the housing reconstruction programme in Bam, a consequence of inconsistencies in the system's formation and implementation. Examining the motivations linked with such vulnerability, it can be concluded that the broader economic and social contexts also contributed, for example, inflation and mistrust between the society and the government in general. Nevertheless, emergent problems during implementation were significantly influential in taking advantage of the system. By the increase of the size of the reconstruction delivery system consequently the number of interactions and the level of vulnerability increase, requiring more attention to system configuration. Taking advantage of the system loopholes in the Bam case became a means to compensate for the consequences of those inconsistencies and shortfalls in the delivery system formation and adjustments during implementation.

References

Curado C (2006) Organisational learning and organisational design. Learn Organ 13(1):25–48

Ekhlaspour R (2009) Bam, review of society and economic conditions of Bam during the last two centuries. Iran History, Tehran. (in Persian)

Fallahi A (2007) Lessons learned from the housing reconstruction following the Bam earthquake in Iran. Aust J Emerg Manage 22:26–35

Gharaati Kopaei, M (2009) Knowledge transfer in post-disaster reconstruction; the problem of post-post-disaster reconstruction. PhD thesis, McGill University, Montreal. http://digitool.library.mcgill.ca/webclient/StreamGate?folder_id=0&dvs=1420020480363~426. Accessed 21 Aug 2014

HFIR (2005) Guidance for reconstruction, residential, retail and public units of Bam, Baravat and surrounding villages, 2nd edn. Housing Foundation Islamic Revolution, Bam. (in Persian)

Morgan G (1989) Creative organization theory: a resource book. Sage, Newbury Park, CA

Morgan G (2006) Images of organization, updated edition (first edition 1997). Sage, Thousand Oaks

Pelling M, High C, Dearing J, Smith D (2008) Shadow spaces for social learning: a relational understanding of adaptive capacity to climate change within organisations. Environ Plann 40(4):867–884. https://doi.org/10.1068/a39148

Quzai U (2010) Pakistan: implementing people-centred reconstruction in urban and rural areas. In: Lyons M, Schilderman T, Boano C (eds) Building back better: delivering people-centred housing reconstruction at scale. Practical Action Pub., Warwickshire, UK, pp 113–134

Roosli R, O'Brien G (2011) Social learning in managing disasters in Malaysia. Disaster Prev Manage Int J 20:386–397. https://doi.org/10.1108/09653561111161716

Sull DN (2007) Closing the gap between strategy and execution. MIT Sloan Manage Rev 48(4). http://sloanreview.mit.edu/the-magazine/articles/2007/summer/48412/closing-the-gap-between-strategy-and-execution/?type=x&reprint=48412. Accessed 19 Jan 2011

The Municipality of Bam (2007) Move again (Report of the Bam Municipality-December 2003 to the End of March 2006). The Municipality of Bam, Kerman. (in Persian)

Wisner B (2004) Chapter 9, Towards a safer environment. At risk: natural hazards, people's vulnerability and disasters. Routledge, London, pp 321–376

Chapter 10
Analysis of the Multi-organisational Nature: The Bam Housing Reconstruction Organisation

Abstract When system elements include organisations, the complexities of their formation and implementation are greater. The sociotechnical delivery system of a reconstruction programme connects system elements, including organisations that in nature, size and internal arrangement might be different, but through the system formation, their activities are (or must be) integrated towards the objectives of the reconstruction. This delivery system is, in fact, a mega system, a concept which has also recently emerged within the literature on construction management. This chapter analyses main considerations and features related to the multi-organisational nature of the sociotechnical system of housing reconstruction programme in the Bam case that influenced the programme performance and its longer-term effects. The research makes a case that auto-adaptation within a multi-organisational delivery system of the reconstruction programme is necessary but not sufficient. The case exemplifies the need for balance between auto-adaptation and overall strategic supervision of the programme as a whole. The workflow as an attribute of organisational configuration had a multiple importance as both the production chain and simultaneously the standardisation force of the coordinating mechanism. Dynamics of power balance between organisations and their internal motivations were influential. The engagement of influential local and national organisations which had role in housing development in normal situations in the programme contributed to sustaining improvements after the reconstruction period, transferring lessons from reconstruction to the normal housing development situation.

10.1 Introduction

A sociotechnical system brings together a number of elements that interact with each other for the greater purpose of the system. These system elements can be individuals, organisations or a mixture of both. The complexities of the formation and implementation of inter-organisational sociotechnical systems are greater than the systems themselves, which bring individuals together (Morgan 2006). Previous literature review showed that reconstruction programmes bring together

© Springer International Publishing AG, part of Springer Nature 2018

F. F. Arefian, *Organising Post-Disaster Reconstruction Processes*,
The Urban Book Series, https://doi.org/10.1007/978-3-319-70911-6_10

organisations for the delivery of the programme. The delivery system is, in fact, a mega system, a concept which has also recently emerged within the literature on construction management that sees emerging megaprojects as construction programmes (Davies and Mackenzie 2013). This (mega) delivery system connects these organisations that in nature, size and internal arrangement might be different, but through the system formation their activities are (or must be) integrated towards the purpose of the system. Similarly, the case of Bam, as shown in Chap. 5, brought together a number of organisations that were either at the higher level of the system formation or at the operational level of the implementation, and which had to work with each other.

10.2 Coordination Mechanism Among the Delivery System Elements

The coordination mechanism in the Bam case was mostly based on the standardisation of the workflow by creating operational linkages between the system elements, and coordinating meetings. The workflow greatly influenced the way the programme performed and the need for adjustments, and the decision making towards undertaking various approaches. Scholars in disaster studies (e.g. Barakat 2003; Maskrey 1994; Christopolos cited in Pelling 2003) emphasise the need for better coordinated activities. Acknowledging the above statement, the Bam case illuminates the links between the coordination mechanism, workflow and organisational configuration for reconstruction programme delivery systems. While within a simple organisation the workflow defines the production chain, within the multi-organisational sociotechnical system, the workflow also connects various organisations together, and it is a manifestation of how various organisations are coordinated toward the delivery of the programme. As in the Bam case, it became the integration mechanism that created unity among the system, shaping the acts undertaken by each organisation and the outcomes of each organisation's activities act as the input for another organisation. This re-emphasises the importance of perceiving workflow process as an organisational attribute and examining it with regards to other attributes within organisational configuration and strategic nature as discussed in Chaps. 7 and 8.

Organisation theory also suggests that achieving mutual adjustment among organisations might result in a coordination mechanism. Some scholars such as Comfort and Kapucu (2006), in their discussions on the sociotechnical systems that are formed by various organisations to act in response to a disaster in extreme situations, suggest that an auto-adaptation model is a way of achieving the sought-after mutual adjustment among participant organisations. They see such a system of interacting units as each performing at its own rate, but adjusting performance to that of its near-neighbours in the system in response to incoming information from the environment. Thus, such a system will become a system of continued learning, which fosters initiatives and actions at all levels through mutual adjustment.

However, the Bam case exemplifies that auto-adaptation is necessary, but not sufficient. For example, there was an in-built flexibility on auto-adaptation for organisational capacity in each participant organisation as described in Chap. 5. This was in response to the unknown trends of the owner-driven workload, but it was not helpful for improving the effectiveness of the coordinated activities at the design phase. It was expected that after granting self-adaptive flexibility to organisations, setting objectives and forming the workflow/working-process, those organisations would automatically adapt themselves to the needs of the programme during implementation based on the received information from the environment. This did not happen. There were (and always are) other influential factors that impacted upon the organisations' decisions on adapting themselves towards the needs of the programme, for example, internal motivations, the broader professional context and their disciplinary narrow-mindedness.

Pure reliance on auto-adaptation and mutual agreement is an overly theoretical ambition with a minimal chance of success in the reality of post-disaster reconstruction in practice. This ambition assumes that organisations always know what is good for the programme as a whole and would therefore self-adapt accordingly. Internal considerations, arrangements and approaches are important when considering adaptive flexibility and auto-adaptation in each organisation as criteria for effective inter-organisational efforts in a post-disaster situation. Moreover, reconstruction as the strategic agent for recovery pursues strategic objectives. Therefore, it is not only that the adaptive flexibility should be aligned with what is needed for each task and against the near-neighbour element, but it should also be aligned with the overall strategic direction and aimed towards considering the big picture of the reconstruction. The case clearly showed that despite having autonomy for flexibility in order to meet the programme needs during implementation; some internal changes were not necessarily in favour of the strategic objectives. Auto-adaptation carries risks of fragmentation of disciplines, narrow-mindedness and hindering the bigger picture (which is the recovery strategy), and such risks are also linked with the unit grouping for organisational configuration.

I suggest that the notion of auto-adaptation for multi-organisational sociotechnical systems working in extreme situations must be revisited and supported by careful examination of organisational configuration and application of strategic management. Even if the space for auto-adaptation exists in a sociotechnical delivery system, it is almost utopian to assume that all these organisations will auto-adapt themselves in the best possible way when needed. Such optimism hinders the importance of organisational configuration that positions all these different organisations in relation to each other and integrates them through the working-process/workflow. The greater number of elements makes reaching harmonised auto-adaptation more difficult, and an increase in the number of participant organisations increases the likelihood of tensions and conflicts between different directions of the auto-adaptations, which can be linked to power dynamics and internal motivations between or within system elements. Exactly for this reason, it is important that the organisational configuration of the delivery system minimises the risk of operational conflicts. The programme also requires strategic umbrella supervision in order to safeguard the strategic nature and the effectiveness of the overall multi-organisational activities within the coordinated mechanism, and see the bigger picture. This supervisor can be a person, a multidisciplinary committee or an umbrella organisation. Regardless of the form taken, the core principle is to apply an integrated understanding of what is the bigger picture for the benefit of the programme. The Bam case lacked such integrated strategic supervision and suffered from fragmented disciplinary narrow-mindedness (Arefian 2016). There is a need to balance between flexibility for internal auto-adaptation in each organisation, and the ability of the strategic superviser to request organisations to adapt themselves for the good of the overall programme performance during implementation. This is unlike waiting for organisations to adjust themselves whenever they wish.

In the Bam case, coordinating meetings at various levels were held in Bam and Tehran, for example, meetings with the BAUC and each organisation or with a group of organisations working with each other. They were to support mutual adjustment and coordination as members of the BAUC, Setads and municipalities explained. While formal meetings are better than nothing for sharing experiences, raising issues and finding solutions, they are not enough for achieving mutual adjustments. Such formal meetings cannot and do not replace the informal mutual adjustments that can happen when a socio-operational space for informal multidisciplinary and multi-organisational interactions exists. For example, at the construction phase, a greater number of system elements managed to collaborate and follow an inexperienced workflow (similar to the design phase), while tougher control mechanism was in place (as compared to the construction phase for housing development practice). But the related workflow did not encounter similar problems to the one for the design phase, and the way of doing things did not change as it did in the design phase. This can be linked to the multi-organisational, multidisciplinary working units that ultimately formed after KEO appointed supervision teams to each unit group zones, working with their allocated Setads. This multi-organisational unit grouping facilitated informal multidisciplinary interactions that helped recognise other perspectives towards mutual adjustments. The design phase lacked such facilitation, as a result of inconsistency in organisational configuration for the control mechanism and the avoidance of auto-adaptation by the controllers, as was explored before.

10.3 Dynamics of Power Balance Between Participant Organisations

Perceiving systems as social entities naturally entails recognition of dynamisms at both individual and organisational levels that might influence how a system is formed or performs. As Morgan (2006) reviews, the importance of power balance in the functionality of a sociotechnical system is recognised in organisation theory. The Bam case also saw dynamics of power balance at both a high-level and operational level that influenced the way the programme performed.

First is the power balance between HFIR and MHUD at the high-level programme formation, starting somehow with an equal power balance, leaning towards HFIR as the programme was put into practice and finally minimising the role of MHUD towards the end of the programme. The influence of such a dynamic in power balance further reduced the importance of the strategic objective on urban architectural identity that was reflected in the diminishing the role of BAUC at the organisational level. It contributed to the BAUC, as the responsible body over the qualitative architecture and urbanism, not influencing the programme performance from the midway point. This was related to the political administrative change in the country. At the stages of setting the core principles for the reconstruction, outlining the role, and forming the sociotechnical delivery system for the programme, the power balance between MHUD and HFIR veered more towards equality. The unique characteristics of the Bam earthquake led to high-profile negotiations between concerned organisations. The Strategic Steering Committee quickly created high-profile collaboration between HFIR and MHUD. The role of MHUD for overall reconstruction was to engage in policy making, and directing the overall project control. Concurrently, the role of HFIR was to be the executive body of the overall reconstruction scheme, including housing reconstruction. By analysing the initial documents we saw that it was even difficult to identify what strategic objective or what consideration was first suggested by whom. The collective inputs were gathered in the reconstruction programme proposed by the HFIR, for example, it was MHUD's input to establish the BAUC for supporting HFIR over the architectural and urbanism aspects of the Bam reconstruction, and inviting and commissioning KEO for the control mechanism.

However, after the programme was put into practice, when practical issues emerged, the balance of power leaned towards HFIR and its more pragmatic approach towards speeding up the reconstruction operation. This imbalance sharply deepened after the political change in the country and administrative change at MHUD that saw change at the middle management level who were in charge of organising and heading the BAUC meetings but did not share the same passion of their for change of the planning system. Since such administrative change did not happen at the ORA, its pragmatic view (mixed with architectural flavour) took over and remained until the end. For example, e.g. when the new promise of reconstruction funds for tenants and newlywed couples was made, its related practical considerations, e.g. the ability to provide a piece of land, and their possible influences on the urban fabric and architectural identity, were not discussed in the BAUC. Additionally, it must be noted that HFIR was the executive body for the operation and Setads were its provincial branches. Setads were responsible for the housing reconstruction programme in their allocated zones. So HFIR had a strong operational power in the field, therefore, even at the early stages of programme formation, operational power of HFIR was predictable. The operational power of MHUD, however, was through the BAUC and Mother Consultancy and to some extent the local consultancies, were all only engaged at the design phase, not in the whole housing reconstruction workflow.

Second was the power balance at the operational level between KEO and Mother Consultancy. It led the KEO's conservative approach on the design phase to overtake some of the architectural design guidance and imposed cube-shaped houses, a tendency to over design structural elements that in turn contributed to the shortage of funds for completing the construction operation that in turn potentially contributed to the number of uncompleted structures. The fact that such conflict was not solved meant that the conservative approach of KEO—a sign of its overtaking power—stayed in place. Such a dynamic is related to influential situational factors, for example, the emotional context after the destructive earthquake in Bam, the blame game towards KEO following the extensive collapse of newly built houses during the earthquake.

The above examples elaborate unwritten and implicit dynamisms of the power balance among participant organisations at both the formation and implementation of reconstruction delivery systems. Given the sociotechnical account of organisational systems, it appears that a dynamic power balance is inevitable in post-disaster reconstruction programmes. However, since organising reconstruction programmes already happens in a complex and dynamic context, attempts are needed to reduce the complexities of such power shift as much as possible. This, as the research suggests, occurs through bringing stability within participant organisations at higher levels of decision making. This might not eradicate the power balance dynamism at the system formation stage, but it reduces the likelihood of power shift during implementation. This, in turn, contributes to a consistent approach and way of doing things during system formation and implementation, nurtured by organisational continuity and opportunities for learning.

The Bam case highlights the importance of stability in key decision makers in organisations that are collaborating for the programme formulation and the formation of its sociotechnical delivery system. The context surrounding each project or programme also has implications for its managerial style and organisational culture (Davies and Mackenzie 2013). Changing key people in strategic collaborations is a shock for the whole delivery system that was formed under a certain condition. For example, ministerial administration change in MHUD changed the pattern of the influential key people in the programme and their priorities. This damage, however, was to some extent limited because the other part of the strategic collaboration, ORA at HFIR, did not change. If ORA, similarly to MHUD, had experienced a major managerial change, the reconstruction programme as a whole, including the housing reconstruction programme, could have encountered further damages. Thus, awareness of the inherited existence of such dynamism within a reconstruction programme is crucial. Avoiding radical changes in participant organisations and awareness of internal motivations potentially is needed. Radical changes add more instability to an already turbulent post-disaster context, which could potentially negatively affect the programme implementation.

10.4 Internal Motivations and Arrangements Within Participant Organisations

Reconstruction programmes, similar to any other project or programme, are defined and formulated in a certain moment in all participant organisations which are part the context. As time goes by, each of these organisations continues to experience their own internal dynamics. As Davies and Mackenzie (2013) highlight, complex construction projects and programmes, regardless of where they are conducted (e.g. London Olympic 2012 construction programme) are inherently uncertain because of the inability to predict how the system components will interact when joined together. Internal sources of such uncertainty deal with the novelty of the project specifications and temporal urgency of the project. External sources of uncertainty refer to the social, political and variety stakeholders which together form the context within which a complex programme is embedded. While the importance of interdependencies within an organisational system and the programme/project context has long been recognised, a radical change in that context adds to the turbulence and increases the uncertainty within the sociotechnical system of the project or programme (Milleti and Gillespie 1976).

The Bam case provides an interesting example of how internal motivations and considerations influence the programme performance and its adjustments during implementation. An example of this is the participation of the KEO that was suggested by the BAUC as a sign of the reconfirmation of the role of the Country's Engineering Organisation (CEO), as it was described in the National Engineering Organisation Law after the Manjil earthquake. Similar to a number of international cases the Bam case also saw newly built houses, which did not comply with the existing seismic regulations, collapse and cause great fatalities. In many international cases, as Twigg and Benfield Greig (2006) point out, building codes were undermined by a combination of careless building practices and lax enforcement of building regulations, like in the Bam case. Taking the blame over the poor quality of those new houses, which were built after the introduction of the Law, KEO became the controller of structural aspects of building operation in the Bam case. Thus, it is no surprise that KEO applied a conservative approach for doing its allocated tasks as explored before. In the design phase, KEO changed the status of a number of secondary seismic codes from optional to compulsory; and in the construction phase, KEO increased the number of key controlling points from five to nine. The previous role of the Engineering Organisation, as the earthquake revealed, was not effective to reassure the application of the compulsory seismic building codes during the construction operation. Hence, having a key role in reconstruction, therefore, was an opportunity for KEO to redeem its reputation and establish its role within the housing development practice.

The above is an example of how internal dynamics in an organisation affects the sociotechnical system formation and implementation and therefore affects the way the programme works. Internal motivation in an organisation as a system element is also linked with individuals in the organisation. Internal motivation in MHUD changed after influential decision makers were replaced, and so did the their overall understanding of the importance of architectural features and urban identity.

Thus, the reconstruction programme itself becomes vulnerable to extreme internal dynamics in high-level participant organisations if it is linked to political positions. The reason is that the political scene in each country has a certain timetable for elections, but disasters can happen at any time, irrelevant to the political timetable. At the programme level, therefore, chances are that a programme's sociotechnical delivery system is formed in a certain political context, but is implemented in a completely different political mindset. The impacts of such internal dynamics can be in favour of or against the strategic objectives of the reconstruction programme or the recovery in general. It damages the continuity of the importance of strategic objectives, strategic approaches towards them, and their practical considerations at the operational level. Such uncertainties, raised from the exposure of a reconstruction programme to internal dynamics of participating organisations and the broader sociopolitical context, affect the programme more if representatives of those organisations in the programme are unstable. If such representation in a reconstruction programme is directly linked with a political administrative position in an organisation, any radical change in that organisation potentially contributes to a reconstruction programme encountering changes. Obviously, the impacts of such exposure are more influential for higher level organisations, e.g. MHUD, HFIR and the BAUC, than operational ones, e.g. Setads and consultancies. Successors of the previous representatives may not share the same mindset that their predecessors had during the programme formulation, as it was in the MHUD. A similar (to some extent) international case is the case of Japan after the Tohoku earthquake and tsunami in March 2011, with the Democratic Party of Japan (DPJ) in power after an initial advisory group for recovery, called the Reconstruction Design Council, was formed. It saw reconstruction as an opportunity for creating more sustainable communities, able to deal with the pressing issues of demographic and economic challenges. Following a political change two years later, the Liberal Democratic Party, a follow up body for the Council, stressed the need for resilient social infrastructure able to mitigate future disaster and self-reliant local societies. These objectives certainly indicate different priorities, even though they both might be worthwhile (Dimmer 2014).

The above issue is potentially also the case for other development and construction programmes, for example, the London Olympic 2012 construction programme started during a time when the Labour Party was in power and was completed while the Conservative Party had taken over the government. However, it must be noted that the social, political, psychological and procurement context of a post-disaster reconstruction programme is already far more complex than other development or construction programmes. Those development programmes and mega projects are planned and implemented in normal situations as part of the development pathway (local, regional or national), which offers a degree of predictability for the sociopolitical context. A disaster, as an interruption of the development pathway in a certain context, already causes an extreme level of uncertainty (e.g. Baas et al. 2008; Wisner 2004; Masurier et al. 2006) for which attempts must be made to reduce complexities and accommodate uncertainties.

This re-emphasises the importance of stability of key decision makers in post-disaster reconstruction programmes. Instability increases uncertainties and internal motivations during implementation and may lead to power shift within the delivery system. Even if there are shortfalls at the programme formulation stages, the suggested stability brings continuity that nourishes organisational learning within the delivery system in and between organisations that is helpful for adaptive correction of the programme. Each system component is an organisation with its own internal arrangements and dynamics that are part of the programme context. Stability of representatives of key decision making organisations in a reconstruction programme is a prominent factor for implementing the programme based on the initially formulated arrangements and mindset. It is associated with organisational continuity and since political positions are usually temporary, it will be helpful to separate reconstruction positions from political roles. This, however, might be difficult to achieve in practice.

10.5 Engaging Main Related National and Local Organisations

The Bam case exemplifies how the engagement of organisations that are involved or influential on normal housing development procedure in reconstruction might positively convey the momentum for improvement to after reconstruction period and indirectly influence the normal housing practice at local and national levels. As we saw the formation of the sociotechnical delivery system of the housing reconstruction programme in Bam was indeed an effort to improve the previous existing practice of housing construction procedure, within the existing understanding of the shortfalls and processes. The reconstruction programme in Bam triggered longer term influences, which are linked to engaging related influential national and local organisations, which play a role in either setting regulations or being in charge of their application within the programme. The case presents two examples of this:

First is the engagement of KEO in general, the provincial branch of CEO, which is a qualifying body for engineers, in charge of the application of the national building codes and holding engineers accountable. It helped to reposition KEO within the normal housing development practice and mainstream a tougher approach towards the construction quality in Bam and Baravat. Although KEO's participation was biased towards structural discipline, and was narrow-minded and conflicted with BAUC's requests and Mother Consultancy's controlling criteria, the KEO's engagement was valuable for future improvements. Inconsistencies in organisational configuration, insufficient attention to the strategic nature of the programme, and examining the interdependencies, as well as internal motivations, were among the contributing factors to the emerged issues and problems, not the KEO's engagement per se.

Second is the engagement of BHRC to control the construction materials at their departure point to the field. BHRC, as the research wing for introducing construction methods and national building codes, was not directly part of the process of delivering houses, but it was a system component that had to prevent non-standard materials from entering the Bam market through the examination of construction materials. It provided a unique opportunity for BHRC to be engaged with a large scale practice and reflect upon lessons on the priorities for the organisation agenda. As it was later documented, BHRC conducted 675 visits in total and sent 54 reports back to the Strategic Steering Committee, HFIR, MHUD and other organisations. Before the reconstruction, only two construction materials (concrete and bars) had to be standard. As the president of BHRC recalls, reconstruction of Bam triggered BHRC to propose to extend this requirement to all construction materials, pursue the proposal, and negotiate it with MHUD, the government and the Institute of Standards and Industrial Research up until its approval (Saemian and Erfanian Daneshvar 2011).

10.6 Conclusion

This chapter analysed main features of the multi-organisational nature of the sociotechnical delivery system of the housing reconstruction programme in the Bam case that influenced the programme performance and its longer-term effects which were not explored in other chapters. Multiple perspectives re-emphasised the importance of workflow as both the production chain and simultaneously as the standardisation force of the coordinating mechanism. The workflow is an attribute of organisational configurations; therefore, it puts more emphasis on the discussion of the organisational configuration and consistencies or inconsistencies.

The analysis makes a case for some research suggestions: First, auto-adaptation is necessary but not sufficient; there is a need for balance between auto-adaptation and overall strategic supervision of the programme as a whole. Auto-adaptation is influenced by power balance, internal motivations and dynamics in participant organisations, especially the ones that perform at higher levels of decision making. Changes in internal dynamics of those organisations and the power balance between participant organisations may or may not be in favour of one or more of the strategic objectives. Second, organisational continuity and personnel stability are important.

Thus, the separation of representative-operational roles in post-disaster reconstruction from the political position-based roles is helpful for reducing radical changes and unforeseen problems. As political positions are temporary and the time frame for change in the political scene of a country usually does not match the time frame for reconstruction but they affect internal motivations and power balance. Third, engaging influential organisations (on housing development in normal situations) as participant organisations in the housing reconstruction programme proved to have positive longer-term influences and was potentially helpful for sustaining improvements and conveying the lessons learned in reconstruction to the everyday situation. This is crucial for linking reconstruction to urban development activities afterwards.

References

Arefian FF (2016) Getting ready for urban reconstruction: organising housing reconstruction in Bam. In: Arefian FF, Moeini SHI (eds) Urban change in Iran: stories of rooted histories and ever-accelerating developments. The urban book series. Springer International Publishing, Cham, Heidelberg, New York, Dordrecht, London, pp 231–257

Baas S, Ramasami S, Dey De Pryck J, Battista F (2008) Disaster risk management systems analysis: a guide book. Volume 13 of Environment and Natural Resources Management. Food and Agriculture Organization of the United Nations, Rome

Barakat S (2003) Housing reconstruction after conflict and disaster. Network Paper, ODI, London. http://www.odihpn.org/documents%5Cnetworkpaper043.pdf. Accessed 23 Aug 2009

Comfort LK, Kapucu N (2006) Inter-organizational coordination in extreme events: The World Trade Center attacks, 11 Sept 2001. Nat Hazards 39(2):309–327. https://doi.org/10.1007/s11069-006-0030-x

Davies A, Mackenzie I (2013) Project complexity and systems integration: constructing the London 2012 Olympics and Paralympics games. Int J Project Manage 32(5):773–790. https://doi.org/10.1016/j.ijproman.2013.10.004

Dimmer C (2014) Evolving place governance innovations and pluralising reconstruction practices in post-disaster Japan. Plann Theor Pract 15(2):260–265

Le Masurier J, Suzanne W, Shestakova Y (2006) An analysis of alliancing procurement method for reconstruction following an earthquake. In: Proceedings of the 8th national conference on earthquake engineering. Presented at the 8th national conference on earthquake engineering, San Francisco, CA, USA, paper 290. http://nisee.berkeley.edu/documents/elib/www/documents/200601/8NCEE/8NCEE-000290.pdf. Accessed 8 Dec 2014

Maskrey A (1994) Disaster mitigation as a crisis of paradigms: reconstructing after the alto mayo earthquake, per. In: Varley A, Wiley J (eds) Disasters, development and environment, Chichester, New York

Milleti DS, Gillespie DF (1976) An integrated formalization of organization-environment interdependencies. Hum Relat 29(1):85–100. https://doi.org/10.1177/001872677602900106

Morgan G (2006) Images of organization, updated edition (first edition 1997). Sage, Thousand Oaks

Pelling M (2003) Chapter 2, Cities as sites of disaster. In: Pelling M (ed) The vulnerability of cities: natural disasters and social resilience. Earthscan Publications, London, pp 19–45

Saemian S, Erfanian Daneshvar A (2011) The process of technical control and supervision in reconstruction of bam. (Unpublished report). (in Persian)

Twigg J, Benfield Greig Hazard Research Centre (2006) Technology, post-disaster housing reconstruction and livelihood security—Practical Action Consulting (Website) Practical Action. http://practicalactionconsulting.org/t4sl_disasterapproaches. Accessed 2 Sept 2011

Wisner B (2004) Chapter 9, Towards a safer environment. In: At Risk: natural hazards, people's vulnerability and disasters. Routledge, London, pp 321–376

Part IV
Concluding Discussion

Chapter 11
Concluding Discussions and Propositions

Abstract This chapter outlines the essence of this research as a coherent whole. It discusses the generalisability of the research, through the strategic choice of the case, where the researcher has the opportunity to access a deeper level understanding of a complex phenomenon and interactions. The Bam case was an attempt to 'urbanise' the previous reconstruction experience in Iran, which had been based on large scale rural reconstruction experiences. The research advances the discourse on organising reconstruction, the organisational configuration and management of post-disaster reconstruction activities, focusing on housing. The research connects to knowledge areas of disaster-development studies in urban contexts, strategic management and contingency view in organisation theory and construction management, assisting to achieve better reconstruction programmes in other national and international cases. Taking the discourse forward in-depth learning opportunities are identified and linked back to theoretical discussions. Through the iterative process, the research proposes a graphically presented analytical framework, which integrates multiple overlapping perspectives for understanding organisation design and management of a post-disaster reconstruction programme, which have developmental objectives in an urban context, offering overlapping theoretical insights on this subject. Building on the learning opportunities the proposed analytical model is supported by complementing propositions.

11.1 What Was It All About?

This research addressed the existing imbalance in disaster literature in understanding the reality of organising reconstruction activities, and to contribute to bridging the theoretical gap on organising reconstruction programmes and processes. Moving towards the above aim, it explored how organisational design and management influenced the approach to and the achievement of the reconstruction objectives in Bam. The case offered instrumental values as well as, logically,

© Springer International Publishing AG, part of Springer Nature 2018

F. F. Arefian, *Organising Post-Disaster Reconstruction Processes*,
The Urban Book Series, https://doi.org/10.1007/978-3-319-70911-6_11

intrinsic ones. The characteristics of the Bam case and my position, as a participant-observer at the real time phenomena and a professional insider of the broader context, characterised the Bam case as a strategic choice for this research. As Flyvbjerg (2006, p. 226) states, in the case of a single case study, 'strategic choice of a case greatly adds to the generalisability of a case study', through rich descriptions and the ability to focus on the depth of a single case study where the researcher has the opportunity to access a deeper level understanding of a complex phenomenon and interactions (e.g. Eisenhardt 1989; Bryman 2008; Yin 2003, 2009).

Since the notion of organisational design in reconstruction is an emerging discourse, the research stood as an exploratory case study research, for which the case became a laboratory for novel investigation, offering learning opportunities for other international cases, as well as those inside Iran. The research synthesised the broader interdisciplinary literature on disaster-development and organisation theory to propose a tentative conceptual model, integrating overlapping insights to the case, which in turn verified, polished and took the study to a deeper level of exploration and understanding. The research examined main organisational and managerial aspects that influenced the way the system, through its certain initial formation and multiple adjustments facilitated the programme to approach its strategic objectives; from multiple overlapping insights of the synthesised proposed model, those learning opportunities emerged. The research propositions build on those emerged learning opportunities and from examining them analytically within the broader multidisciplinary literature.

The core principles of the Bam reconstruction that were employed directly by the housing reconstruction programme as the strategic objectives. Those were as below:

- Safeguarding historical urban architectural identity
- Building earthquake resistant buildings
- Beneficiaries to be mobilised and encouraged to participate

The research examined how the introduction of these core reconstruction principles was related to the broader existing urban planning system in the country that was requested to respect the historical urban identities and qualitative approach; the existing housing development procedures and its shortfalls for ensuring the application of the seismic regulations during the construction; and the previous post disaster reconstructions experiences that saw people's participation as a win-win situation for people and the government. Such overlapping, underlying broader issues met each other in the context of the Bam reconstruction. The core principles of the Bam reconstruction as a whole were directly translated as the strategic objectives for the housing reconstruction. Extensive previous experience in both post-disaster and post-war reconstruction, in addition to the organisational continuity in dealing with reconstruction activities, enabled Iran to apply lessons learned from previous reconstructions to the new cases. Bringing all lessons in one massive experience in Manjil led to the establishment of broad policies and know-how on dealing with reconstruction. However, the Bam case—given its unprecedented characteristics—was different. It was the first large-scale earthquake, after the Manjil one that had its epicentre underneath a relatively dense urban area, which was famous for its historical urban identity in the form of the garden city urban fabric and unique architectural styles. It destroyed 85% of the built environment in the city and its surrounding area.

The housing reconstruction programme in Bam therefore had to facilitate achieving these objectives for the scale of an initially estimated 20,000 to 25,000 houses in an urban area. The programme plan directly addressed each of these objectives and also specified the broader financial and administrative established policies in the Bam case. The programme saw the housing reconstruction as a portfolio of 25,000 individual reconstruction cases. In order to organise the reconstruction, an innovative sociotechnical multi-organisational system for delivery of the housing reconstruction programme was formed.

The practical considerations of the following aspects provided the basis for forming the delivery system: dividing the area to districts and creating district level administration; inviting and selecting construction consortiums and contractors; reaching a broad agreement with contractors; contracting between contractors and people; creating the BAUC for providing, directing and supervising design codes; organising local consultancies in Bam; establishment of a one-stop-shop compound (construction Bazaar) for provision of technical-professional services and exhibition of construction materials and techniques; supervising the construction of houses by HFIR, Engineering Organisation and BHRC; and financial assistance.

The programme delivery system was implemented in two main phases, design and construction, after entering the reconstruction programme. The design phase led to the issuance of the planning permission for starting the construction operations, similar to the housing development procedure found in a normal situation. These phases embedded practical considerations for approaching the strategic objectives on safeguarding historical urban identity and building earthquake resistant houses. Bam was divided into zones; Setads and consultancies were appointed to each zone. The order of activities for each phase was defined as a standard workflow that included control mechanisms by the Mother Consultancy and KEO in order to assure the considerations for approaching the strategic objectives were met.

Organisations as socio-technical systems are not still, as they evolve (formally or informally) over the time, and the same applies to reconstruction programmes as temporary organisations. The research therefore explored how the delivery system evolved through its adjustments, and examined the potential contributors to the emergent needs of adjustments, and how the results of such adjustments acted as intermediary contributors to the system and facilitated approaching the strategic objectives. It showed that the initial system had to go through adjustments in order to deal with the unfolding problems and emerging issues that were related to the inconsistencies within the initial system formation, i.e. unit grouping, accountability, decentralisation, workflow and the purposeful approach, rising prices, insufficiency of initial funds at the middle of the programme, and reflecting new political promises that allowed new entrants to the reconstruction programme.

11.2 The Analytical Framework and Learning Opportunities from the Bam Case

In my view, it can be concluded that the Bam case was an attempt to '*urbanise*' the previous reconstruction experience in Iran, which had been based on large scale rural reconstruction experiences. Simultaneously, it was an attempt at incorporating improvements in housing development procedure, which was reflected in the introduction of the strategic objectives. Learning opportunities from this attempt built on analytical discussions on both achievements (e.g. on the technical issues for building earthquake resistant buildings and also public knowledge on this) and underachievement. Although the objective of safeguarding the historical urban identity cannot be labelled as an absolute failure of the housing reconstruction programme, compared to initial expectations, it shows a clear setback that can be interpreted as underachievement. The Bam case also had longer-term influences. At the local level this influence took the form of sustaining a control mechanism over the seismic structural regulations, even after the reconstruction period. At a broader level, the influence took the form of attempts for institutional development aimed at updating and suggesting operational frameworks for the application of the national building codes and standardisation of the construction industry in a normal situation.

Building on previous analytical pathway presented in Chap. 1 and in-depth analyses of the Bam case, Fig. 11.1 presents a more detailed multi-perspective conceptual framework for understanding reconstruction organisations. The framework synthesises the learning opportunities for organisational design and management of reconstruction programmes requires for organising processes, especially the housing reconstruction that have emerged from this research. The following discussion is arranged around explaining the framework, which in turn addresses these learning opportunities.

11.2.1 Innovative, yet Consistent Organisational Configuration: Influential Attributes

Organising reconstruction has been concluded to be an organisational design problem (Davidson 2009). Despite this acknowledgement there is no theoretical model as a point of reference for understanding 'how' to design the delivery system a reconstruction programme, especially for the housing recnstruction. From the organisational design perspective, the research conceptual model suggests considering reconstruction programmes as temporary organisations that theoretically must be able to work in a complex and highly dynamic environment. Attempts to apply improvements in the previous status quo require innovative organisational formation and organisational configuration. However, based on a contingency view, which is the manifestation of systems theory in organisational studies, the best

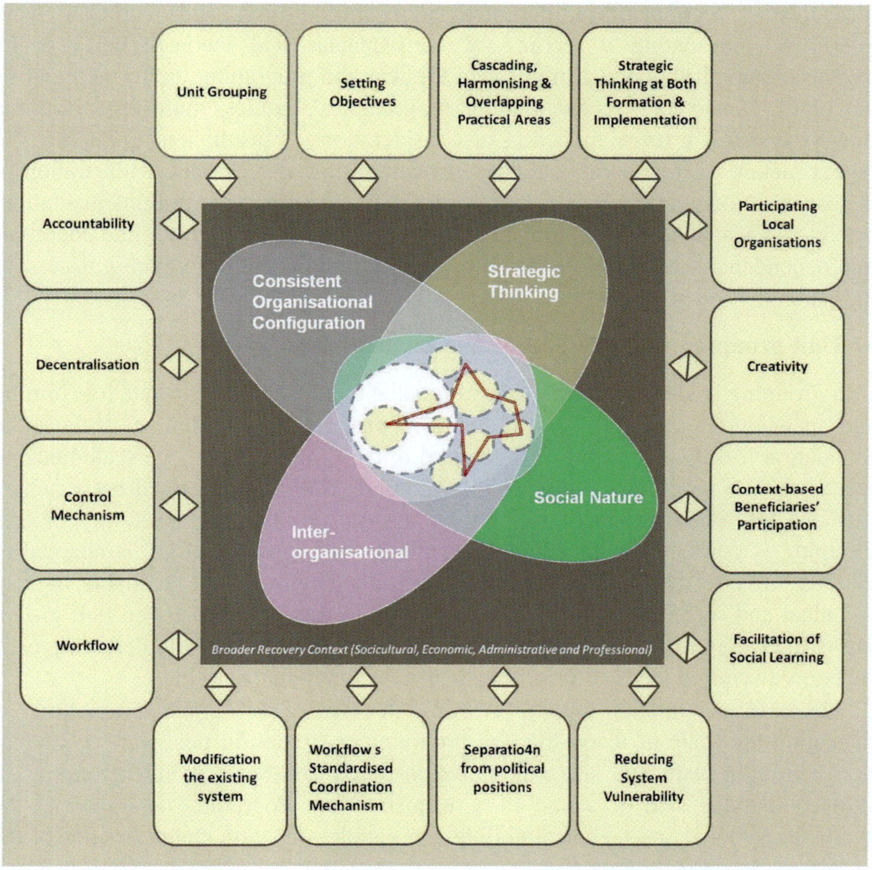

Fig. 11.1 The research learning opportunities related to the proposed tentative conceptual model for organisation design and management for post-disaster reconstruction programmes

organisation design is the contingent one. In other words, effective configuration and consistency during the design of the organisational configuration for the system formation is crucial. The case study verified the Bam housing programme delivery system as an innovative organisation, as suggested by my synthesis in the literature review. The Bam case brought together established practical considerations of previous reconstructions and demands for improvements in the form of strategic objectives, within a complex situation of a large scale devastating destruction in an historic urban fabric.

I examined the case on deeper level. The research identified and examined organisational attributes in the delivery system of the housing reconstruction programme in Bam, and determined that their consistencies/inconsistencies influenced the way in which the programme performed, its outcomes and/or longer-term impacts. Those organisational attributes, namely, unit grouping, control mechanisms, accountability, decentralisation mechanisms, and workflow were interlinked, presenting a mixture of consistencies and inconsistencies in its organisational attributes that influenced the way the programme performed and its outcomes. Since the reconstruction programmes require multi-organisational (mega) systems, a mere collection of considerations is insufficient to capture the bigger picture of the operational complexities in practice. Graphical illustration of the organisational configuration and workflow in advance, prior to implementation, provides a visual analytical tool to understand the bigger picture, and identifies interdependencies between the participant organisations and introducing necessary measures if needed.

- **Unit grouping as a key organisational attribute**

Unit grouping is identified as a major influential organisational attribute for forming and managing a delivery system for a reconstruction programme. It lays the foundation for organisational configuration of the programme. Large scale housing reconstruction programmes require unit grouping on an end-focused basis, which includes the whole production chain for the operational unit. This way of unit grouping is consistent with vertical decentralisation, and therefore, is adapted to various sizes of reconstruction programmes, depending on the available human, technical and organisational capacity. Contrary to this, function-based unit grouping is likely to lead towards competing priorities, as each unit's undertaking each function/discipline tends to prioritise their own function/discipline.

The case examination tapped on Iran's decades of developing know-how on managing the scale of reconstruction programmes (rooted in post-war reconstruction) through customer, size, and location-based end-focused unit grouping. It showed that the size of a unit in a reconstruction programme is linked to the available personal and technical abilities; in parallel, location characteristics of the reconstruction area (compact or dispersed) are influential factors for operational unit grouping. The combination of those qualities interactively influences the need for unit grouping in organising post-disaster reconstruction programmes. Logically, larger scale programmes require unit grouping to divide the workload and to create smaller-size operational units. Such location characteristics also influence the basis for unit grouping. For example, where the overall housing reconstruction programme covers dispersed locations, e.g. villages in rural areas, it naturally indicates geographic-based end-focused unit grouping. For a large scale reconstruction in a compact area (dense urban area), although both end-focused and function-focused unit grouping might be possible, the research suggests that, again, geographic-based end-focused unit grouping inherently allows for greater potential for facilitating mutual adjustments and understanding within system elements. Function-based unit grouping in reconstruction programmes has a tendency towards narrow-mindedness. Therefore, introducing greater numbers of such function-based unit groupings requires more attempts to create harmony, facilitating shared understanding and an integrated approach between units in critical points. Function-based unit grouping for organising reconstruction programmes inherits the danger of focusing more on specialised functions, as a means, rather than an end of the programme, therefore it may even add problems to the implementation of the programme whilst the traumatic situation of the post-disaster reconstruction is also concerned with other social and financial situations.

However, although the end-focused unit grouping might present itself as advantageous for organisational configuration, it is important that this approach includes all system elements of the workflow in the production chain. Unit grouping as a major organisational attribute is linked with other attributions and considerations, e.g. job description, internal arrangements and control mechanisms, and so on. Harmonising other attributes and practical considerations towards the purpose of the programme for each attribute, and between them, as well as providing a clear understanding of what is the common good for the sake of the programme, rather than for the sake of the individual organisations, is crucial.

- **Integrated approach to control mechanisms**

The Bam case showed divergent approaches in control mechanisms as one of the parameters to be configured (Mintzberg et al. 2003). The control mechanism was directly built around objectives and was also linked with delegation of power and decentralisation. Each of the controllers assumed its related disciplinary strategic objective had priority over every other. Growing divergence led to a bottleneck with physical, financial and social consequences, as discussed before. Such divergence was linked to discipline/function-based unit grouping (in contrast with the main end-focused unit grouping), which has the tendency towards narrowing perspectives and rigidity (Mintzberg et al. 2003). Other contributors were the lack of prioritisation of objectives and facilitating mutual understanding on objectives.

Organisational configuration should promote an integrated approach and prevent divergence in designing the control mechanism of the reconstruction programme if needed a all. If dividing the control mechanism by function between numbers of participant organisations is an absolute necessity, then awareness about potential narrow-mindedness during the implementation is crucial. Therefore, efforts must be made to harmonise various function-based control mechanisms, to prioritise objectives and reach mutual understanding of those priorities. Any duality and parallel activities have the hidden potential of growing diversity during implementation, if mutual understanding and adjustments do not exist.

- **Delegating accountability**

Within the case study, handling accountability emerged as an important factor when the size of the reconstruction programme increases. This is simply because the increase in the size of the programme means a greater number of engaged teams and individuals. Managing accountability for numerous individuals and teams accordingly is a way of managing the scale of human resources involved in reconstruction operations, reducing the risk of misbehaviour.

The Bam case showed both consistency and inconsistency in this regard. The lack of attention to managing accountability for contractors, in combination with the lack of unit grouping for them, linked with inconsistent job specification and procurement, contributed to the unforeseen situation in which a number of cowboy contractors deceived some local people. This had a domino-effect that led to a major adjustment in the way the socio-technical system of reconstruction worked in terms of beneficiaries participation. Beneficiaries' could become the contractor of their houses if they wanted to, moving from beneficiary-applicant to beneficiary-contractor. This contributed to various phenomena, for example, speeding up the physical progress and the increase of changes during construction phase. In contrast (and fortunately), the accountability for other operational teams, from design to administrative and engineering teams, was established through consultancy companies, Setads and the KEO.

- **Decentralisation and its prerequisites**

Decentralisation is seen important while organising reconstruction activities (Lyons et al. 2010); it is also advocated in guiding principles suggested by international organisations, for example, the Developing Recovery Framework by GFDRR (2014). However, the Bam case brought another dimension to light in this regard. Positioning the case analysis against the literature on organisation theory, the research argues that decentralisations for the sake of decentralisation per se might even be harmful. Decentralisation should be seen and decided on in light of its contribution to approaching the programme strategic direction and objective(s) and promoting mutual understanding of the bigger picture. Such decentralisation benefits from a clear workflow that eliminates competing functionalities, and reduces the number of interlocked interdependencies that integrate participant organisations' tasks.

Compared to the highly centralised attitude of the government, the housing reconstruction programme in Bam was relatively decentralised. The research identified two types of decentralisation in the Bam case, recognised in organisation theory as vertical and horizontal decentralisation (Mintzberg et al. 2003). They were linked with the basis for unit grouping and managing accountability. First was the vertical decentralisation that was correspondent with the customer-, size-, location-based end-focused operational unit grouping. The second was horizontal decentralisation that was correspondent with the control mechanisms. While the vertical decentralisation was a way of managing the scale and had the potential of replication, contrarily, the horizontal decentralisation divided the approval decision making power equally between two parallel organisations, which had divergent understanding of the strategic objectives and priorities. This all led to a nearly deadlock situation, that in turn had a domino-effect, as previously discussed, on how the programme performed and the outcomes of the programme.

Differences in vertical and horizontal decentralisation are linked to the following: shared organisational culture; previous experience of working in post-disaster reconstruction and working with each other; competing or non-competing nature of functions which are interconnected in the production chain; shared understanding of the bigger picture of reconstruction objectives and priorities that might go beyond specific functional disciplines and operational interconnections; and interdependencies that determine the sequence of performed tasks within the workflow. Thus, not all kinds of decentralisation are good for reconstruction programmes.

While the research favours vertical decentralisation for organising reconstruction, any kind of decentralisation requires its prerequisites, such as shared understanding of priorities, attempts to reach mutual adjustments, reducing competing functionalities, and reducing the number of interdependencies. Decentralisation without such prerequisites probably creates more harm than good, because it eliminates the ultimate decision making power, which is crucial in that posttraumatic situation.

- **Modification of the existing system**

In Bam, given the size and geographical extent of the housing reconstruction programme, (the whole urban area), the workflow was treated as a modified version of the previous housing development procedure towards improving its existing shortfalls on enforcing safe construction and application of construction supervision and appreciating the traditional architectural identity in urban development practice, including housing. This modified improved workflow created a temporary housing development process for the area. Such an approach reduced the complexities of the initial system formation and influenced the housing development after the reconstruction period. The delivery system of the reconstruction programme in Bam was in fact based on such modification and the integration of specific measures and mechanisms for approaching the objectives. This supports quicker decision making for identifying critical areas for improvements based on available human and technical resources and capacities instead of working from scratch. It also provides some familiarity for beneficiaries, and the improved system naturally fits within the socio-political context, avoids challenges and shows spaces for improvement within the existing system that is crucial for longer-term improvements. However, the research also discussed that such an approach is inherently non-radical, as modifications will be improvements not radical changes of the overall housing process, e.g. changing the existing separation of design construction phases towards turn-key arrangements.

Building on the above and agreeing with Wisner (2004) that within every system there is a space for improvement, I take an optimistic but realistic approach, to avoid the idealistic 'illusive window of opportunity' stated by Christoplos et al. (2006) in recovery and reconstruction. The modification of the existing system is a relatively realistic way for the application of improvements in the form of new regulations, mechanisms or techniques in construction, and has potential benefits as it focuses on resolving the previous shortfalls instead of starting from scratch, and also has the potential to mainstream and sustain improvements to some extent. Through such an approach, the programme benefits from the momentum for improving the existing practice and experiencing another ways of doing things. Simultaneously it builds on the realities on the ground and has some broad familiarity, as well as shared elements, with the previous practice. Therefore, it has the potential to contribute to mainstreaming some of the improvements made possible during the reconstruction period.

This however, emphasises the importance of thoughtful formation of this improved workflow and examining it against potential issues as much as possible before implementation. Achieving consistency of the organisational configuration of this modified system and the multi organisational nature of the programme in the form of engaging organisations which are influential on the housing development procedures in everyday situations. In Bam, despite creating a modified version of the housing development practice, the programme implementation could not benefit from its full potential. This was a result of an unforeseen bottleneck in the workflow and a conflict between two decision making participant organisations.

- **A Fluent workflow: a production chain, coordinating mechanism and integrating system process**

The above interrelated organisational attributes for the reconstruction programme come to life through the workflow, which, as discussed before, is not only a production chain but also the manifestation of the coordination mechanism among the system elements—an important organisational attribute which connects all participants, and defines the order of activities and their interdependencies (Mintzberg et al. 2003; Goold 2002). The Bam case showed how the workflow became the connective integrating process for the mega system at the programme level that brought participant organisations together (each having their own systems), similar to what Davies and Mackenzie (2013) found recently in construction programmes. The research stressed the importance of workflow as the system integration process, suggesting the advantages of modifying and improving the existing housing procedure, and the need of special attention to be given to the fluency of the workflow, examining it before programme implementation.

The workflow must not be underestimated as a collection of practical considerations. Modification of the existing workflows and procedures for approaching strategic objectives for organising reconstruction is advantageous, benefits from creativity, and necessitates sketching out the workflow for its examination against various potential scenarios and from the perspective of participant organisations. This might not eliminate the possibility of unforeseen issues arising during implementation, but it more likely reduces the number of unforeseen issues.

In Bam, the operational bottleneck was a manifestation of various inconsistencies between organisational attributes and size, duality and conflict of priorities towards strategic objectives. The lack of attention to operational interconnections and internal adjustments in the control mechanisms with the operational unit grouping created a major bottleneck. Creative problem solving in Bam highlighted the need for creativity in handling the situation at both the initial design and implementation phases. Any collection of practical considerations offers more than one option for organisational configuration, and its workflow requires exploring and examining from the perspective of participant organisations.

11.2.2 *Reflecting on the Strategic Nature of the Reconstruction Organisation, Its Delivery System Formation and Implementation*

Synthesising the multidisciplinary literature, the research suggests that whether it is declared publicly (such the Bam case) or not, reconstruction programmes within the contemporary perception of reconstruction must be strategic to serve multidimensional developmental recovery. This, as recognised in organisation theory, requires strategic thinking and management, associated in essence with vertical cascading from higher levels of strategies to the operational level of practical considerations, harmonising levels of activities, as well as creativity (Armstrong 2009; Morgan 2006). In Bam, by declaring three strategic objectives in reconstruction activities, including the housing reconstruction programme, the case declared itself as a purposeful strategic programme. The Bam case presents signs of strategic thinking however it had also shortfalls in the application of a strategic management (Arefian 2016).

The case study showed achievements or underachievement of the reconstruction programme were also directly related to how the strategic nature of the reconstruction was handled at both the system formation and implementation phases. This can be summarised as: on mobilising beneficiaries' participation in an owner-driven reconstruction programmes quantitatively, beneficiaries participated as it was the objective of the programme. Qualitatively, the beneficiaries' role after the system adjustment became similar to the people's role in housing development procedure in everyday situations that is tented by the overall socio-political context of the country. Building earthquake resistant buildings included tougher control mechanisms and improved supervisory services that, as was expressed by interviewees and locals, created a trust and confidence in achieving this objective. Those improvements and control mechanisms became mainstreamed within the post reconstruction housing development process. Additionally public local knowledge on structural aspects of safe construction exists. However, performance towards the objective on safeguarding historical urban identity that on its own was idealistic and tried to capture a broader opportunity at national level that was not as initially expected. The Bam reconstruction influenced long-term institutional improvements in different construction industry-related aspects that were introduced by the national level participant organisations (Arefian 2016).

- **Cascading, harmonising and attention to overlapping areas**

The research built on the literature in organisation theory and more recent project/programme management on cascading higher level strategy in an iterative way from higher levels of strategy to the programme and operational considerations levels, as well as harmonising activities at horizontal levels (Ansoff 1990; Morris and Jamieson 2005).

Bringing this to the reconstruction field, the Bam case during the initial system formation vertical cascading was addressed by translating each objective to operational considerations accordingly. This helped in managing the complexity of the organisational formation, defining participant organisations, their roles and responsibilities. However it missed identifying horizontal interconnections of different groups of practical considerations that aimed to address different strategic objectives. This hindrance in turn contributed to the creating of the aforementioned conflict and bottleneck with its domino-effect consequences. This happened because different objectives were treated in isolation from each other, but all indeed were related to each other in the provision of housing. By increasing the number of objectives for reconstruction activities, the delivery system for a reconstruction programme experiences more complexities in form of the more tasks, greater numbers of participant organisations that must perform those tasks, and more interconnections and overlapping operational considerations. The increase of complexities can easily hinder the strategic nature of the reconstruction programme by forcing the system in favour of one strategic objective and against the other one, especially during the implementation. Creating horizontal harmonisation is more difficult when the number of objectives increases. Given the above, the research confirms the applicability of the notion of vertical cascading from higher level strategy to lower level policies and practical considerations, as well as the need for harmonising those at each level towards the strategic directions provided by the objectives.

It is crucial to identify overlapping areas between practical considerations for different objectives. All objectives of a programme are related to the same product, for example, a house, and there will inevitably be overlapping areas within the workflow in the field even if they are not recognised in advance. Graphical illustration of the vertical cascading and horizontal alignments on objectives individually and related to each other in different levels (e.g. policy formulation and practical considerations), prior to implementation, provides a visual analytical tool to understand the bigger picture and to identify overlapping areas in advance, and to reduce the risks of unforeseen emerging problems in practice.

- **Objective setting**

Although having more objectives adds to the complexities of organising the programme, it seems to be inevitable for a multidimensional recovery. Within such a situation, the case exemplified the importance of the prioritisation of objectives when there is more than one objective. As the Chartered Management Institute (CMI n.d.) state, such a prioritisation is one of the basics of managing strategically. In Bam, the lack of clarity and mutual understanding on the priorities combined with the function-based unit grouping fuelled a damaging conflict between KEO and Mother Consultancy. Different individual interpretations by participant organisations in favour of their own professional disciplines contributed to the emergent of the aforementioned bottleneck in the workflow.

- **Contextual ceiling on beneficiaries' participation**

Peoples' role in development procedures in normal situations is an indication of their level of beneficiaries' participation in a reconstruction programme. This level might be slightly manipulated but the modified role cannot challenge the established socio-political system in the country. Hence, beneficiaries' participation in reconstruction activities has a ceiling related to the country's socio-political and planning system. A radical change to the context-based acceptable level of people' participation in any given context, logically, reduces the chance of success for its implementation. Therefore, instead of talking of people's participation as an abstract notion in the reconstruction field, it is helpful and realistic to see and position it within the particular socio-political context within which reconstruction operations are conducted.

The overall approach to beneficiaries' participation in Bam, similar to a report by Lyons et al. (2010) on other international cases, had to become more liberal from its initial limiting approach. This was to reflect the unforeseen problem of dodgy contractors. The new liberal approach allowed beneficiaries to become their own contractors if they wanted to, similar to the normal housing development procedure, linked with the approach on modification and improvement of the existing systems in practice, which in turn are overarched by the broader national socio-political system in the country.

- **Creativity: exploring new ways**

The aforementioned shortcomings in organising the reconstruction programme in Bam that at a certain point blocked the system from effectively working was solved for by innovative thinking. Creativity is seen as a required element in strategic thinking as suggested in the literature on strategic management and recently in construction management (Harpham 2009; Morris and Jamieson 2005). Interestingly, the case showed how managerial creativity helped the programme to deal with the consequences of existing inconsistencies in the organisational configuration and during its implementation which was even debated at the national Parliament.

- **Strategic thinking at both delivery system formation and implementation**

In the Bam case, the attention to the strategic nature of the housing reconstruction programme was mainly at the system formation stage. During the implementation period, the system had to be adjusted, and it evolved to reflect the realities of the system formation with its certain initial organisational configuration (with all consistencies and inconsistencies). The system formation was based on three core directions, but the system adjustments that introduced new practical considerations for the workability of the system were not examined against those core directions. The objective of safeguarding historical urban identity was mostly absorbed in the reflections of such system adjustments. Also the interconnections of the proposed adjustments to deal with emerged problems and issues during the programme implementation were not seen against this objective, and therefore negatively influenced approaching this objective.

Reflecting on the strategic nature of a reconstruction programme must occur at the system formation stage, the initial organisational design stage, as well as during its implementation in the field. It is simplistic to assume that the initial reconstruction programme delivery system formation will not encounter difficulties or some pre-assumptions will not prove wrong; this is linked to the highly dynamic and complex post-disaster context of a reconstruction programme combined with social nature of the reconstruction programmes. The formation of a reconstruction programme delivery system takes place at the exact time that limited information on how the realities will be unfolded and new information will be emerged (Sull 2007). The strategic nature of the programme, and the application of corrective reflective adjustments in a re-examination of the new practical considerations to deal with the unfolded realities and solve problems, was not considered in the Bam case.

11.2.3 Recognising and Reflecting on the Social Nature of Reconstruction Organisations

Epistemological advances in organisational theory perceive organisations as socio-technical systems -a combination of both social and technical aspects. Dynamic interactions among organisational societies influence the way an organisation performs, and through this they influence the outcomes of the implementation of that system (Morgan 2006; Perrow 1973). Since participatory reconstruction programmes blur the boundaries between organisational society and society as a whole, dynamics of society as a whole (or its representatives) mixes with the dynamics of interactions within the system. Also the way of doing things, and the culture of all other participant organisations towards influence the way the socio-technical delivery system of the reconstruction programme works. The major social dynamics of the housing reconstruction programme in Bam influenced the programme performance through influencing the system formation and implementation are as follows.

- **Recognition of underlying social dynamics towards facilitating social learning during reconstruction**

Organisational society, people and culture are soft tissues which as recognised in organisation theory, organisational design must also deal with them in addition to the formal organisational configuration (Goold 2002). A reconstruction programme delivery system creates hardware towards a given strategic objective; the underlying social dimension of the system is the software of the programme. The system works in full capacity when the hardware and software are harmonised and work with together.

In the Bam case, there is an interesting and on the existing public knowledge and awareness about the main structural elements of building earthquake-resistant buildings among groups of people who are not usually expected to know such things (Gharaati Kopaei 2009). It was linked with the existing common place (construction site) for social interactions that facilitated informal knowledge transfer at a very operational level, a way of organisational learning that touched the broader society, as discussed before. Informal social interactions, which happen in common places, play an important role for knowledge transfer at the individual level and public awareness that in turn plays a role in the programme achievements, under-achievement and longer term impacts. Such individual learning, which is conditioned by the social environment, is an interpretation of social learning (Pelling et al. 2008). There is a link between social learning and the existence of common places that facilitate social interactions, despite such intrinsic uncertainty in human interactions, as noted by Curado (2006) and the Bam case strongly presents this. Social learning is linked with organisational learning and social systems, and addresses the areas of local knowledge, the social context and social position for information flow. It is however unexplored within the post-disaster recovery literature (Wisner 2004).

Social learning as an informal knowledge transfer is able to strengthen or undermine the formal performance of the delivery system for improving the previous status quo towards objectives of the reconstruction and therefore the recovery strategy. The Bam programme did not recognise social dynamism and the flow of social learning, therefore missed to direct or benefit from them for the good of the programme performance. In the absence of such attention any influence of such social interactions on the programme will be accidental. The ignorance of this dynamism means that outcomes of such informal social learning cannot be directed towards the reconstruction strategic objectives and/or longer-term influences. This very interesting phenomena of knowledge transfer in Bam was unplanned since the existing and emerging opportunities for benefiting from social learning, to culturally strengthen the efforts on achieving the objectives, were not noted. The programme did not take into account the existence of a common social place for family members and construction teams, as well as the new social initiative, Shora-Yar, and its potentials for enhancing reconstruction programme performance.

Crucially, social dynamics exist and social learning happens, even if the reconstruction programme does not take them into account. By acquiring a social learning insight towards the reconstruction programme and its delivery system formation reconstruction managers indeed help themselves. Although social systems in reconstruction programmes are complex and social interactions cannot be confidently predicted, organisational attributes and practical considerations can encourage and facilitate common places for informal social interactions and this must be recognised. This requires awareness of underlying social dynamism. Figure 11.2 illustrates the link between organisational formation and social learning towards approaching strategic objectives.

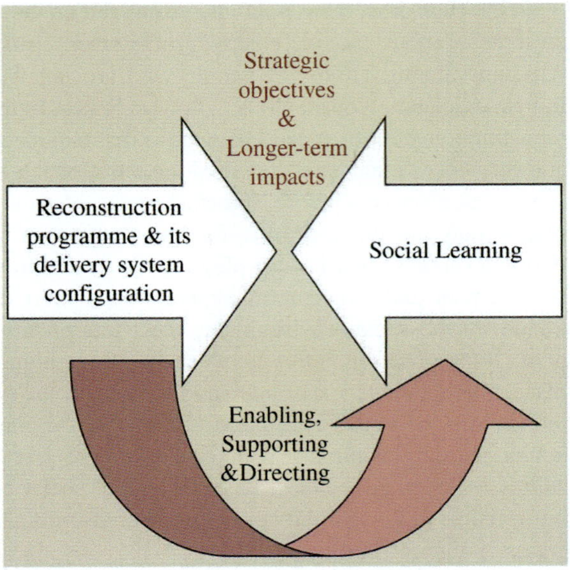

Fig. 11.2 The schematic correlation between the socio-technical system of the reconstruction programme and social learning towards approaching reconstruction strategic objectives

- **Reducing system vulnerability through organisational configuration**

Increase of the size, number of participant organisations and interactions of a programme ultimately increases complexities and, as recognised in the organisation field, it increases risks of internal system loopholes (Bonabeau 2007). Examinations on how the reconstruction programme delivery system in Bam was vulnerable, brought together a variety of contributors: the scale of the human loss, rising prices, organisational, socio-political and economic issues that might even be related to deeper issues beyond the capacity of a housing reconstruction programme, e.g. historical mistrust between the government and the people. The system vulnerability and taking advantage of the system loopholes was greatly influenced by the combination of a) the above issues which might be beyond the scale of reconstruction activities; and b) the inconsistencies within the organisational configuration that is within the scope of a reconstruction programme. The extent of the uncompleted structures in Bam is an approximate manifestation of these.

While the scope of reconstruction activities cannot address broader issues, necessary actions are needed to address the issues that are in the capacity of the reconstruction programme, such as organisational considerations, and providing a social lens to understand and examine the system. For instance, within that historical mistrust, the problem of dodgy contractors was harmful for the reputation of HFIR as the representative of the government.

The minimum benefit of recognition of the existing and emerging social opportunities for transparency and linkages between top-down and bottom-up communication through innovative initiatives, is that it provides shared understandings of the complexities and problems for delivering large scale reconstruction programmes. System loopholes can always be found in complex systems; but the risk of abusing a system increases when in a post-disaster reconstruction situation construction material prices are rising, beneficiaries are deceived by cowboy contractors, and they are trapped within a lengthy process due to a bottleneck in the workflow. In such a situation, inconsistencies in organisational configuration and the lack of attention to its strategic nature are more responsible for the system vulnerability than the existing historical mistrust between people and governments.

11.2.4 Reflecting on the Multi-organisational Nature of Reconstruction Programmes' Delivery Systems

Synthesising the literature on reconstruction and organisation theory, the multi-organisational nature of reconstruction programmes characterised the delivery system for the programme as a mega system, within which each component is a system. The difficult task of forming such integration is highlighted both in the reconstruction field and the construction management of megaprojects identified as construction programmes (Comfort and Kapucu 2006; Davies and Mackenzie 2013). The Bam case showed the major dynamics of the multi-organisational nature of the housing programme, was indeed interrelated with all the organisational configuration, strategic nature and social nature of the programme, and influenced the programme performance through its system formation and implementation.

- **Auto-adaptation is necessary but not sufficient**

Examining the features of the multi-organisation that influences approaches to the strategic objectives, the Bam examination made a case to show that auto-adaptation is necessary but not sufficient. Auto-adaptation of the participant organisations must be complemented with overall supervision of the performance of the system as a whole accordingly. This challenges what Comfort and Kapucu (2006) suggest, and adds a strategic dimension to their discussion. They suggest the auto-adaptation model for the formation of this delivery system as a way of achieving mutual adjustment among organisations towards coordination mechanism suggested in the organisation theory.

The Bam case showed that the notion of auto-adaptation although required to make the system flexible and adjustable, is influenced by internal motivations and key decision making in participated organisations, especially ones at higher levels. Participant organisations have internal dynamics that affect the programme over the reconstruction period. The notion of auto-adaptation on its own might lead to divergent directions and deviation of the strategic nature direction of the programme. This reinforces the importance of attention to the strategic nature of the reconstruction programme during system formation and implementation as discussed earlier. The research therefore suggests that the notion of auto-adaptation must be complemented by strategic supervision over the bigger picture and the overall direction of the programme.

- **Engaging relevant organisations, advantageous for a strategic approach**

The Bam case exemplifies how engaging organisations which are normally involved or influential in housing development practice in post-disaster reconstruction will positively convey the momentum for improvement to the post-reconstruction situation and influence the normal housing practice and/or construction industry. Two examples elaborated on local and national levels of suck influences. As a result of the engagement of KEO, the local branch of a national professional organisation for construction industry, related measures and practical considerations acted as a catalyst for more permanent improvements in the structural control mechanisms after the reconstruction period, when the task of structural controls was permanently delegated with the Bam branch of KEO. Another example was the engagement of BHRC that further advanced the introduction of seismic codes for traditional construction materials and the standardisation of more construction materials across the country.

Engaging relevant national and local organisations in reconstruction programmes is beneficial for the strategic approach towards longer-term impacts, at both national and local levels. It will be advantageous for mainstreaming the improved measures and methods through facilitating organisational learning, exposing room for negotiation, and retaining the new knowledge, measures and mechanisms after reconstruction. The application of a newly improved process over time by local organisations leads to organisational learning (Argate and Miron-Spektor 2011), seizes the existing spaces for future negotiation, and keeps the knowledge and new way of doing things within the locality which influence future experiences after the reconstruction period. Thus reconstruction activities become pilot projects and programmes and nevertheless part of the overall development of the area. Understandably engaging more organisations requires more attention to organisational configuration to prevent system vulnerability.

- **Workflow as the standardisation of coordination mechanism among participant organisations**

Approaching strategic objectives through a multi-organisation reconstruction programme reinforces the importance of a fluent consistent system configuration and workflow as an integrating process, as discussed earlier, and was also confirmed recently by Davies and Mackenzie (2013) in the context of construction programmes. In multi-organisation delivery systems, different organisations with different natures, internal structure, working cultures, and motivations are brought together for the delivery of strategic objectives. The coordination mechanism between different organisations, to bring unity among them, is manifested by a standard workflow, which is part of organisational configuration that was concluded earlier in Sect. 11.2.1.

- **Power balance and political positions**

Dynamics of power balance influence the programme performance in unpredicted directions, which might or might not be in favour of the strategic objectives of the programme and the recovery strategy. This occurs regardless of how equal the inputs are from different organisations at the stage of the programme formulation and the delivery system formation. What happens in practice is also influenced by the operational power/links that decision making organisations have in the field. The operational capacity is one of the influential factors of the dynamic of power balance.

The case also exemplifies the importance of the power balance in a reconstruction programme formation and implementation, especially during conflicts and disputes. It showed that internal motivations were linked with the previous role of the organisation in the normal urban development activities before the disaster in KEO and MHUD, HFIR and the BAUC. In turn it affects the formulation, implementations and adjustments of a reconstruction programme. Seeing an opportunity in reconstruction attempts to redeem organisational reputation and reposition organisational standards in normal urban development activities.

11.3 Taking the Discourse Forward: The Propositions

Identifying opportunities for learning from the Bam case and positioning it within the multidisciplinary literature, the main contribution of this research sits within the interconnected areas of reconstruction and strategic management, and the contingency view in organisation theory and urban development, and construction management. This, in my view, advances the discourse on organising reconstruction, which takes the form of overlapping the above theoretic insights. This builds on both achievement and underachievement and provides instrumental and intrinsic values of the research with the potential to inform other reconstruction cases. The instrumental value comes from understanding the underlying contributors to achievements and underachievement towards informing the practice for international cases that share some issues with Iran in one way or another, while for Iran itself it will be dealing with underlying contributors to underachievement. Building on all above we can take the previous analytical framework (Fig. 11.1) to a more detailed level that emerged, and provide propositions. Figure 11.3 presents a more detailed level of the analytical framework. This builds on the learning opportunities, which in turn relate to the multidisciplinary literature. This multi-perspective conceptual framework can be explained in the form of four overlapping propositions. Organising reconstruction processes sits in their overlapping area. The following elaborates on these four research propositions:

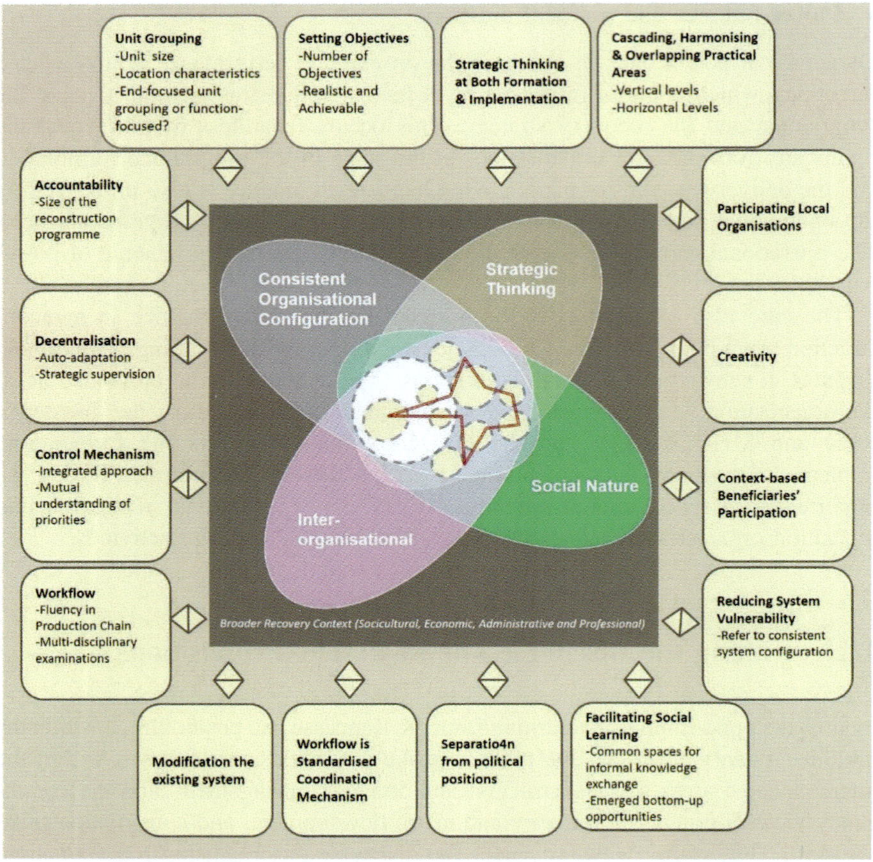

Fig. 11.3 Developed conceptual model for organisation design and management for post-disaster reconstruction programmes, proposed by this research

11.3.1 Proposition A. Reconstruction Programmes Require Innovative yet Consistent and Strategic Delivery Systems

Organisational configuration of the delivery system must be examined during the system formation. Consistencies and inconsistencies among the organisational attributes and the size and location of characteristics of the post-disaster reconstruction programmes are the most important organisational factors for achievement or underachievement of a reconstruction programme regarding a strategic objective. Unit grouping, setting control mechanisms, delegating accountability, (de)centralisation and workflow are of the main organisational attributions that must be seen in relation to each other and to the size and location of the characteristics of a reconstruction programme

- End-focused customer-based unit grouping can be seen as a flexible modular approach for managing the scale of the reconstruction; it however needs to be complemented with a consistent approach in other elements and job specification for participants.
- The number of functional units depends on the size and location characteristics of the programme and the assumed capacity for each unit.
- Decentralisation requires prerequisites in the form of avoiding competing situations and facilitating mutual understanding of objectives and priorities.
- The larger the size of the programme, the more the need for clear delegation of accountability increases. Therefore, larger-sized programme require middle layers between the numerous operational workforce and the high-level reconstruction management system.
- For control mechanisms, duality and discipline-based parallel control mechanisms must be avoided as much as possible. Instead on integrated control mechanisms, for example, creating multidisciplinary working teams are suggested.
- The workflow brings the reconstruction system to life and exposes hidden inconsistencies during implementation in practice. The structure of the workflow must be examined against the possibilities through critical points of interconnections before the implementation of the programme. A workflow is the standardisation of the coordination mechanism between participant organisations. Participant organisations in the programmes should contribute to examining the workflow before its implementation.
- Modification of the existing system is a realistic approach for integrating improvements and measures for achieving strategic objectives within the reconstruction process. If conducted through consistent organisational configuration and by engaging local organisations, it paves the way to mainstreaming those improvements, at least to some extent.

11.3.2 Proposition B. Reconstruction Programmes Require Strategic Management at All Levels

Reconstruction activities are strategic and require strategic thinking during their formation and during the implementation levels of activities, including programmes:

- Strategic approach to reconstruction programmes starts with the setting of objectives.
- Strategic Objectives must be realistic and take the post-traumatic situation, time and finances into account. Expectations for improvements can be challenging

but they should be achievable. Idealist and ambitious strategic objectives potentially will end up with underachievement.

- In case of having two or more objectives for a reconstruction programme, clear mutual understanding of priorities is needed from the outset.
- Vertical cascading between various levels (from high level strategy to policy formulation and practical operational considerations) and horizontal alignment at each level are crucial to be examined and safeguarded. They are the essence of strategic thinking at all times for achieving objectives. Each participant organisation or each operational task is in place to serve the strategic objectives.
- The number of strategic objectives must be limited. A greater number of objectives potentially require a greater number of participants and practical considerations. These increase complexities in the formation and implementation of the programme. Thus, it increases the likelihood of deviation from the core direction towards the strategic objectives.
- In case of having two or more strategic objectives, they must be seen in relation to each other, and practical considerations for one objective should be examined against their potential influences on approaching other objectives as well. A greater number of objectives creates more interconnections and overlaps between different sets of practical considerations. Such overlaps and interconnections require special attention.
- Creativity and the art of balancing are important for exploring new ways of doing things within the existing availabilities and limitations.

11.3.3 Proposition C. Delivery System Configuration Can Facilitate Informal Social Learning Through Understanding Underlying Social Dynamics

Acquiring a social lens for understanding the existing social context and underlying dynamisms beyond formal urban processes has the potential to support the reconstruction programme to be impactful and increase the chances of success:

- Reconstruction programmes must address and facilitate informal knowledge exchange and social learning in order to enhance the reconstruction programme's approach towards its strategic objectives. To do this, awareness of the existing social dynamism is necessary.
- When thinking about organisational configuration must take into account how it creates common physical places (for example, as a work place, meeting point or social gatherings) so that they facilitate informal social interactions too. If such common places already exist practical considerations and supportive measures are needed to maximise the impact of the reconstruction programme.
- A post-disaster situation often sees emergence of social initiatives. They are emerging opportunities for a two-way top-down and bottom-up communication,

- Inconsistent organisational configuration potentially increases the vulnerability of the reconstruction delivery system as a result of its complications and negative effects during implementation.
- Instead of talking about people's participation in abstract, the contextual people's participation that reconstruction programmes must consider based on the certain socio-political situation that is a tentative ceiling for any kind of people's participation. Radical changes endanger the realism and deliverability of the programme.

11.3.4 Proposition D. Delivery System Must Maximise Its Inherent Potentials and Minimise Its Hidden Threats in a Multi-Organisation Reconstruction Programme

Reconstruction programmes have inherent potential opportunities for longer-term impacts and threats of divergence and conflict. A consistent and strategic delivery system during formation and implementation must maximise the potentials and minimise threats:

- Workflow in a multi-organisational programme is not only a production chain; it is also a coordination mechanism among participant organisations.
- Auto-adaptation must be balanced with strategic supervision for the performance of the whole system.
- Participating organisations that are influential in housing development in normal situations (for regulation, standards and so on) are potentially helpful to trigger and mainstream improvements made during the reconstruction period and after.
- Creating operational linkages between reconstruction programmes and political positions in participant institutions inherits the likelihood of instability of decision making positions and of change in reconstruction approaches.

11.4 Conclusion

This chapter outlined the essence of this research as a whole. It presented learning opportunities from the Bam case and demonstrated how those advance the existing knowledge on organising reconstruction within the overlapping knowledge areas of disaster-development studies, strategic management and contingency view in organisation theory. Accordingly, the research proposed an analytical model for understanding organisation design and management of a post-disaster reconstruction programme that offers overlapping theoretical insights on this subject. Through the iterative process of the application of the model, the research advanced the existing knowledge within the reconstruction field. In brief, the research suggested innovative, yet consistent and strategic delivery systems for organisational design and management of post-disaster reconstruction programmes are required. The research also proposed the notion of contextual people's participation, the importance of social learning within reconstruction programmes, and maximising the potentials for mainstreaming improvements. It proposed an analytical framework which brings together all required insights to organise reconstruction activities.

Throughout this research, I treated the case as an experimental laboratory in order to develop the current state-of-the-art theories on this subject. At the end of this research, what appears is that the Bam case for housing reconstruction programmes in fact was again a laboratory for Iran to examine how to deal with a large scale urban reconstruction, as a result of the permanent risk of large scale urban disaster. From the outset, the case showed that post-disaster reconstruction in a historic landscape adds further layer of expectations to what already expected from the reconstruction. It means that the great complexities of urban reconstruction are even greater when the city had previously historic urban characteristics. Attention to historic urban contexts has been of the new urban agenda of the Habitat 3 conference in 2016 and logically adds another objective to reconstruction programmes, including housing reconstruction, in historic cities which the Bam case exemplified.

Regardless of the way the system worked, I see such an attempt on its own as one step forward for dealing with one of the biggest challenges that Iran, similar to many other earthquake-prone countries, progressively faces: a combination of rapid urbanisation and likelihood of destructive earthquakes in urban areas.

Finally, the research, interestingly, found that the latest advances on construction management for megaprojects deals to some extent with same concerns about forming mega systems for construction programmes (Davies and Mackenzie 2013). Although in my view the complexities of participatory post-disaster reconstruction exceed such construction programmes, further research is needed to identify how these two fields can inform each other to deal with the same concern.

References

Ansoff, H (1990) Implanting strategic management, 2nd ed. Prentice Ha, New York

Arefian FF (2016) Getting ready for urban reconstruction: organising housing reconstruction in Bam. In: Arefian FF, Moeini SHI (eds) Urban change in Iran: stories of rooted histories and ever-accelerating developments. The urban book series. Springer International Publishing, Cham, Heidelberg, New York, Dordrecht, London, pp 231–257

Argate L, Miron-Spektor E (2011) Organisational learning: from experience to knowledge. https://server1.tepper.cmu.edu/Seminars/docs/ARGOTE_SPEKTOR_OS_%20in-press.pdf. Accessed 3 Sep 2012

Armstrong M (2009) Armstrong's handbook of management and leadership: a guide to managing for results, 2nd edn. Kogan

Bonabeau E (2007) Understanding and managing complexity risk. MIT Sloan Management Review, 48

Bryman A (2008) Social research methods, 3rd edn. Oxford University Press, Oxford, New York

Chartered Management Institute (CMI), (n.d.) Igor Ansoff father of corporate strategy. Thinker 052. http://www.managers.org.uk for CMI members. Accessed 19 Jan 2011

Christopolos I, ALNAP, Glemminge Development Research (2006) Provention consortium forum 2006. Strengthening global collaboration in disaster risk reduction. The elusive 'Window of Opportunity' for risk reduction in post-disaster recovery. http://ipcc-wg2.gov/njlite_download.php?id=5282. Accessed 20 Dec 2010

Comfort LK, Kapucu N (2006) Inter-organizational coordination in extreme events: the world trade center attacks, 11 Sept 2001. Nat Hazards 39(2):309–327. https://doi.org/10.1007/s11069-006-0030-x

Curado C (2006) Organisational learning and organisational design. Learn Organ 13(1):25–48

Davidson C (2009) Multi-actor arrangements and project management. In: Lizarralde G, Johnson C, Davidson C (eds) Rebuilding after disasters: from emergency to sustainability. Spon Press, London, New York

Davies A, Mackenzie I (2013) Project complexity and systems integration: constructing the London 2012 Olympics and Paralympics games. Int J Project Manage 32(5):773–790. https://doi.org/10.1016/j.ijproman.2013.10.004

Eisenhardt KM (1989) Building Theories from case study research. Acad Manage Rev 14(4):532–550

Flyvbjerg B (2006) Five misunderstandings about case-study research. Qual Inq 12:219–245. https://doi.org/10.1177/1077800405284363

Gharaati Kopaei M (2009) Knowledge transfer in post-disaster reconstruction; the problem of post-post-disaster reconstruction. PhD thesis, McGill University, Montreal. http://digitool.library.mcgill.ca/webclient/StreamGate?folder_id=0&dvs=1420020480363~426. Accessed 21 Aug 2014

Global Facility for Disaster Reduction and Facility. GFDRR (2014) Guide to developing disaster recovery framework (World Reconstruction Conference Version), GFDRR, Washington DC

Goold M (2002) Designing effective organizations: how to create structured networks. Jossey-Bass; Wiley (distributor), San Francisco Calif

Harpham A (2009) Bridging the Gap between Corporate Strategy and Project, Introduction to Programme Management. www.pm4success.com. Accessed 21 Aug 2009

Lyons M, Schilderman T, Boano C, D'Urzo S (2010) Conclusions. In: Lyons M, Schilderman T, Boano C (eds) Building back better: delivering people-centred housing reconstruction at scale. Practical Action Pub, Warwickshire UK, pp 345–359

Mintzberg H, Ghoshal S, Lampel J, Quinn JB (2003) The strategy process: concepts, contexts, cases, 4th ed. Pearson Education

Morgan G (2006) Images of organization, updated edition (first edition 1997). Sage, Thousand Oaks

Morris PWG, Jamieson A (2005) Moving from corporate strategy to project strategy. Proj Manage J 36(4):5–18

Pelling M, High C, Dearing J, Smith D (2008) Shadow spaces for social learning: a relational understanding of adaptive capacity to climate change within organisations. Environ Plan 40(4):867–884. https://doi.org/10.1068/a39148

Perrow C (1973) The short and glorious history of organizational theory. Org Dyn 2(1):3–15. https://doi.org/10.1016/0090-2616(73)90007-7

Sull DN (2007) Closing the gap between strategy and execution. MIT Sloan Manage Rev 48(4). http://sloanreview.mit.edu/the-magazine/articles/2007/summer/48412/closing-the-gap-between-strategy-and-execution/?type=x&reprint=48412. Accessed 19 Jan 2011

Wisner B (2004) Chapter 9, Towards a safer environment. At risk: natural hazards. People's vulnerability and disasters. Routledge, London, pp 321–376

Yin RK (2003) Case study research: design and methods. Sage

Yin RK (2009) Case study research: design and methods. Sage, London, New Delhi, Singapore

Printed by Printforce, the Netherlands